Mechanics and Analysis of Advanced Materials and Structures

Mechanics and Analysis of Advanced Materials and Structures

Editors

Sanichiro Yoshida
Giovanni Pappalettera

MDPI • Basel • Beijing • Wuhan • Barcelona • Belgrade • Manchester • Tokyo • Cluj • Tianjin

Editors
Sanichiro Yoshida
Southeastern Louisiana University
USA

Giovanni Pappalettera
Politecnico di Bari
Italy

Editorial Office
MDPI
St. Alban-Anlage 66
4052 Basel, Switzerland

This is a reprint of articles from the Special Issue published online in the open access journal *Materials* (ISSN 1996-1944) (available at: https://www.mdpi.com/journal/materials/special_issues/Mech_Anal_Adv_Mater_Struct).

For citation purposes, cite each article independently as indicated on the article page online and as indicated below:

LastName, A.A.; LastName, B.B.; LastName, C.C. Article Title. *Journal Name* **Year**, *Volume Number*, Page Range.

ISBN 978-3-0365-7774-6 (Hbk)
ISBN 978-3-0365-7775-3 (PDF)

© 2023 by the authors. Articles in this book are Open Access and distributed under the Creative Commons Attribution (CC BY) license, which allows users to download, copy and build upon published articles, as long as the author and publisher are properly credited, which ensures maximum dissemination and a wider impact of our publications.

The book as a whole is distributed by MDPI under the terms and conditions of the Creative Commons license CC BY-NC-ND.

Contents

Preface to "Mechanics and Analysis of Advanced Materials and Structures" vii

Sanichiro Yoshida and Giovanni Pappalettera
Mechanics and Analysis of Advanced Materials and Structures
Reprinted from: *Materials* 2023, 16, 2123, doi:10.3390/ma16052123 1

Chiung-Shiann Huang and S. H. Huang
Analytical Solutions Based on Fourier Cosine Series for the Free Vibrations of Functionally Graded Material Rectangular Mindlin Plates
Reprinted from: *Materials* 2020, 13, 3820, doi:10.3390/ma13173820 5

Luis Mercedes, Christian Escrig, Ernest Bernat-Masó and Lluís Gil
Analytical Approach and Numerical Simulation of Reinforced Concrete Beams Strengthened with Different FRCM Systems
Reprinted from: *Materials* 2021, 14, 1857, doi:10.3390/ma14081857 25

Leonid P. Obrezkov, Taija Finni and Marko K. Matikainen
Modeling of the Achilles Subtendons and Their Interactions in a Framework of the Absolute Nodal Coordinate Formulation
Reprinted from: *Materials* 2022, 15, 8906, doi:10.3390/ma15248906 43

Panos Tsakiropoulos
On the Stability of Complex Concentrated (CC)/High Entropy (HE) Solid Solutions and the Contamination with Oxygen of Solid Solutions in Refractory Metal Intermetallic Composites (RM(Nb)ICs) and Refractory Complex Concentrated Alloys (RCCAs)
Reprinted from: *Materials* 2022, 15, 8479, doi:10.3390/ ma15238479 55

Sungjong Cho, Hyunjo Jeong and Ik Keun Park
Optimal Design of Annular Phased Array Transducers for Material Nonlinearity Determination in Pulse–Echo Ultrasonic Testing
Reprinted from: *Materials* 2020, 13, 5565, doi:10.3390/ma13235565 97

Hyunjo Jeong, Sungjong Cho, Shuzeng Zhang and Xiongbing Li
Absolute Measurement of Material Nonlinear Parameters Using Noncontact Air-Coupled Reception
Reprinted from: *Materials* 2021, 14, 244, doi:10.3390/ma14020244 117

Gianni Niccolini, Stelios M. Potirakis, Giuseppe Lacidogna and Oscar Borla
Criticality Hidden in Acoustic Emissions and in Changing Electrical Resistance during Fracture of Rocks and Cement-Based Materials
Reprinted from: *Materials* 2020, 13, 5608, doi:10.3390/ma13245608 135

Yuma Murata, Tomohiro Sasaki and Sanichiro Yoshida
Stress Dependence on Relaxation of Deformation Induced by Laser Spot Heating
Reprinted from: *Materials* 2022, 15, 6330, doi:10.3390/ma15186330 153

Shun Takahashi, Sanichiro Yoshida, Tomohiro Sasaki and Tyler Hughes
Dynamic ESPI Evaluation of Deformation and Fracture Mechanism of 7075 Aluminum Alloy
Reprinted from: *Materials* 2021, 14, 1530, doi:10.3390/ma14061530 169

Caroline Kopfler, Sanichiro Yoshida and Anup Ghimire
Application of Digital Image Correlation in Space and Frequency Domains to Deformation Analysis of Polymer Film
Reprinted from: *Materials* 2022, 15, 1842, doi:10.3390/ma15051842 191

**Claudia Barile, Caterina Casavola, Giovanni Pappalettera
and Vimalathithan Paramsamy Kannan**
Damage Progress Classification in AlSi10Mg SLM Specimens by Convolutional Neural Network and k-Fold Cross Validation
Reprinted from: *Materials* **2022**, *15*, 4428, doi:10.3390/ma15134428 207

Alexander A. Pavlovskii, Konstantin Pushnitsa, Alexandra Kosenko, Pavel Novikov and Anatoliy A. Popovich
Organic Anode Materials for Lithium-Ion Batteries: Recent Progress and Challenges
Reprinted from: *Materials* **2023**, *16*, 177, doi:10.3390/ma16010177 223

Preface to "Mechanics and Analysis of Advanced Materials and Structures"

Modern materials and structures differ significantly from conventional ones. Often, their sizes are orders of magnitude smaller, and their compositions are so sophisticated as to render traditional characteristic methods inapplicable. This Special Issue aims to discuss these problems by inviting authors from various fields of science and technology.

We hope that the articles collected in this Special Issue will allow readers to conceive new ideas in an interdisciplinary fashion.

Sanichiro Yoshida and Giovanni Pappalettera
Editors

Editorial

Mechanics and Analysis of Advanced Materials and Structures

Sanichiro Yoshida [1,*] and Giovanni Pappalettera [2]

1 Department of Chemistry & Physics, Southeastern Louisiana University, Hammond, LA 70402, USA
2 Dipartimento di Meccanica, Matematica e Management, Politecnico di Bari, Via Orabona 4, 70125 Bari, Italy
* Correspondence: sanichiro.yoshida@selu.edu

Citation: Yoshida, S.; Pappalettera, G. Mechanics and Analysis of Advanced Materials and Structures. *Materials* 2023, *16*, 2123. https://doi.org/10.3390/ma16052123

Received: 7 February 2023
Accepted: 10 February 2023
Published: 6 March 2023

Copyright: © 2023 by the authors. Licensee MDPI, Basel, Switzerland. This article is an open access article distributed under the terms and conditions of the Creative Commons Attribution (CC BY) license (https://creativecommons.org/licenses/by/4.0/).

Modern technological development has made the designing and characterization of materials sophisticated. The design concepts used for centuries for macroscopic objects do not necessarily apply to micro or nano-scale counterparts. The elastic modulus of a nano-scale wire can be significantly different from that of the same material at the macroscopic scale. Composite materials comprising multiple materials possessing completely different properties can exhibit unexpected behaviors.

This situation requires the advancement of diagnostic methods in characterizing material properties. Great efforts are necessary to extend the type of materials and the conditions to make analysis convenient and to improve the figure of merits such as the sensitivity, resolution, non-destructiveness, and rapidness of the measurement. Such advancement can be an extension or a combination of existing techniques or the development of entirely new methodologies. Often, analytical or numerical modeling is necessary to expedite material characterization. Recent software development provides various computational tools, including machine learning and artificial intelligence algorithms, for numerical analysis and modeling.

The present Special Issue is a collection of articles that shed light on the above issues. It has 11 original research papers and one review article. Below we take a quick look at the gist of these articles.

The first four papers in the reference list discuss the mechanical properties of composite materials.

The vibrational behavior of plates is complex. It has many modes, and the dynamics depend on the boundary conditions and the material's characteristics. Huang et al. [1] discuss an analytical solution based on the Mindlin plate theory that describes the free vibration of rectangular plates of functionally graded material. They analyzed various combinations of boundary conditions and validated the results via comparison with published results. They tabulated data from this study so that researchers could use it to judge the accuracy of numerical methods.

Fabric reinforcement of materials is another complex process. While the fabric material increases strength, controlling the interface with the matrix material is challenging. Mercedes et al. [2] developed an analytical approach and a numerical simulator to evaluate the strength of concrete beams with various types of fabrics. They successfully parameterized complex strengthening behaviors with a single parameter called the reduction coefficient. They conclude that this strategy simplifies the analytical design of fiber-reinforced cementitious matrices.

In a slightly different context from the above two works, Obrezkov et al. [3] performed numerical modeling. They developed a numerical method to analyze a biomedical composite material. They modeled the elongation-induced deformation of the human Achilles tendon by assuming that it consists of beam elements using absolute nodal coordinate formulation. They report that this method reduces the degree of freedom, benefiting cost reductions in computation. They conclude that with some limitations, this model is feasible.

The aerospace industry requires the development of ultra-high-temperature materials. In ref. [4], Tsakiropoulos analyzed the stability of complex-concentrated and high-entropy

Niobium-based solid solutions for ultra-high temperature materials by considering oxygen concentrations. The author discusses the increase in hardness and Young's modulus with oxygen concentrations under various conditions in detail.

The following two papers in the reference list discuss nonlinear acoustic probing.

Acoustic probing is a well-established technique. Acoustic waves pass through most materials in a wide range of frequencies. This feature is contrastive to electromagnetic waves. One drawback is its wavelength. The typical acoustic frequency used for probing is several GHz, whereas the acoustic velocity in most materials is less than ten km/s. The corresponding acoustic wavelength is of the order of microns. This situation limits the size of anomalies to detect. A solution to this problem is the use of the nonlinear response of the material to the probing signal. The following two papers discuss nonlinear acoustic probing.

Cho et al. [5] used the second harmonic of an acoustic emitter signal in their probing system operating in the pulse-echo mode. A challenge in such nonlinear techniques is the smallness of the signal received by the sensor. To overcome this issue, the authors of this paper considered intensifying the acoustic signal by focusing the acoustic beam from the emitter. In this paper, they propose an annular phased-array acoustic transducer and demonstrate the increase in the nonlinear sensitivity via numerical analyses.

High acoustic coupling between the acoustic transducer and the specimen is a significant factor for successful probing. Distilled water and acoustic gel are common materials for coupling. Usually, these materials increase the coupling efficiency so that the acoustic signal detected by the receiver is reasonably high. However, the receiver signal depends on the contact pressure between the transducer and the specimen, compromising the repeatability of data.

In ref. [6], Jeong et al. propose a method to evaluate the nonlinear parameter with air-coupled acoustic receivers. By developing a two-layer acoustic model consisting of the solid specimen and air, considering diffraction and attenuation, they constructed an algorithm to evaluate the nonlinear parameter relative to a reference specimen. In this paper, they verified the algorithm via comparisons with the experiment. This paper also discusses factors that affect the accuracy of the method.

Multiple methods applied to characterize the same specimen always provide extra information. In the following two papers, the authors take advantage of it.

Niccolini et al. [7] investigated the critical behavior of fracture processes in rocks and cement-based materials by acquiring acoustic emission and electrical resistance time series. They conducted compression experiments using rod specimens of Luserna stone and cement mortar. The experimental results indicate that the change in electric resistance precedes the acoustic emission in all tested specimens. The authors ascribe this observation to the fact that the acoustic emission signal is related to larger cracks than the electric resistance signal. In this paper, they discuss other findings in terms of the fracturing process.

Residual stresses are difficult to analyze because they hide in the material. In many situations, it is necessary to alter their status for the observer to evaluate them. On the other hand, too much alteration would change the stress, compromising the measurement. This nature makes nondestructive estimation of residual stress challenging. Murata et al. [8] demonstrated a nondestructive method of residual stress analysis. They applied laser spot heating to residually stressed specimens and probed the relaxation to estimate residual stress. Using a thermal camera, they monitored the thermal process. In addition, by mechanically applying initial loads to the specimen, they made quantitative analyses. The authors verified the results via comparisons with strain gauge measurements.

The recent advancement in software, generally called machine learning algorithms, facilitates the existing technology of deformation analyses. Regardless of the choice of hardware, the output signal representing deformation is complicated. Often, the information is buried in noise, and its spatiotemporal characteristics are complex. This situation causes human operators to make errors in diagnosis. Computers are capable of analyzing a large volume of data and extracting features.

In the following three papers, the authors utilize the advantages of computer-based algorithms to deal with complex data.

Takahashi et al. [9] used an optical interferometric method called Electronic Speckle-Pattern Interferometry (ESPI) to discuss the dynamics of plastic deformation and fracture under tensile loading. The ESPI technique reveals the contour of differential displacements (the displacement occurring during a short time interval) as dark fringes in a full-field optical image.

For the analysis of deformation dynamics, it is necessary to analyze numerous fringe images. An algorithm to automatically locate the fringes is essential. Using speckles, the optical image containing the differential displacement contour is noisy. Takahashi et al. applied Gaussian filtering to reduce the noise and located the fringes continuously until the specimen fractured. With this method, they found that in the transition from late plastic deformation to the fracturing stage, the rotational mode of deformation plays a significant role in the process of stress concentration.

Kopfler et al. [10] use optical speckles for another purpose. They utilized the randomness of the speckle pattern to uniquely identify local areas of a transparent specimen (polyethylene films) and analyzed nonuniform deformation under tensile loading. With the numerical algorithm known as Digital Image Correlation, they found the nonuniform displacement field of the specimen. In the cases where the deformation is more uniform, they analyzed the deformation in the frequency domain using Fast Fourier Transform.

Barile et al. [11] propose an acoustic emission-based analysis of the damage progression stage in Sintered Laser Melting (SLM) materials. Characterization of materials processed by SLM is of great interest but also challenging due to the intrinsic orthotropy of these materials. When tested by acoustic emission methods, the orthotropic structure of the material causes the spectral content of the resulting waveforms to depend on the nature of the source and the building direction of the materials. In this view, the authors have taken advantage of machine learning tools and specifically of a convolutional neural network to classify detected signals based on the damage stage and the building direction. In particular, by adopting k-fold cross-validation, they have demonstrated drastic improvement in classification accuracy.

Finally, Pavlovskii et al. [12] reviewed the recent progress and challenges in organic anode materials for Lithium-Ion Batteries (LIB), one of the most demanding technologies of the modern era. This review compares the electrochemical performances of different organic anode materials, discussing the advantages and disadvantages of each class of organic materials in research and commercial applications. After addressing the practical applications of some organic anode materials, the paper discusses some techniques to address significant issues, including low discharge voltages and the undesired dissolution of the anode material into electrolytes.

Author Contributions: Writing—Original Draft Preparation, S.Y.; writing—review and editing, S.Y. and G.P.; supervision, S.Y. All authors have read and agreed to the published version of the manuscript.

Funding: This research received no external funding.

Conflicts of Interest: The authors declare no conflict of interest.

References

1. Huang, C.S.; Huang, S.H. Analytical Solutions Based on Fourier Cosine Series for the Free Vibrations of Functionally Graded Material Rectangular Mindlin Plates. *Materials* **2020**, *13*, 3820. [CrossRef] [PubMed]
2. Mercedes, L.; Escrig, C.; Bernat-Masó, E.; Gil, L. Analytical Approach and Numerical Simulation of Reinforced Concrete Beams Strengthened with Different FRCM Systems. *Materials* **2021**, *14*, 1857. [CrossRef] [PubMed]
3. Obrezkov, L.P.; Finni, T.; Matikainen, M.K. Modeling of the Achilles Subtendons and Their Interactions in a Framework of the Absolute Nodal Coordinate Formulation. *Materials* **2022**, *15*, 8906. [CrossRef] [PubMed]
4. Tsakiropoulos, P. On the Stability of Complex Concentrated (CC)/High Entropy (HE) Solid Solutions and the Contamination with Oxygen of Solid Solutions in Refractory Metal Intermetallic Composites (RM(Nb)ICs) and Refractory Complex Concentrated Alloys (RCCAs). *Materials* **2022**, *15*, 8479. [CrossRef] [PubMed]

5. Cho, S.; Jeong, H.; Park, I.K. Optimal Design of Annular Phased Array Transducers for Material Nonlinearity Determination in Pulse–Echo Ultrasonic Testing. *Materials* **2020**, *13*, 5565. [CrossRef] [PubMed]
6. Jeong, H.; Cho, S.; Zhang, S.; Li, X. Absolute Measurement of Material Nonlinear Parameters Using Noncontact Air-Coupled Reception. *Materials* **2021**, *14*, 244. [CrossRef] [PubMed]
7. Niccolini, G.; Potirakis, S.; Lacidogna, G.; Borla, O. Criticality Hidden in Acoustic Emissions and in Changing Electrical Resistance during Fracture of Rocks and Cement-Based Materials. *Materials* **2020**, *13*, 5608. [CrossRef] [PubMed]
8. Murata, Y.; Sasaki, T.; Yoshida, S. Stress Dependence on Relaxation of Deformation Induced by Laser Spot Heating. *Materials* **2022**, *15*, 6330. [CrossRef] [PubMed]
9. Takahashi, S.; Yoshida, S.; Sasaki, T.; Hughes, T. Dynamic ESPI Evaluation of Deformation and Fracture Mechanism of 7075 Aluminum Alloy. *Materials* **2021**, *14*, 1530. [CrossRef] [PubMed]
10. Kopfler, C.; Yoshida, S.; Ghimire, A. Application of Digital Image Correlation in Space and Frequency Domains to Deformation Analysis of Polymer Film. *Materials* **2022**, *15*, 1842. [CrossRef] [PubMed]
11. Barile, C.; Casavola, C.; Pappalettera, G.; Kannan, V.P. Damage Progress Classification in AlSi10Mg SLM Specimens by Convolutional Neural Network and k-Fold Cross Validation. *Materials* **2022**, *15*, 4428. [CrossRef] [PubMed]
12. Pavlovskii, A.A.; Pushnitsa, K.; Kosenko, A.; Novikov, P.; Popovich, A.A. Organic Anode Materials for Lithium-Ion Batteries: Recent Progress and Challenges. *Materials* **2023**, *16*, 177. [CrossRef] [PubMed]

Disclaimer/Publisher's Note: The statements, opinions and data contained in all publications are solely those of the individual author(s) and contributor(s) and not of MDPI and/or the editor(s). MDPI and/or the editor(s) disclaim responsibility for any injury to people or property resulting from any ideas, methods, instructions or products referred to in the content.

Article

Analytical Solutions Based on Fourier Cosine Series for the Free Vibrations of Functionally Graded Material Rectangular Mindlin Plates

Chiung-Shiann Huang * and S. H. Huang

Department of Civil Engineering, National Chiao Tung University, 1001 Ta-Hsueh Rd., Hsinchu 30050, Taiwan; mark071819@gmail.com
* Correspondence: cshuang@mail.nctu.edu.tw

Received: 7 July 2020; Accepted: 25 August 2020; Published: 29 August 2020

Abstract: This study aimed to develop series analytical solutions based on the Mindlin plate theory for the free vibrations of functionally graded material (FGM) rectangular plates. The material properties of FGM rectangular plates are assumed to vary along their thickness, and the volume fractions of the plate constituents are defined by a simple power-law function. The series solutions consist of the Fourier cosine series and auxiliary functions of polynomials. The series solutions were established by satisfying governing equations and boundary conditions in the expanded space of the Fourier cosine series. The proposed solutions were validated through comprehensive convergence studies on the first six vibration frequencies of square plates under four combinations of boundary conditions and through comparison of the obtained convergent results with those in the literature. The convergence studies indicated that the solutions obtained for different modes could converge from the upper or lower bounds to the exact values or in an oscillatory manner. The present solutions were further employed to determine the first six vibration frequencies of FGM rectangular plates with various aspect ratios, thickness-to-width ratios, distributions of material properties and combinations of boundary conditions.

Keywords: analytical solution; Fourier cosine series; vibrations; FGM rectangular plates; Mindlin plate theory

---

1. Introduction

Functionally graded materials (FGMs) were first produced in the mid-1980s [1]. An FGM is composed of varying mixtures of different materials, such as ceramics and metals. The material properties of FGMs smoothly and continuously vary, in contrast to conventional laminated composite materials. Consequently, FGMs do not comprise stress singularities formed due to discontinuities in the material properties. FGMs can be designed to possess the high heat resistance and corrosion resistance of ceramics as well as the high mechanical strength of metals. Over the previous three decades, FGMs have been extensively explored in various fields including aerospace, energy, electronics, optics, biomedicine, and mechanical engineering.

Plates are employed in a wide range of mechanical and structural system components in civil, mechanical and aeronautical engineering. The behaviors of FGM plates have attracted research attention. Different reviews [2–6] have provided exhaustive summaries of the studies published on the free vibrations and buckling of FGM plates according to various plate theories and the three-dimensional elasticity theory.

Numerous studies have investigated the free vibrations of FGM rectangular plates, and most of these studies have employed various numerical methods. For example, on the basis of the classical plate theory, Abrate [7] and Zhang and Zhou [8] reported that an FGM plate behaves similar to a homogeneous plate if a suitable reference plane is adopted, while Yang and Shen [9] employed a one-dimensional differential quadrature approximation and the Galerkin procedure to determine the frequencies of initially stressed plates. According to the first-order shear deformation plate theory, Zhao et al. [10] applied the element-free kp-Ritz method to analyze the vibrations of square and skew plates under different combinations of boundary conditions. Fu et al. [11] employed the Ritz method with admissible functions consisting of double Fourier cosine and several closed-form auxiliary functions to study the vibrations of orthotropic FGM plates with general boundary restraints. Ferreira et al. [12] investigated the vibrations of square plates employing the first-order and third-order shear deformation plate theories and the collocation method with multiquadric radial basis functions. Huang et al. [13] adopted the third-order shear deformation plate theory and Ritz method for analyzing the vibrations of rectangular plates with and without side cracks. Hong [14] investigated the thermal vibrations of plates via the generalized differential quadrature method and the third-order shear deformation plate theory. Using the higher-order shear deformation plate theories, Qian et al. [15] applied the Petrov–Galerkin meshless method and Roque et al. [16] adopted the multiquadric radial basis function method to find the vibration frequencies of thick plates. Using the Ritz method and three-dimensional elasticity theory, Uymaz and Aydogdu [17] studied the vibrations of plates with various combinations of boundary conditions, and Cui et al. [18] performed vibration analysis of an FGM sandwich rectangular plate resting on an elastic foundation using admissible trigonometric functions. Huang and his coworkers [19,20] proposed a set of admissible functions, which can accurately describe the behaviors of a crack, for examining the vibrations of cracked FGM rectangular plates and also showed the natural frequencies of intact plates. Burlayenko et al. [21] employed the commercial finite element package ABAQUS to analyze the vibrations of thermally loaded FGM sandwich plates.

Only a few studies have been devoted to analytical solutions for the free vibrations of FGM rectangular plates based on various plate theories. The solutions in these studies consider rectangular plates with two opposite edges or four simply supported edges (faces). Using the first-order shear deformation plate theory, Hosseini-Hashemi et al. [22,23] introduced new potential and auxiliary functions to construct exact closed-form solutions for the vibrations of rectangular plates having two opposite edges simply supported with and without considering the in-plane displacement components, respectively, while Ghashochi-Bargh and Razavi [24] proposed analytical solutions for the vibrations of orthotropic FGM rectangular plates without considering the in-plane displacement components. Hosseini-Hashemi et al. [25] extended their studies by using a third-order shear deformation theory. Considering simply supported conditions on four edge surfaces, Matsunaga [26] and Sekkal et al. [27] developed solutions for FGM rectangular plates and sandwich plates, respectively, according to the higher-order shear deformation plate theories. Based on three-dimensional elasticity theory, Vel and Batra [28] used the power series method to construct solutions for the vibrations of FGM rectangular plates, while Reddy and Cheng [29] employed an asymptotic approach along with a transfer matrix. Huo et al. [30] employed the recursive matrix method to develop the solutions for the vibrations of FGM sandwich plates.

According to plate theories, there are 21 distinct combinations of boundary conditions (i.e., free, simply supported and clamped) for rectangular plates. The literature review found that except for the six cases in which two opposite edges are simply supported, no analytical solution for the vibrations of FGM rectangular plates with various combinations of boundary conditions exists. The present study aims to fill a gap in the literature and proposes analytical solutions based on the Mindlin plate theory for the vibrations of FGM rectangular plates with 21 combinations of boundary conditions. The proposed solutions are established using the Fourier cosine series with polynomial supplementary functions, which eliminate the validity requirement for the term-wise differentiation of the Fourier sine series of a function to accurately represent the differential of the function [31]. The validity of the present solutions is confirmed through comprehensive convergence studies for plates with four combinations of boundary conditions and by comparing the obtained vibration frequencies with those published in the literature. The material properties along the thickness of an FGM plate are estimated using the power-law or the Mori–Tanaka scheme. The solutions are further applied to determine the vibration frequencies of Al/Al$_2$O$_3$ FGM plates with nine combinations of boundary conditions. The obtained analytical results can serve as a benchmark for the evaluation of other solutions obtained through various approximate or numerical approaches.

2. Methodology

2.1. Material Models

Depicted in Figure 1 is a rectangular FGM plate with a length of a, width of b and thickness of h. The material properties of FGM plates are assumed to vary along their thickness (z) according to the power-law or Mori–Tanaka scheme, which are two popular material models used in the literature. The FGM plates under consideration are made of aluminum (Al) and ceramic (zirconia (ZrO$_2$) or alumina (Al$_2$O$_3$)), the material properties of which are given in Table 1.

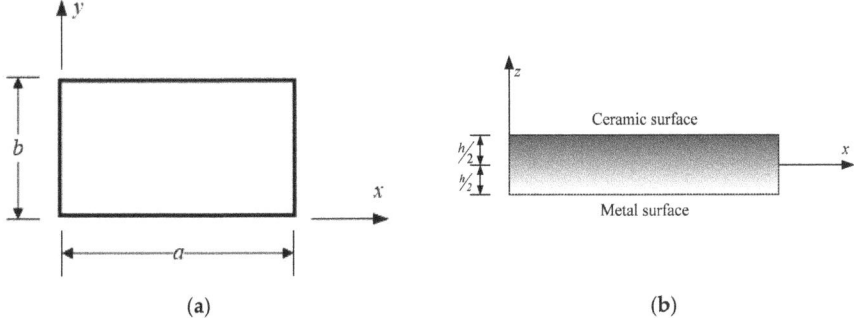

Figure 1. Geometry of a functionally graded material (FGM) plate and coordinates: (**a**) top view, (**b**) side view.

Table 1. Material properties of the FGM ingredients.

Material	Properties		
	E (GPa)	Poisson's Ratio (ν)	ρ (kg/m^3)
Aluminum (Al)	70.0	0.3	2702
Alumina (Al$_2$O$_3$)	380	0.3	3800
Zirconia (ZrO$_2$)	200	0.3	5700

A power-law distribution of material properties is often assumed for FGMs. The material properties (i.e., Young's modulus ($E = E(z)$), Poisson's ratio ($\nu(z)$), and mass density ($\rho = \rho(z)$)) along the thickness of an FGM plate are given as follows:

$$P(z) = P_b + V(z)(P_t - P_b) = P_b + V(z)\Delta P \tag{1}$$

where $V(z) = \left(\frac{z}{h} + \frac{1}{2}\right)^{\overline{m}}$; P_b and P_t denote the material properties at the bottom face ($z = -h/2$) and top face ($z = h/2$), respectively; ΔP is the difference between P_b and P_t, and $\overline{m}$ is the material property gradient index that governs the material variation profile in the thickness direction. Equation (1) indicates that if $P_b = P_t$ or $\overline{m} = 0$, $P(z)$ is constant. FGM plates consisting of Al and ZrO$_2$ (or Al$_2$O$_3$) exhibit a constant Poisson's ratio because Al and ZrO$_2$ (or Al$_2$O$_3$) have the same Poisson's ratio. Figure 2 illustrates the distributions of $E(z)$ and $\rho(z)$ along the thickness of the Al/Al$_2$O$_3$ plates when $\overline{m} = 0.5, 2,$ and 5, where E_c and ρ_c are the Young's modulus and density of Al$_2$O$_3$, respectively.

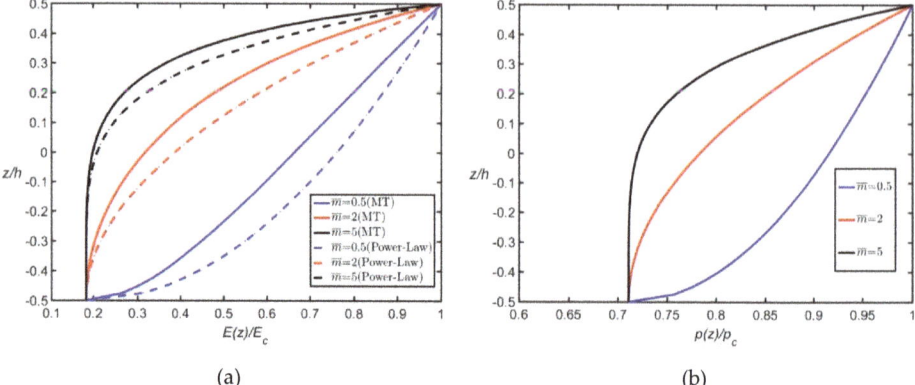

Figure 2. Variations of $E(z)$ and $\rho(z)$ for Al/Al$_2$O$_3$ through the thickness: (a) $E(z)$; (b) $\rho(z)$.

The Mori–Tanaka scheme is also frequently used to describe the material properties of FGMs. The effective mass density along the thickness of an FGM plate is given by

$$\rho(z) = \rho_1 V_1(z) + \rho_2 V_2(z) \tag{2}$$

$$V_1(z) + V_2(z) = 1 \tag{3}$$

$$V_1(z) = V_1^b + (V_1^t - V_1^b)\left(\frac{z}{h} + \frac{1}{2}\right)^{\overline{m}} \tag{4}$$

where subscripts 1 and 2 indicate materials 1 and 2, respectively, and V_1^t and V_1^b are the volume fractions of material 1 on the top and bottom surfaces of the plate, respectively. The effective local bulk modulus K and the shear modulus G are given by

$$\frac{K(z) - K_1}{K_2 - K_1} = \frac{V_2(z)}{1 + \frac{(K_2 - K_1)V_1(z)}{K_1 + (4/3)G_1}}, \quad \frac{G(z) - G_1}{G_2 - G_1} = \frac{V_2(z)}{1 + \frac{(G_2 - G_1)V_1(z)}{G_1 + f_1}} \tag{5}$$

where $f_1 = \frac{G_1(9K_1 + 8G_1)}{6(K_1 + 2G_1)}$. After the effective moduli K and G are estimated, the effective Young's modulus and Poisson's ratio are obtained using the following equation:

$$E(z) = \frac{9K(z)G(z)}{3K(z) + G(z)} \quad \text{and} \quad \nu(z) = \frac{3K(z) - 2G(z)}{2(3K(z) + G(z))} \tag{6}$$

In the Mori–Tanaka scheme, the Poisson's ratio is a function of z even when materials 1 and 2 have the same Poisson's ratio. Equation (2) can be converted to the form of Equation (1), so the density distribution based on the Mori–Tanaka scheme is the same as that described by Equation (1). The distributions of $E(z)$ and $\rho(z)$ along the thickness of Al/Al$_2$O$_3$ plates with $\overline{m}$ = 0.5, 2, and 5 are illustrated in Figure 2 and denoted by "M–T".

2.2. Governing Equations and Boundary Conditions

In the Mindlin plate theory [32], the displacement components of a plate are expressed as follows:

$$\overline{u}(x,y,z,t) = u_0(x,y,t) + z\psi_x(x,y,t), \quad \overline{v}(x,y,z,t) = v_0(x,y,t) + z\psi_y(x,y,t),$$
$$\overline{w}(x,y,z,t) = w_0(x,y,t) \tag{7}$$

where $\bar{u}$, $\bar{v}$, and $\bar{w}$ are the displacement components in the x-, y-, and z-directions, respectively; u_0, v_0, and w_0 are the displacements on the mid-plane, and ψ_x and ψ_y are the rotations of the mid-plane normal in the x- and y-directions, respectively. The stress resultants are defined as follows:

$$Q_\beta = \int_{-h/2}^{h/2} \sigma_{\beta z} dz, \quad \begin{Bmatrix} N_{\beta\beta} \\ M_{\beta\beta} \end{Bmatrix} = \int_{-h/2}^{h/2} \sigma_{\beta\beta} \begin{Bmatrix} 1 \\ z \end{Bmatrix} dz, \quad \begin{Bmatrix} N_{xy} \\ M_{xy} \end{Bmatrix} = \int_{-h/2}^{h/2} \sigma_{xy} \begin{Bmatrix} 1 \\ z \end{Bmatrix} dz \tag{8}$$

where the subscript β represents x or y, and σ_{ij} represents the stress components. The equations of motion are given as follows:

$$N_{xx,x} + N_{xy,y} = I_0 \ddot{u}_0 + I_1 \ddot{\psi}_x, \; N_{xy,x} + N_{yy,y} = I_0 \ddot{v}_0 + I_1 \ddot{\psi}_y, \; Q_{x,x} + Q_{y,y} = I_0 \ddot{w}_0,$$
$$M_{xx,x} + M_{xy,y} - Q_x = I_1 \ddot{u}_0 + I_2 \ddot{\psi}_x, \; M_{xy,x} + M_{yy,y} - Q_y = I_1 \ddot{v}_0 + I_2 \ddot{\psi}_y \tag{9}$$

where $I_l = \int_{-h/2}^{h/2} \rho(z) z^l dz$ ($l = 0$, 1 and 2). The subscript comma denotes the partial derivative with respect to the coordinates defined by the variable after the comma.

Substituting linear strain–displacement (Equation (10)) and stress–strain relationships (Equation (11)) into Equation (8) yields the expressions for stress resultants in terms of displacement related components (Equation (12)).

$$\varepsilon_{xx} = \frac{\partial \bar{u}}{\partial x}, \; \varepsilon_{yy} = \frac{\partial \bar{v}}{\partial y}, \; \varepsilon_{xy} = \frac{1}{2}(\frac{\partial \bar{v}}{\partial x} + \frac{\partial \bar{u}}{\partial y}), \; \varepsilon_{yz} = \frac{1}{2}(\frac{\partial \bar{w}}{\partial y} + \frac{\partial \bar{v}}{\partial z}), \; \varepsilon_{zx} = \frac{1}{2}(\frac{\partial \bar{u}}{\partial z} + \frac{\partial \bar{w}}{\partial x}) \tag{10}$$

$$\begin{Bmatrix} \sigma_{xx} \\ \sigma_{yy} \\ \sigma_{xy} \\ \sigma_{yz} \\ \sigma_{zx} \end{Bmatrix} = \begin{bmatrix} \frac{E}{1-\nu^2} & \frac{\nu E}{1-\nu^2} & 0 & 0 & 0 \\ \frac{\nu E}{1-\nu^2} & \frac{E}{1-\nu^2} & 0 & 0 & 0 \\ 0 & 0 & 2G & 0 & 0 \\ 0 & 0 & 0 & 2G & 0 \\ 0 & 0 & 0 & 0 & 2G \end{bmatrix} \begin{Bmatrix} \varepsilon_{xx} \\ \varepsilon_{yy} \\ \varepsilon_{xy} \\ \varepsilon_{yz} \\ \varepsilon_{zx} \end{Bmatrix} \tag{11}$$

$$N_{xx} = \bar{E}_0 u_{0,x} + \bar{E}_1 \psi_{x,x} + \bar{D}_0 v_{0,y} + \bar{D}_1 \psi_{y,y}, \; N_{yy} = \bar{D}_0 u_{0,x} + \bar{D}_1 \psi_{x,x} + \bar{E}_0 v_{0,y} + \bar{E}_1 \psi_{y,y},$$
$$N_{xy} = \bar{G}_0(u_{0,y} + v_{0,x}) + \bar{G}_1(\psi_{x,y} + \psi_{y,x}), \; M_{xx} = \bar{E}_1 u_{0,x} + \bar{E}_2 \psi_{x,x} + \bar{D}_1 v_{0,y} + \bar{D}_2 \psi_{y,y},$$
$$M_{yy} = \bar{D}_1 u_{0,x} + \bar{D}_2 \psi_{x,x} + \bar{E}_1 v_{0,y} + \bar{E}_2 \psi_{y,y}, \; M_{xy} = \bar{G}_1(u_{0,y} + v_{0,x}) + \bar{G}_2(\psi_{x,y} + \psi_{y,x}),$$
$$Q_x = \kappa \bar{G}_0(w_{0,x} + \psi_x), \; Q_y = \kappa \bar{G}_0(w_{0,y} + \psi_y) \tag{12}$$

where

$$\bar{G}_i = \int_{-h/2}^{h/2} G z^i dz, \; \bar{E}_i = \int_{-h/2}^{h/2} \frac{E}{1-\nu^2} z^i dz \text{ and } \bar{D}_i = \int_{-h/2}^{h/2} \frac{\nu E}{1-\nu^2} z^i dz$$

The parameter κ is the transverse shear correction coefficient and is taken as 5/6 in the following analyses.

By substituting Equation (12) into Equation (9), the governing equations are obtained in terms of the displacement functions as follows:

$$\bar{E}_0 u_{0,xx} + \bar{E}_1 \psi_{x,xx} + \bar{D}_0 v_{0,xy} + \bar{D}_1 \psi_{y,xy} + \bar{G}_0(u_{0,yy} + v_{0,xy}) + \bar{G}_1(\psi_{x,yy} + \psi_{y,xy}) = I_0 \ddot{u}_0 + I_1 \ddot{\psi}_x \tag{13}$$

$$\bar{D}_0 u_{0,xy} + \bar{D}_1 \psi_{x,xy} + \bar{E}_0 v_{0,yy} + \bar{E}_1 \psi_{y,yy} + \bar{G}_0(u_{0,xy} + v_{0,xx}) + \bar{G}_1(\psi_{x,xy} + \psi_{y,xx}) = I_0 \ddot{v}_0 + I_1 \ddot{\psi}_y \tag{14}$$

$$\kappa \bar{G}_0 \left[(w_{0,xx} + \psi_{x,x}) + (w_{0,yy} + \psi_{y,y}) \right] = I_0 \ddot{w}_0 \tag{15}$$

$$\overline{E}_1 u_{0,xx} + \overline{E}_2 \psi_{x,xx} + \overline{D}_1 v_{0,xy} + \overline{D}_2 \psi_{y,xy} + \overline{G}_1(u_{0,yy} + v_{0,xy})$$
$$+ \overline{G}_2(\psi_{x,yy} + \psi_{y,xy}) - \kappa \overline{G}_0(w_{0,x} + \psi_x) = I_1 \ddot{u}_0 + I_2 \ddot{\psi}_x \quad (16)$$

$$\overline{D}_1 u_{0,xy} + \overline{D}_2 \psi_{x,xy} + \overline{E}_1 v_{0,yy} + \overline{E}_2 \psi_{y,yy} + \overline{G}_1(u_{0,xy} + v_{0,xx})$$
$$+ \overline{G}_2(\psi_{x,xy} + \psi_{y,xx}) - \kappa \overline{G}_0(w_{0,y} + \psi_y) = I_1 \ddot{v}_0 + I_2 \ddot{\psi}_y \quad (17)$$

Each edge of a rectangular plate is simply supported (S), clamped (C) or free (F). For the edge with y = constant, the S, C and F boundary conditions are defined as follows:

Simply supported: $u_0 = w_0 = \psi_x = N_{yy} = M_{yy} = 0$;
Clamped: $u_0 = v_0 = w_0 = \psi_x = \psi_y = 0$, and
Free: $N_{yy} = N_{xy} = Q_y = M_{yy} = M_{xy} = 0$.

Similar definitions of boundary conditions are also applicable for the edge with x = constant.

2.3. Series Solutions

To establish the Fourier cosine series solutions for the vibrations of plates, let

$$u_0(x, y, t) = U_0(x, y) \cdot e^{i\omega t}, \ v_0(x, y, t) = V_0(x, y) \cdot e^{i\omega t}, \ w_0(x, y, t) = W_0(x, y) \cdot e^{i\omega t},$$
$$\psi_x(x, y, t) = \Psi_x(x, y) \cdot e^{i\omega t}, \ \psi_y(x, y, t) = \Psi_y(x, y) \cdot e^{i\omega t} \quad (18)$$

And

$$U_0(x,y) = \sum_{m=0}^{M} \sum_{n=0}^{N} A_{mn}^{(1)} \cos\alpha_m x \cos\beta_n y + \sum_{l=1}^{2} \xi_l(x) \sum_{n=0}^{N} B_{ln}^{(1)} \cos\beta_n y + \sum_{l=1}^{2} \eta_l(y) \sum_{m=0}^{M} C_{lm}^{(1)} \cos\alpha_m x \quad (19)$$

$$V_0(x,y) = \sum_{m=0}^{M} \sum_{n=0}^{N} A_{mn}^{(2)} \cos\alpha_m x \cos\beta_n y + \sum_{l=1}^{2} \xi_l(x) \sum_{n=0}^{N} B_{ln}^{(2)} \cos\beta_n y + \sum_{l=1}^{2} \eta_l(y) \sum_{m=0}^{M} C_{lm}^{(2)} \cos\alpha_m x \quad (20)$$

$$W_0(x,y) = \sum_{m=0}^{M} \sum_{n=0}^{N} A_{mn}^{(3)} \cos\alpha_m x \cos\beta_n y + \sum_{l=1}^{2} \xi_l(x) \sum_{n=0}^{N} B_{ln}^{(3)} \cos\beta_n y + \sum_{l=1}^{2} \eta_l(y) \sum_{m=0}^{M} C_{lm}^{(3)} \cos\alpha_m x \quad (21)$$

$$\Psi_x(x,y) = \sum_{m=0}^{M} \sum_{n=0}^{N} A_{mn}^{(4)} \cos\alpha_m x \cos\beta_n y + \sum_{l=1}^{2} \xi_l(x) \sum_{n=0}^{N} B_{ln}^{(4)} \cos\beta_n y + \sum_{l=1}^{2} \eta_l(y) \sum_{m=0}^{M} C_{lm}^{(4)} \cos\alpha_m x \quad (22)$$

$$\Psi_y(x,y) = \sum_{m=0}^{M} \sum_{n=0}^{N} A_{mn}^{(5)} \cos\alpha_m x \cos\beta_n y + \sum_{l=1}^{2} \xi_l(x) \sum_{n=0}^{N} B_{ln}^{(5)} \cos\beta_n y + \sum_{l=1}^{2} \eta_l(y) \sum_{m=0}^{M} C_{lm}^{(5)} \cos\alpha_m x \quad (23)$$

where $\alpha_m = m\pi/a$, $\beta_n = n\pi/b$, and $\xi_l(x)$ and $\eta_l(y)$ are supplementary functions.

Tolstov [31] showed the following theorem on the differentiation of the Fourier series of a function:

Theorem 1. *Let f(x) be a continuous function and have an absolutely integrable derivative on [0, L]. When f(x) is expanded as*

$$f(x) = \sum_{n=1}^{\infty} \widetilde{b}_n \sin \lambda_n x, \text{ where } \lambda_n = n\pi/L, \quad (24)$$

$$f'(x) = \frac{f(L) - f(0)}{L} + \sum_{n=1}^{\infty} \{\frac{2}{L}[(-1)^n f(L) - f(0)] + \lambda_n \widetilde{b}_n\} \cos \lambda_n x \quad (25)$$

When f(x) is expanded as

$$f(x) = \widetilde{a}_0 + \sum_{n=1}^{\infty} \widetilde{a}_n \cos \lambda_n x, \ f'(x) = -\sum_{n=1}^{\infty} \lambda_n \widetilde{a}_n \sin \lambda_n x \quad (26)$$

The theorem indicates that the Fourier cosine series can be differentiated term-by-term, while such an operation can be applied to the Fourier sine series only if $f(0) = f(L) = 0$. To remedy such shortcoming of the sine series, Li [33] proposed to add some supplementary functions into the cosine series in Equation (26) and to determine $f''(x)$ and $f^{(iv)}(x)$ via term-by-term differential.

According to Li [33,34], the supplementary functions are typically determined by satisfying the following conditions:

$$\xi_{1,x}(0) = 1,\ \xi_{1,x}(a) = 0,\ \xi_{2,x}(0) = 0,\ \xi_{2,x}(a) = 1,\ \eta_{1,y}(0) = 1,\ \eta_{1,y}(b) = 0,\ \eta_{2,y}(0) = 0,$$
$$\eta_{2,y}(b) = 1,\ \int_0^a \xi_l(x)dx = 0 \text{ and } \int_0^b \eta_l(y)dy = 0\ (l = 1,\ 2). \tag{27}$$

If polynomial functions are used, Equation (27) leads to

$$\xi_1(x) = -\frac{x^2}{2a} + x - \frac{a}{3},\ \xi_2(x) = \frac{x^2}{2a} - \frac{a}{6},\ \eta_1(y) = -\frac{y^2}{2b} + y - \frac{b}{3},\ \eta_2(y) = \frac{y^2}{2b} - \frac{b}{6}. \tag{28}$$

Substituting Equations (18)–(23) and (28) into the boundary conditions yields a set of linear algebraic equations for the coefficients $A_{mn}^{(i)}$, $B_{ln}^{(i)}$ and $C_{ln}^{(i)}$. For instance, when $U_0(a, y) = 0$, which is one of the fixed boundary conditions at $x = a$, the following equation is obtained:

$$\sum_{n=0}^{N}[\sum_{m=0}^{M} A_{mn}^{(1)} \cos\alpha_m a + \sum_{l=1}^{2} \xi_l(a)B_{ln}^{(1)} + \sum_{l=1}^{2}\bar{c}_{ln}\sum_{m=0}^{M} C_{lm}^{(1)} \cos\alpha_m a]\cos\beta_n y = 0 \tag{29}$$

In establishing Equation (29), $\eta_l(y)$ is expressed in its Fourier cosine series as follows:

$$\eta_l(y) = \sum_{n=0}^{N} \bar{c}_{ln} \cos\beta_n y \tag{30}$$

where $\bar{c}_{ln} = \int_0^b \eta_l(y)\cos\beta_n y dy / \int_0^b (\cos\beta_n y)^2 dy$. Equation (29) includes $N + 1$ functions of $\cos\beta_n y$, and each coefficient of $\cos\beta_n y$ must equal zero in order to satisfy the equation. Consequently, Equation (29) provides $N + 1$ linear algebraic equations for the coefficients $A_{mn}^{(1)}$, $B_{ln}^{(1)}$ and $C_{ln}^{(1)}$,

$$\sum_{m=0}^{M} A_{mn}^{(1)} \cos\alpha_m a + \sum_{l=1}^{2} \xi_l(a)B_{ln}^{(1)} + \sum_{l=1}^{2}\bar{c}_{ln}\sum_{m=0}^{M} C_{lm}^{(1)} \cos\alpha_m a = 0\ (n = 0,\ 1,\ 2,\ \cdots,\ N) \tag{31}$$

Similarly, one can establish 10 $(M + N + 2)$ linear algebraic homogeneous equations for $A_{mn}^{(i)}$, $B_{ln}^{(i)}$ and $C_{ln}^{(i)}$ from the boundary conditions along the four edges of a rectangular plate. Such set of equations can be further expressed in the matrix form as follows:

$$\mathbf{B_p\ P} = \mathbf{B_A\ A} \tag{32}$$

where $\mathbf{P} = (B_{ln}^{(1)}\ C_{lm}^{(1)}\ B_{ln}^{(2)}\ C_{lm}^{(2)}\cdots B_{ln}^{(5)}\ C_{lm}^{(5)})^T$ and $\mathbf{A} = (A_{mn}^{(1)}\ A_{mn}^{(2)}\cdots A_{mn}^{(5)})^T$ $(l = 1, 2; m = 0, 1, \cdots M; n = 0, 1, \cdots N)$.

To satisfy the governing equations, substituting Equations (18)–(23) and (28) into Equations (13)–(17) also yields a set of linear algebraic equations for the coefficients $A_{mn}^{(i)}$, $B_{ln}^{(i)}$ and $C_{lm}^{(i)}$. For example, Equation (13) yields the following equation:

$$
\begin{aligned}
& \sum_{m=0}^{M}\sum_{n=0}^{N}\left\{(-\alpha_m^2\overline{E}_0 - \beta_n^2\overline{G}_0)A_{mn}^{(1)} + \sum_{l=1}^{2}(-\beta_n^2\overline{G}_0\widetilde{C}_{lm} + \overline{E}_0\widetilde{C}_{lm}^{(2)})B_{ln}^{(1)} + \sum_{l=1}^{2}(-\alpha_m^2\overline{E}_0\widetilde{J}_{ln} + \overline{G}_0\widetilde{J}_{ln}^{(2)})C_{lm}^{(1)} + \right. \\
& \sum_{i=0}^{M}\sum_{j=0}^{N}\alpha_i\beta_j(\overline{D}_0 + \overline{G}_0)\widetilde{S}_{mn}^{(ij)}A_{ij}^{(2)} + \sum_{j=0}^{N}\sum_{l=1}^{2}-\beta_j(\overline{D}_0 + \overline{G}_0)\widetilde{C}_{lm}^{(1)}\widetilde{S}_{yn}^{(j)}B_{lj}^{(2)} + \\
& \sum_{i=0}^{M}\sum_{l=1}^{2}-\alpha_i(\overline{D}_0 + \overline{G}_0)\widetilde{J}_{ln}^{(1)}\widetilde{S}_{xn}^{(i)}C_{li}^{(2)} + (-\alpha_m^2\overline{E}_1 - \beta_n^2\overline{G}_1)A_{mn}^{(4)} + \sum_{l=1}^{2}(\overline{E}_1\widetilde{C}_{lm}^{(2)} - \beta_n^2\overline{G}_1\widetilde{C}_{lm})B_{ln}^{(4)} + \\
& \sum_{l=1}^{2}(-\alpha_m^2\overline{E}_1\widetilde{J}_{ln} + \overline{G}_1\widetilde{J}_{ln}^{(2)})C_{lm}^{(4)} + \sum_{i=0}^{M}\sum_{j=0}^{N}\alpha_i\beta_j(\overline{D}_1 + \overline{G}_1)\widetilde{S}_{mn}^{(ij)}A_{ij}^{(5)} + \\
& \left. \sum_{j=0}^{N}\sum_{l=1}^{2}-\beta_j(\overline{D}_1 + \overline{G}_1)\widetilde{C}_{lm}^{(1)}\widetilde{S}_{yn}^{(j)}B_{lj}^{(5)} + \sum_{i=0}^{M}\sum_{l=1}^{2}-\alpha_i(\overline{D}_1 + \overline{G}_1)\widetilde{J}_{ln}^{(1)}\widetilde{S}_{xn}^{(i)}C_{li}^{(5)} \right\}\cos\alpha_m x \cos\beta_n y = \\
& -\omega^2 \sum_{m=0}^{M}\sum_{n=0}^{N}\left\{I_0 A_{mn}^{(1)} + I_1 A_{mn}^{(4)} + \sum_{l=1}^{2}\widetilde{C}_{lm}(I_0 B_{ln}^{(1)} + I_1 B_{ln}^{(4)}) + \sum_{l=1}^{2}\widetilde{J}_{ln}(I_0 C_{lm}^{(1)} + I_1 C_{lm}^{(4)})\right\}\cos\alpha_m x \cos\beta_n y.
\end{aligned}
\tag{33}
$$

In establishing Equation (33), the following functions are expressed in terms of the Fourier cosine series to factor out $\cos\alpha_m x \cos\beta_n y$:

$$
\begin{aligned}
& \xi_l(x) = \sum_{m=0}^{M}\widetilde{C}_{lm}\cos\alpha_m x, \quad \xi_{l,x}(x) = \sum_{m=0}^{M}\widetilde{C}_{lm}^{(1)}\cos\alpha_m x, \quad \xi_{l,xx}(x) = \sum_{m=0}^{M}\widetilde{C}_{lm}^{(2)}\cos\alpha_m x \\
& \eta_l(y) = \sum_{n=0}^{N}\widetilde{J}_{ln}\cos\beta_n y, \quad \eta_{l,y}(y) = \sum_{n=0}^{N}\widetilde{J}_{ln}^{(1)}\cos\beta_n y, \quad \eta_{l,yy}(y) = \sum_{n=0}^{N}\widetilde{J}_{ln}^{(2)}\cos\beta_n y, \\
& \sin\alpha_i x = \sum_{m=0}^{M}\widetilde{S}_{xm}^{(i)}\cos\alpha_m x, \quad \sin\beta_j y = \sum_{n=0}^{N}\widetilde{S}_{yn}^{(j)}\cos\beta_n y, \\
& \sin\alpha_i x \sin\beta_j y = \sum_{m=0}^{M}\sum_{n=0}^{N}\widetilde{S}_{mn}^{(ij)}\cos\alpha_m x \cos\beta_n y.
\end{aligned}
\tag{34}
$$

Similarly, one can establish $5(M+1)(N+1)$ linear algebraic homogeneous equations for $A_{mn}^{(i)}$, $B_{ln}^{(i)}$ and $C_{lm}^{(i)}$ from the governing equations (Equations (13)–(17)). Such set of equations can be further expressed in the following matrix form:

$$(\hat{\mathbf{K}}\mathbf{A} + \widetilde{\mathbf{K}}\mathbf{P}) - \omega^2(\hat{\mathbf{M}}\mathbf{A} + \widetilde{\mathbf{M}}\mathbf{P}) = 0 \tag{35}$$

Equation (32) gives

$$\mathbf{P} = (\mathbf{B}_p^{-1}\mathbf{B}_A)\mathbf{A} = \mathbf{\Gamma}\mathbf{A} \tag{36}$$

Substituting Equation (36) into Equation (35) yields

$$(\hat{\mathbf{K}} + \widetilde{\mathbf{K}}\mathbf{\Gamma})\mathbf{A} = \omega^2(\hat{\mathbf{M}} + \widetilde{\mathbf{M}}\mathbf{\Gamma})\mathbf{A} \tag{37}$$

which forms an eigenvalue problem.

3. Convergence Studies and Comparisons

The boundary conditions for a rectangular plate at the edges $x = 0$, $y = 0$, $x = a$ and $y = b$ are specified by four letters in a respective series. For example, CSFF boundary conditions mean a clamped boundary condition at $x = 0$, a simply supported boundary condition at $y = 0$ and a free boundary condition at $x = a$ and $y = b$. To validate the proposed solutions, comprehensive convergence studies were carried out for the nondimensional vibration frequencies Ω ($=\omega(a^2/h)\sqrt{\rho_c/E_c}$, where the subscript "c" indicates ceramic material properties) of the first six modes of square plates with $h/b = 0.1$ and with SSSS, SCSC, CFFF and FFFF boundary conditions. The obtained results were compared with the published results from the literature. In the following, the solution terms M and N in Equations (19)–(23) are set equal, and $\overline{M}$ $(= M + 1)$ mainly equals 5, 10, 15, 25, 30 and 35 in the convergence studies. An Al/Al$_2$O$_3$ FGM, whose material properties are described by Equation (1) (the power-law model), is mainly considered.

Table 2 presents the convergence studies for Al/Al$_2$O$_3$ and Al/ZrO$_2$ FGM ($\overline{m} = 1$) plates with SSSS boundary conditions. Notably, the material properties of the Al/ZrO$_2$ FGM are determined by the Mori–Tanaka scheme. In the following tables, the mode denoted by "*" is the in-plane displacement-dominated mode. The results determined from simple exact closed-form solutions (see Appendix A) are given, and the superscripts "(m, n)" denote the wave numbers in the x-direction and y-direction, respectively. Some published results based on the Mindlin plate theory are also given in Table 2 for comparison. The published results presented in the table include (1) the results provided by Hosseini-Hashemi et al. [23], who proposed exact analytical solutions for FGM rectangular plates having two simply supported opposite edges; (2) the numerical results of Zhao et al. [10], who obtained solutions by using an element-free kp-Ritz method with shape functions constructed based on the kernel particle concept; (3) the numerical results of Ferreira et al. [12], who used the global collocation method with multi-quadric radial basis functions.

Table 2. Convergence of $\Omega = \omega(a^2/h)\sqrt{\rho_c/E_c}$ for SSSS FGM square plates with $h/b = 0.1$ and $\overline{m} = 1$.

Material Model	Material Ingredient	Mode	$\overline{M}=M+1$						Exact Closed-Form Sol.	Published
			5	10	15	25	30	35		
Power-Law	Al/Al$_2$O$_3$	1	4.510	4.433	4.422	4.419	4.418	4.418	4.419 $^{(1,1)}$	<4.420> (4.347)
		2	11.03	10.63	10.60	10.60	10.58	10.58	10.59 $^{(1,2)}$	<10.59> (10.42)
		3	11.03	10.63	10.60	10.56	10.58	10.58	10.59 $^{(2,1)}$	</> (10.42)
		4*	16.22	16.20	16.20	16.20	16.20	16.20	16.20 $^{(1,0)}$	<x> (15.94)
		5*	16.22	16.20	16.20	16.20	16.20	16.20	16.20 $^{(0,1)}$	<x> (/)
		6	16.90	16.34	16.31	16.30	16.30	16.30	16.31 $^{(2,2)}$	<16.31> (/)
M-T	Al/ZrO$_2$	1	5.288	5.205	5.193	5.190	5.190	5.190	5.192 $^{(1,1)}$	{5.096}
		2	12.90	12.45	12.42	12.41	12.41	12.41	12.41 $^{(1,2)}$	{12.30}
		3	12.90	12.45	12.42	12.41	12.41	12.41	12.41 $^{(2,1)}$	{12.30}
		4*	18.10	18.09	18.08	18.08	18.08	18.08	18.08 $^{(1,0)}$	{17.49}
		5*	18.10	18.09	18.08	18.08	18.08	18.08	18.08 $^{(0,1)}$	{17.49}
		6	19.74	19.12	19.08	19.07	19.06	19.06	19.09 $^{(2,2)}$	{18.87}

Note: 'x' denotes data missed; '/' denotes data not available; < > denotes results of Hosseini-Hashemi et al. [23]; ( ) denotes results of Zhao et al. [10]; { } denotes results of Ferreira et al. [12]; "*" denotes the in-plane displacement-dominated model; the superscripts "( )" denote the wave numbers in the x-direction and y-direction.

The present results converge from the upper-bounds of solutions as the number of solution terms increases. The results obtained using $\overline{M} = 15$ show the agreement of three significant figures with the results determined from the simple exact closed-form solutions given in Appendix A, and the differences are less than 0.1%. It is interesting to observe that although the solutions of Hosseini-Hashemi et al. [23] provided accurate results for out-of-plane displacement-dominated modes, they did not provide the results for a wave number of zero in the x-direction or y-direction, which corresponded to the in-plane displacement-dominated modes. The first six modal shapes are depicted in Figure 3. In this figure, the contours of out-of-plane displacement (W_0) (represented by solid lines) and the nodal lines (represented by dashed lines) are depicted for the out-of-plane flexural modes. Moreover, the in-plane modal deformations are displayed for the in-plane displacement-dominated modes. Compared with the results obtained from the exact closed-form solutions, the results of Zhao et al. [10] exhibited a difference of approximately 1.6% and the results of Ferreira et al. [12] exhibited a difference between 0.9% and 3.3%. Consequently, the present results obtained using $\overline{M} = 15$ are more accurate than those of Zhao et al. [10] and Ferreira et al. [12].

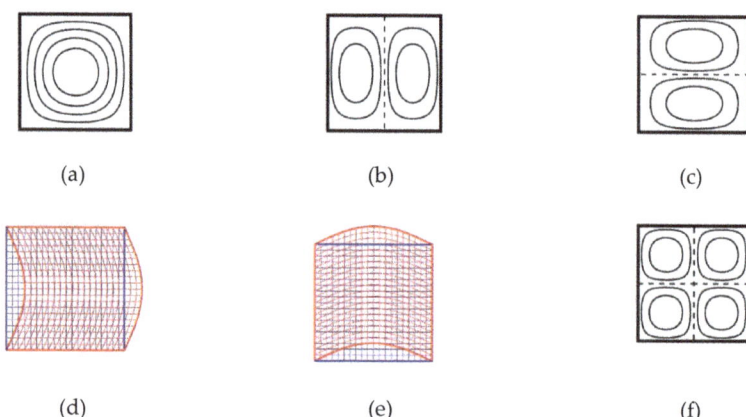

Figure 3. Mode shapes for an FGM square plate with SSSS boundary conditions: (a) mode 1; (b) mode 2; (c) mode 3; (d) mode 4; (e) mode 5; (f) mode 6.

Table 3 lists the non-dimensional natural frequencies Ω obtained using $\overline{M}$ equal to 5, 10, 15, 25, 30 and 35 for FGM ($\overline{m} = 0$ and 1) square plates with SCSC boundary conditions. As $\overline{M}$ increases, the natural frequencies converge from the upper-bounds of solutions. The results obtained using $\overline{M} = 15$ and 35 show the consistency of three significant figures. The present results for the homogeneous plate ($\overline{m} = 0$) obtained using $\overline{M} = 35$ show excellent agreement of four significant figures with those of Liew et al. [35] and Du et al. [36], who investigated the vibrations of homogeneous rectangular plates. Liew et al. [34] determined the frequencies of out-of-plane modes via the conventional Ritz method with polynomial admissible functions, while Du et al. [36] established series solutions for the in-plane vibrations of rectangular plates. The present results for out-of-plane displacement-dominated modes obtained using $\overline{M} \geq 25$ for the FGM ($\overline{m} = 0$ and 1) square plates are consistent with the results of Hosseini-Hashemi et al. [23] for at least three significant figures. Again, the solutions of Hosseini-Hashemi et al. [23] neglected the in-plane mode for a wave number of zero in the x-direction.

Table 4 presents the convergence of the non-dimensional natural frequencies Ω of cantilevered FGM ($\overline{m} = 0$ and 5) square plates with various numbers of solution terms. Interestingly, the results indicate different convergence trends from those observed from Tables 1 and 2. The values of Ω for the homogeneous plate ($\overline{m} = 0$) reveal convergence from the upper-bounds of solutions for the first and third modes, convergence from the lower-bounds of solutions for the second, fourth and sixth modes, and oscillatory convergence for the fifth mode. These findings are not applied to the results for the FGM plate with $\overline{m} = 5$. For example, oscillatory convergence is observed for the first and fifth modes. The present results obtained using $\overline{M} \geq 25$ exhibit excellent agreement with those of Liew et al. [35] with the differences less than 0.1%. The differences between the present convergent results and those of Zhao et al. [10] can be larger than 1%.

Table 3. Convergence of Ω for SCSC FGM square plates with $h/b = 0.1$.

$\overline{m}$	Mode	$\overline{M}=M+1$						Published
		5	10	15	25	30	35	
0	1	8.183	8.079	8.073	8.071	8.071	8.070	{8.070} <8.070>
	2	15.37	14.91	14.88	14.87	14.86	14.86	{14.86} <14.86>
	3	18.25	17.95	17.93	17.92	17.92	17.92	{17.92} <17.92>
	4 *	19.50	19.49	19.48	19.48	19.48	19.48	[19.48] < × >
	5	24.49	23.91	23.87	23.85	23.85	23.85	{23.85} <23.85>
	6	27.72	26.44	26.32	26.29	26.29	26.28	{26.28} </>
1	1	6.320	6.228	6.223	6.221	6.221	6.221	<6.220>
	2	11.89	11.51	11.48	11.47	11.47	11.47	<11.47>
	3	14.17	13.94	13.92	13.91	13.91	13.91	<13.92>
	4 *	16.22	16.20	16.20	16.20	16.20	16.20	<x>
	5	19.05	18.57	18.54	18.53	18.53	18.53	<18.54>
	6	21.62	20.48	20.38	20.35	20.35	20.35	</>

Note: 'x' denotes data missed; '/' denotes data not available; { } denotes results of Liew et al. [35]; [] denotes results of Du et al. [36]; < > denotes results of Hosseini-Hashemi et al. [23]; "*" denotes the in-plane displacement-dominated mode.

Table 4. Convergence of Ω for CFFF FGM square plates with $h/b = 0.1$.

$\overline{m}$	Mode	$\overline{M}=M+1$						Published
		5	10	15	25	30	35	
0	1	1.039	1.038	1.038	1.038	1.038	1.038	{1.038} (1.030)
	2	2.399	2.428	2.435	2.438	2.439	2.439	{2.440} (2.391)
	3	6.134	6.082	6.079	6.079	6.079	6.079	{6.080} (6.005)
	4 *	6.548	6.576	6.578	6.580	6.581	6.581	{/} (7.636)
	5	7.742	7.702	7.712	7.715	7.716	7.716	{7.716} (/)
	6	8.417	8.518	8.533	8.544	8.545	8.546	{8.548} (/)
5	1	0.6833	0.6826	0.6827	0.6828	0.6828	0.6828	(0.6768)
	2	1.575	1.594	1.599	1.601	1.601	1.601	(1.568)
	3	4.017	3.983	3.981	3.981	3.981	3.981	(3.927)
	4 *	4.253	4.272	4.273	4.274	4.274	4.275	(4.263)
	5	5.065	5.039	5.045	5.047	5.047	5.047	(/)
	6	5.510	5.577	5.586	5.593	5.594	5.594	(/)

Note: '/' denotes data not available; { } denotes results of Liew et al. [35]; ( ) denotes results of Zhao et al. [10]; "*" denotes the in-plane displacement-dominated mode.

Similar to Table 4, Table 5 considers the plates with FFFF boundary conditions. Notably, using supplementary functions given in Equation (28) yields singular $\mathbf{B}_p$ in Equation (32), and its inverse cannot be found for Equation (36). To overcome such numerical difficulties, in addition to the conditions presented in Equation (27), the following conditions are proposed to establish the polynomial supplementary functions:

$$\xi_i(0) = \xi_i(a) = \eta_i(0) = \eta_i(b) = 0 \text{ (for } i = 1 \text{ and 2)}. \tag{38}$$

Satisfying Equations (27) and (38) yields

$$\xi_1(x) = -\frac{5x^4}{2a^3} + \frac{6x^3}{a^2} - \frac{9x^2}{2a} + x, \; \xi_2(x) = \frac{5x^4}{2a^3} - \frac{4x^3}{a^2} + \frac{3x^2}{2a},$$
$$\eta_1(y) = -\frac{5y^4}{2b^3} + \frac{6y^3}{b^2} - \frac{9y^2}{2b} + y, \; \eta_2(y) = \frac{5y^4}{2b^3} - \frac{4y^3}{b^2} + \frac{3y^2}{2b} \tag{39}$$

Table 5 lists the results obtained using $\overline{M}$ = 5, 15, 25, 35, 40 and 45. Notably, six rigid body modes with zero frequencies are not considered in the table. The convergence of the numerical results is slower than that of the results presented in Tables 2–4. When the results of the homogenous plate ($\overline{m} = 0$) are under consideration, oscillatory convergence is found for the third to sixth modes, while convergence from lower-bounds and upper-bounds is observed for the first and second modes, respectively. The results obtained using $\overline{m} \geq 35$ are consistent with those of Liew et al. [35] with the differences less than 0.1%.

Table 5. Convergence of Ω for FFFF FGM square plates with $h/b = 0.1$

$\overline{m}$	Mode	$\overline{M}=\overline{N}$						Published
		5	15	25	35	40	45	
0	1	3.823	3.842	3.846	3.847	3.849	3.849	{3.849}
	2	6.921	5.794	5.745	5.737	5.736	5.736	{5.733}
	3	7.821	7.091	7.064	7.059	7.060	7.060	{7.058}
	4	10.08	9.665	9.656	9.655	9.660	9.660	{9.660}
	5	10.08	9.665	9.656	9.655	9.660	9.660	{9.660}
	6	16.93	16.76	16.74	16.74	16.75	16.75	{16.75}
5	1	2.508	2.521	2.523	2.524	2.524	2.524	(2.512)
	2	4.516	3.790	3.759	3.753	3.752	3.752	(3.746)
	3	5.111	4.640	4.623	4.620	4.620	4.619	(4.608)
	4	6.579	6.314	6.309	6.308	6.308	6.308	(6.270)
	5	6.579	6.314	6.309	6.308	6.308	6.308	(6.270)
	6	11.03	10.92	10.91	10.91	10.91	10.91	(/)

Note: '/' denotes data not available, { } denotes results of Liew et al. [35]; ( ) denotes results of Huang et al. [20].

The convergence behaviors of the results for the FGM plate with $\overline{m}$ = 5 and FFFF boundary conditions are different from those for the homogeneous plate. The convergence of natural frequencies for different modes is monotonic from the upper- or lower-bounds. The results obtained using $\overline{M} \geq 35$ are in good agreement with those obtained from the Ritz method based on three-dimensional elasticity [20] with the differences less than 0.6%.

To simply demonstrate the convergence rates of the present solutions for other combinations of boundary conditions, Table 6 shows the average relative differences in the Ω values of the first six modes obtained using $\overline{M} = \overline{N} = 15$ and $\overline{M} = \overline{N} = 35$ for Al/Al$_2$O$_3$ FGM square plates with $h/a = 0.1$ and $\overline{m}$ = 5 at 16 combinations of boundary conditions. All the differences are less than 0.1%, which indicates that the solutions derived in this study provide accurate results even when $\overline{M} = \overline{N} = 15$ is used. Differences larger than 0.08% occur in the results for SFSF and SFFF boundary conditions, while the differences are less than 0.03% when CSSF, CSCF, CCSF and CCCF boundary conditions are under consideration.

Table 6. Average differences of $\left|\frac{\Omega(\overline{M}=\overline{N}=35)-\Omega(\overline{M}=\overline{N}=15)}{\Omega(\overline{M}=\overline{N}=35)}\right|$ of the first six modes.

Case	SFSF	SSSF	SCSF	SCSS	SFFF	SSFF	CSFF	CSSF
Ave. Differences (%)	0.080	0.045	0.040	0.054	0.088	0.045	0.056	0.030
Case	CFSF	CFCF	CSCF	CCFF	CCSF	CCSS	CCCF	CCCS
Ave. Differences (%)	0.056	0.044	0.020	0.055	0.028	0.045	0.024	0.049

4. Numerical Results

After validating the proposed analytical solutions through convergence studies, we employed the solutions to determine the first six nondimensional frequencies (Ω) of Al/Al$_2$O FGM plates with various aspect ratios ($b/a = 1$ and 2), thickness-to-length ratios ($h/a = 0.02$ and 0.1), power-law index values ($\overline{m} = 0$, 0.5, 2 and 5), and combinations of boundary conditions (CCCC, FFFF, CFFF, CFSF, SSFF, CSFF, CSSF, CCFF and CCCF). The results are summarized in Tables 7–10, in which "*" denotes the in-plane displacement-dominated modes. Figures 4 and 5 depict the variations in Ω with $\overline{m}$ for Al/Al$_2$O FGM rectangular plates ($b/a = 2$ and $h/a = 0.1$) with CFFF and CFSF boundary conditions, respectively. These results were obtained using $\overline{M} = \overline{N} = 35$, except for FFFF plates, whose natural frequencies were determined by using $\overline{M} = \overline{N} = 45$. Since exact closed-form solutions exist for plates with two simply supported opposite edges, such boundary conditions are not under consideration in this section. These tabulated results can serve as benchmark data for evaluating numerical approaches.

Table 7. Nondimensional natural frequencies Ω of CCCC Al/Al$_2$O FGM rectangular plates.

b/a	h/a	$\overline{m}$	Mode					
			1	2	3	4	5	6
1	0.02	0	10.84	22.03	22.03	32.36	39.29	39.49
		0.5	9.184	18.67	18.67	27.44	33.32	33.48
		2	7.527	15.30	15.30	22.49	27.30	27.44
		5	7.133	14.49	14.49	21.29	25.84	25.97
	0.1	0	9.842	18.77	18.77	26.31	31.00	31.30
		0.5	8.409	16.11	16.11	22.64	26.73	26.98
		2	6.902	13.23	13.23	18.58	21.94	22.15
		5	6.451	12.27	12.27	17.15	20.18	20.38
2	0.02	0	7.413	9.593	13.48	19.05	19.23	21.34
		0.5	6.281	8.128	11.43	16.14	16.29	18.09
		2	5.148	6.662	9.363	13.23	13.35	14.83
		5	4.879	6.313	8.872	12.53	12.65	14.04
	0.1	0	6.897	8.815	12.16	16.64	16.75	18.30
		0.5	5.882	7.523	10.39	14.27	14.33	15.70
		2	4.827	6.173	8.521	11.71	11.75	12.88
		5	4.526	5.779	7.960	10.88	10.95	11.95

Table 8. Nondimensional natural frequencies Ω of FFFF Al/Al$_2$O FGM rectangular plates.

b/a	h/a	$\overline{m}$	Mode					
			1	2	3	4	5	6
1	0.02	0	4.038	5.976	7.361	10.44	10.44	18.41
		0.5	3.421	5.067	6.238	8.849	8.849	15.59
		2	2.804	4.153	5.112	7.252	7.252	12.78
		5	2.657	3.932	4.843	6.869	6.869	12.11
	0.1	0	3.849	5.735	7.060	9.660	9.660	16.75
		0.5	3.269	4.861	5.988	8.213	8.213	14.25
		2	2.677	3.969	4.893	6.711	6.711	11.63
		5	2.525	3.752	4.621	6.312	6.312	10.91

Table 8. Cont.

b/a	h/a	$\bar{m}$	Mode 1	2	3	4	5	6
2	0.02	0	1.656	1.997	4.389	4.511	6.690	7.604
		0.5	1.405	1.693	3.719	3.821	5.670	6.442
		2	1.152	1.387	3.048	3.132	4.647	5.280
		5	1.090	1.315	2.888	2.969	4.401	5.004
	0.1	0	1.610	1.927	4.196	4.382	6.419	7.176
		0.5	1.364	1.636	3.563	3.717	5.443	6.099
		2	1.117	1.341	2.918	3.042	4.446	4.991
		5	1.058	1.267	2.753	2.875	4.201	4.700

Table 9. Nondimensional natural frequencies Ω of CFFF and CFSF FGM square plates; "*" denotes the in-plane displacement-dominated mode.

BC	h/a	$\bar{m}$	Material Model	Mode 1	2	3	4	5	6
CFFF	0.02	0		1.049	2.552	6.418	8.180	9.273	16.17
		0.5	Power-Law	0.8888	2.162	5.437	6.930	7.857	13.70
		2		0.7284	1.772	4.456	5.678	6.439	11.23
		5		0.6907	1.680	4.224	5.383	6.102	10.63
	0.1	0	Power-Law or M-T	1.038	2.439	6.079	6.581 *	7.716	8.546
		0.5	Power-Law	0.8000	2.072	5.168	5.907 *	6.555	7.280
			M-T	0.8089	1.960	4.879	5.606 *	6.186	6.877
		2	Power-Law	0.7211	1.698	4.230	4.946 *	5.359	5.962
			M-T	0.6973	1.643	4.087	4.650 *	5.176	5.759
		5	Power-Law	0.6828	1.601	3.981	4.275 *	5.047	5.594
			M-T	0.6666	1.564	3.886	4.073 *	4.926	5.457
CSFF	0.02	0		4.591	6.191	11.91	14.90	16.90	23.15
		0.5		3.664	4.950	9.511	11.89	13.50	18.48
		2		3.084	4.162	8.000	10.00	11.36	15.55
		5	Power-Law	2.950	3.981	7.652	9.569	10.86	14.87
	0.1	0		4.401	5.820	10.89	13.45	15.05	15.57 *
		0.5		3.526	4.678	8.742	10.83	12.14	13.27 *
		2		2.963	3.922	7.316	9.071	10.16	10.99 *
		5		2.818	3.724	6.935	8.563	9.569	9.641 *

Table 10. Nondimensional natural frequencies Ω of FGM square plates ($h/a = 0.1$) with SSFF, CSFF, CSSF, CCFF and CCCF boundary conditions; "*" denotes the in-plane displacement-dominated mode.

BC	$\bar{m}$	Mode 1	2	3	4	5	6
SSFF	0	0.9943	5.011	5.600	10.63	12.16 *	14.10
	0.5	0.8432	4.254	4.754	9.044	10.92 *	12.00
	2	0.6910	3.481	3.891	7.398	9.134 *	9.806
	5	0.6538	3.286	3.672	6.951	7.896 *	9.204
CSFF	0	1.571	5.472	6.977	8.176 *	11.71	14.51
	0.5	1.333	4.648	5.936	7.340	9.984	12.35
	2	1.093	3.804	4.860	6.148 *	8.171	10.10
	5	1.033	3.585	4.569	5.313 *	7.653	9.470
CSSF	0	4.837	8.710	13.92	16.36 *	17.23	17.83
	0.5	4.113	7.416	11.89	14.63 *	14.74	15.25
	2	3.372	6.072	9.747	11.98	12.31 *	12.49
	5	3.175	5.703	9.106	10.60 *	11.23	11.64

Table 10. *Cont.*

BC	$\overline{m}$	Mode					
		1	2	3	4	5	6
CCFF	0	2.019	6.700	7.481	12.71	15.41 *	16.62
	0.5	1.714	5.703	6.369	10.85	13.80 *	14.22
	2	1.405	4.668	5.217	8.883	11.42 *	11.78
	5	1.327	4.385	4.900	8.300	9.983 *	10.87
CCCF	0	6.685	10.78	16.41	19.75	20.31	23.79 *
	0.5	5.701	9.207	14.01	16.90	17.43	21.36 *
	2	4.678	7.546	11.55	13.83	14.29	17.88 *
	5	4.385	7.050	10.72	12.85	13.25	15.46 *

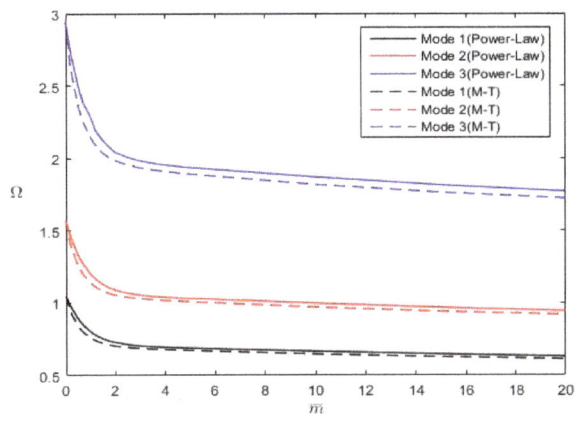

(a)

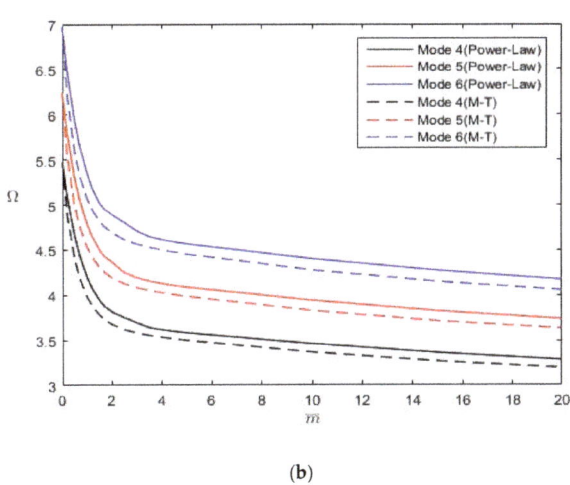

(b)

Figure 4. Variations of Ω with $\overline{m}$ for a CFFF FGM rectangular plate ($b/a = 2$, $h/a = 0.1$): (**a**) for modes 1–3; (**b**) for modes 4–6.

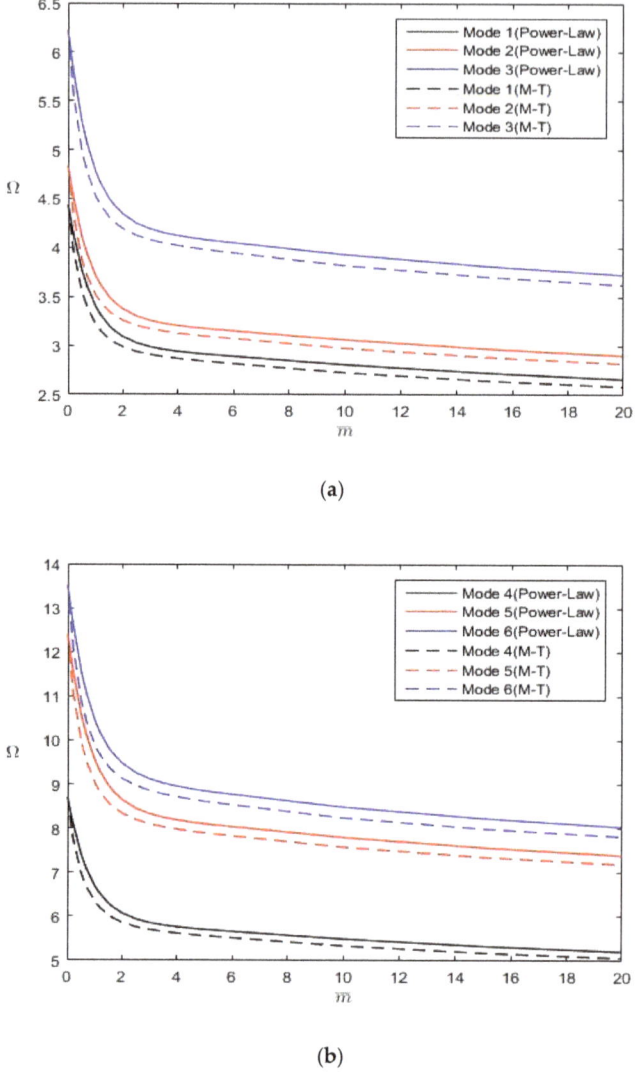

Figure 5. Variations of Ω with $\overline{m}$ for a CFSF FGM rectangular plate ($b/a = 2$, $h/a = 0.1$): (**a**) for modes 1–3; (**b**) for modes 4–6.

The following inferences are drawn from Tables 7–10 and Figures 4 and 5:

1. The constraint increases when a free boundary condition changes to a simply supported boundary condition. The constraint further increases in a clamped boundary condition. Higher constraint results in higher plate stiffness and larger natural frequencies. Therefore, $\Omega_{CCCC} > \Omega_{CSSF} > \Omega_{CSFF} > \Omega_{SSFF}$ and $\Omega_{CCCC} > \Omega_{CCCF} > \Omega_{CCFF} > \Omega_{CSFF} > \Omega_{CFFF} > \Omega_{FFFF}$ (where the subscripts indicate the boundary conditions) if the first six rigid body modes with zero frequencies are considered for plates with FFFF boundary conditions.

2. The Mori–Tanaka material model provides a larger Young's modulus than the power-law material model does; however, both models yield the same density distribution (Figure 2). Consequently,

FGM plates following the Mori-Tanaka material model have larger natural frequencies than those following the power-law material model.
3. An increase in $\overline{m}$ results in a decrease in Ω. Furthermore, as displayed in Figures 4 and 5, the change rate of Ω with $\overline{m}$ gradually decreases with an increase in $\overline{m}$. Notably, an increase in $\overline{m}$ leads to a decrease in the plate stiffness and mass (Figure 2).
4. No in-plane displacement-dominated mode exists in the first six modes for thin square plates with $h/a = 0.02$; however, such a mode may exist for moderately thick plates with $h/a = 0.1$.
5. The nondimensional frequencies (Ω) of plates with $h/a = 0.1$ are less than those of plates with $h/a = 0.02$ because h/a is involved in the definition of Ω. When converting Ω to ω, one finds that the trend is opposite for ω because the plate rigidity increases with h/a.

5. Concluding Remarks

In this study, analytical solutions based on the Mindlin plate theory were developed for the vibrations of FGM rectangular plates with various combinations of boundary conditions. The solutions were established using the Fourier cosine series with polynomial supplementary functions. Fourth-order polynomial supplementary functions were adopted in the solutions for FFFF boundary conditions, and second-order polynomial supplementary functions were adopted in the solutions for the other boundary conditions. The present solutions were validated through comprehensive convergence studies as well as comparisons with published results and the exact closed-form solutions for plates with SSSS boundary conditions. When increasing the number of solution terms, the trends of convergence in vibration frequencies (i.e., monotonous convergence from upper-bounds or lower-bounds or convergence in an oscillatory manner) varied according to the vibration modes, distributions of material properties, and boundary conditions. The vibration frequencies of the first six modes obtained using $\overline{M} = \overline{N} = 15$ were in good agreement with those obtained using $\overline{M} = \overline{N} = 35$, and the average differences were less than 0.1% for FGM square plates with $h/a = 0.1$ and $\overline{m} = 5$ under 17 combinations of boundary conditions, excluding FFFF boundary conditions. The average difference was about 0.6% for FFFF boundary conditions.

The present solutions were also applied to determine the vibration frequencies of Al/Al$_2$O FGM plates with CCCC, FFFF, CFFF, CFSF, SSFF, CSFF, CSSF, CCFF and CCCF boundary conditions. The effects of the plate thickness, material model (Mori–Tanaka and power-law models), and power-law index, $\overline{m}$ on the vibration frequencies were investigated. With a fixed $\overline{m}$, the Mori–Tanaka model yielded higher plate stiffness and larger vibration frequencies for plates than the power-law model. An increase in $\overline{m}$ caused a decrease in the vibration frequencies of Al/Al$_2$O FGM plates. The tabulated data in this study can be used as a standard to judge the accuracy of numerical methods. The present solutions can be simply modified to determine the buckling loads of an FGM rectangular plate under uniform initial stresses and to perform linear static and dynamic analyses of an FGM rectangular plate.

Author Contributions: C.-S.H. initiated the idea, developed the solutions and wrote the manuscript, while S.H.H. wrote parts of the computer codes and prepared the numerical results and figures. All authors have read and agreed to the published version of the manuscript.

Funding: This research was funded by the Ministry of Science and Technology, Taiwan through research grant no. MOST 108-2221-E-009-005.

Acknowledgments: This work reported herein was supported by the Ministry of Science and Technology, Taiwan through research grant no. MOST 108-2221-E-009-005. This support is gratefully acknowledged.

Conflicts of Interest: The authors declare no conflict of interest

Appendix A

The exact closed-form solutions for vibrations of SSSS FGM rectangular plates are given herein. To satisfy SSSS boundary conditions, let

$$u_0(x,y,t) = \sum_{\overline{m}=0}^{\infty} \sum_{\overline{n}=0}^{\infty} a_{\overline{mn}} \cos(\tfrac{\overline{m}\pi}{a}x) \sin(\tfrac{\overline{n}\pi}{b}y)e^{i\omega t},$$

$$v_0(x,y,t) = \sum_{\overline{m}=0}^{\infty} \sum_{\overline{n}=0}^{\infty} b_{\overline{mn}} \sin(\tfrac{\overline{m}\pi}{a}x) \cos(\tfrac{\overline{n}\pi}{b}y)e^{i\omega t},$$

$$w_0(x,y,t) = \sum_{\overline{m}=0}^{\infty} \sum_{\overline{n}=0}^{\infty} c_{\overline{mn}} \sin(\tfrac{\overline{m}\pi}{a}x) \sin(\tfrac{\overline{n}\pi}{b}y)e^{i\omega t},$$

$$\psi_x(x,y,t) = \sum_{\overline{m}=0}^{\infty} \sum_{\overline{n}=0}^{\infty} d_{\overline{mn}} \cos(\tfrac{\overline{m}\pi}{a}x) \sin(\tfrac{\overline{n}\pi}{b}y)e^{i\omega t}$$

$$\psi_y(x,y,t) = \sum_{\overline{m}=0}^{\infty} \sum_{\overline{n}=0}^{\infty} e_{\overline{mn}} \sin(\tfrac{\overline{m}\pi}{a}x) \cos(\tfrac{\overline{n}\pi}{b}y)e^{i\omega t}$$

where $a_{\overline{mn}}$, $b_{\overline{mn}}$, $c_{\overline{mn}}$, $d_{\overline{mn}}$ and $e_{\overline{mn}}$ are coefficients to be determined. Substituting are coefficients to be determined. Substituting the above equations into Equations (13)–(17) yields

$$(\mathbf{K} - \omega^2 \mathbf{M}) \begin{Bmatrix} a_{\overline{mn}} \\ b_{\overline{mn}} \\ c_{\overline{mn}} \\ d_{\overline{mn}} \\ e_{\overline{mn}} \end{Bmatrix} = 0$$

where

$$\mathbf{K} = \begin{bmatrix} (\alpha^2 \overline{E}_0 + \beta^2 \overline{G}_0) & \alpha\beta(\overline{D}_0 - \overline{G}_0) & 0 & (\alpha^2 \overline{E}_1 + \beta^2 \overline{G}_1) & \alpha\beta(\overline{D}_1 - \overline{G}_1) \\ \alpha\beta(\overline{D}_0 - \overline{G}_0) & (\beta^2 \overline{E}_0 + \alpha^2 \overline{G}_0) & 0 & \alpha\beta(\overline{D}_1 - \overline{G}_1) & (\beta^2 \overline{E}_1 + \alpha^2 \overline{G}_1) \\ 0 & 0 & (-\alpha^2 - \beta^2)\kappa\overline{G}_0 & \alpha\kappa\overline{G}_0 & \beta\kappa\overline{G}_0 \\ (\alpha^2 \overline{E}_1 + \beta^2 \overline{G}_1) & \alpha\beta(\overline{D}_1 - \overline{G}_1) & \alpha\kappa\overline{G}_0 & (\alpha^2 \overline{E}_2 + \kappa\overline{G}_0 + \beta^2 \overline{G}_2) & \alpha\beta(\overline{D}_2 - \overline{G}_2) \\ \alpha\beta(\overline{D}_1 - \overline{G}_1) & (\beta^2 \overline{E}_1 + \alpha^2 \overline{G}_1) & \beta\kappa\overline{G}_0 & \alpha\beta(\overline{D}_2 - \overline{G}_2) & (\beta^2 \overline{E}_2 + \kappa\overline{G}_0 + \alpha^2 \overline{G}_2) \end{bmatrix},$$

$$\mathbf{M} = \begin{bmatrix} I_0 & 0 & 0 & I_1 & 0 \\ 0 & I_0 & 0 & 0 & I_1 \\ 0 & 0 & I_0 & 0 & 0 \\ I_1 & 0 & 0 & I_2 & 0 \\ 0 & I_1 & 0 & 0 & I_2 \end{bmatrix}, \alpha = \tfrac{\overline{m}\pi}{a} \text{ and } \beta = \tfrac{\overline{n}\pi}{b}.$$

References

1. Niino, M.; Maeda, S. Recent development status of functionally gradient materials. *ISIJ Int.* **1990**, *30*, 699–703. [CrossRef]
2. Jha, D.K.; Kant, T.; Singh, R.K. A critical review of recent research on functionally graded plates. *Compos. Struct.* **2013**, *96*, 833–849. [CrossRef]
3. Gupta, A.; Talha, M. Recent development in modeling and analysis of functionally graded materials and structures. *Prog. Aerosp. Sci.* **2015**, *79*, 1–14. [CrossRef]
4. Swaminathan, K.; Naveenkumar, D.T.; Zenkour, A.M.; Carrera, E. Stress, vibration and buckling analyses of FGM plates—A state-of-the-art review. *Compos. Struct.* **2015**, *120*, 10–31. [CrossRef]
5. Zhang, N.; Khan, T.; Guo, H.; Shi, S.; Zhong, W.; Zhang, W. Functionally graded materials: An overview of stability, buckling, and free vibration analysis. *Adv. Mater. Sci. Eng.* **2019**, *2019*, 1–18. [CrossRef]
6. Liew, K.M.; Pan, Z.; Zhang, L.W. The recent progress of functionally graded CNT reinforced composites and structures. *Sci. China Phys. Mech. Astron.* **2020**, *63*, 234601. [CrossRef]
7. Abrate, S. Functionally graded plates behave like homogeneous plates. *Compos. Part B Eng.* **2008**, *39*, 151–158. [CrossRef]

8. Zhang, D.G.; Zhou, Y.H. A theoretical analysis of FGM thin plates based on physical neutral surface. *Comput. Mater. Sci.* **2008**, *44*, 716–720. [CrossRef]
9. Yang, Y.; Shen, H.S. Dynamic response of initially stressed functional graded rectangular thin plates. *Compos. Struct.* **2001**, *54*, 497–508. [CrossRef]
10. Zhao, X.; Lee, Y.Y.; Liew, K.M. Free vibration analysis of functionally graded plates using the element-free kp-Ritz method. *J. Sound Vib.* **2009**, *319*, 918–939. [CrossRef]
11. Fu, Y.; Yao, J.; Wan, Z.; Zhao, G. Free vibration analysis of moderately thick orthotropic functionally graded plates with general boundary conditions. *Materials* **2018**, *11*, 273–292. [CrossRef] [PubMed]
12. Ferreira, A.J.M.; Batra, R.C.; Roque, C.M.C.; Qian, L.F.; Jorge, R.M.N. Natural frequencies of functionally graded plates by a meshless method. *Compos. Struct.* **2006**, *75*, 593–600. [CrossRef]
13. Huang, C.S.; McGee, O.G.; Chang, M.J. Vibrations of cracked rectangular FGM thick plates. *Compos. Struct.* **2011**, *93*, 1747–1764. [CrossRef]
14. Hong, C.C. GDQ computation for thermal vibration of thick FGM plates by using fully homogeneous equation and TSDT. *Thin-Walled Struct.* **2019**, *135*, 78–88. [CrossRef]
15. Qian, L.F.; Batra, R.C.; Chen, L.M. Static and dynamic deformations of thick functionally graded elastic plates by using higher-order shear and normal deformable plate theory and meshless local Petrov-Galerkin method. *Compos. Part B* **2004**, *35*, 685–697. [CrossRef]
16. Roque, C.M.C.; Ferreira, A.J.M.; Jorge, R.M.N. A radial basis function approach for the free vibration analysis of functionally graded plates using a refined theory. *J. Sound Vib.* **2007**, *300*, 1048–1070. [CrossRef]
17. Uymaz, B.; Aydogdu, M. Three-dimensional vibration analysis of functionally graded plates under various boundary conditions. *J. Reinf. Plast. Compos.* **2007**, *26*, 1847–1863. [CrossRef]
18. Cui, J.; Zhou, T.R.; Ye, R.C.; Gaidai, O.; Li, Z.C.; Tao, S.H. Three-dimensional vibration analysis of a functionally graded sandwich rectangular plate resting on an elastic foundation using a semi-analytical method. *Materials* **2019**, *12*, 3401. [CrossRef]
19. Huang, C.S.; Yang, P.J.; Chang, M.J. Three-dimensional vibrations of functionally graded material cracked rectangular plates with through internal cracks. *Compos. Struct.* **2012**, *94*, 2764–2776. [CrossRef]
20. Huang, C.S.; McGee, O.G.; Wang, K.P. Three-dimensional vibrations of cracked rectangular parallelepipeds of functionally graded material. *Int. J. Mech. Sci.* **2013**, *70*, 1–25. [CrossRef]
21. Burlayenko, V.N.; Sadowski, T.; Dimitrova, S. Three-dimensional free vibration analysis of thermally loaded FGM sandwich plates. *Materials* **2019**, *12*, 2377. [CrossRef] [PubMed]
22. Hosseini-Hashemi, S.; Rokni Damavandi Taher, H.; Akhavan, H.; Omidi, M. Free vibration of functionally graded rectangular plates using first-order shear deformation plate theory. *Appl. Math. Modeling* **2009**, *34*, 1276–1291. [CrossRef]
23. Hosseini-Hashemi, S.; Fadaee, M.; Atashipour, S.R. A new exact analytical approach for free vibration of Reissner–Mindlin functionally graded rectangular plates. *Int. J. Mech. Sci.* **2011**, *53*, 11–22. [CrossRef]
24. Ghashochi-Bargh, H.; Razavi, S. A simple analytical model for free vibration of orthotropic and functionally graded rectangular plates. *Alex. Eng. J.* **2018**, *57*, 595–607. [CrossRef]
25. Hosseini-Hashemi, S.; Fadaee, M.; Atashipour, S.R. Study on the free vibration of thick functionally graded rectangular plates according to a new exact closed-form procedure. *Compos. Struct.* **2011**, *93*, 722–735. [CrossRef]
26. Matsunaga, H. Free vibration and stability of functionally graded plates according to a 2-D higher-order deformation theory. *Compos. Struct.* **2008**, *82*, 499–512. [CrossRef]
27. Sekkal, M.; Fahsi, B.; Tounsi, A.; Mahmoud, S.R. A novel and simple higher order shear deformation theory for stability and vibration of functionally graded sandwich plate. *Steel Compos. Struct.* **2017**, *25*, 389–401.
28. Vel, S.S.; Batra, R.C. Three-dimensional exact solution for the vibration of functionally graded rectangular plates. *J. Sound Vib.* **2004**, *272*, 703–730. [CrossRef]
29. Reddy, J.N.; Cheng, Z.Q. Frequency of functionally graded plates with three-dimensional asymptotic approach. *J. Eng. Mech. ASCE* **2003**, *129*, 896–900. [CrossRef]
30. Huo, R.L.; Liu, W.Q.; Wu, P.; Zhou, D. Analytical solutions for sandwich plates considering permeation effect by 3-D elasticity theory. *Steel Compos. Struct.* **2017**, *25*, 127–139.
31. Tolstov, G.P. *Fourier Series*; Prentice-Hall: Englewood Cli's, NJ, USA, 1965.
32. Mindlin, R.D. Influence of rotary inertia and shear on flexural motions of isotropic, elastic plates. *J. Appl. Mech. ASME* **1951**, *18*, 31–38.

33. Li, W.L. Free vibrations of beams with general boundary conditions. *J. Sound Vib.* **2000**, *237*, 709–725. [CrossRef]
34. Li, W.L.; Zhang, X.; Du, J.; Liu, Z. An exact series solution for the transverse vibration of rectangular plates with general elastic boundary supports. *J. Sound Vib.* **2009**, *321*, 254–269. [CrossRef]
35. Liew, K.M.; Xiang, Y.; Kitipornchai, S. Transverse vibration of thick rectangular plates-I. Comprehensive sets of boundary conditions. *Comput. Struct.* **1993**, *49*, 1–29. [CrossRef]
36. Du, J.; Li, W.L.; Jin, G.; Yang, T.; Liu, Z. An analytical method for the in-plane vibration analysis of rectangular plates with elastically restrained edges. *J. Sound Vib.* **2007**, *306*, 908–927. [CrossRef]

© 2020 by the authors. Licensee MDPI, Basel, Switzerland. This article is an open access article distributed under the terms and conditions of the Creative Commons Attribution (CC BY) license (http://creativecommons.org/licenses/by/4.0/).

Article

Analytical Approach and Numerical Simulation of Reinforced Concrete Beams Strengthened with Different FRCM Systems

Luis Mercedes *, Christian Escrig, Ernest Bernat-Masó and Lluís Gil

Department of Strength of Materials and Structures in Engineering, Universitat Politècnica de Catalunya UPC, ESEIAAT, Building TR45, c/ Colom 11, 08222 Terrassa, Spain; christian.escrig@upc.edu (C.E.); ernest.bernat@upc.edu (E.B.-M.); lluis.gil@upc.edu (L.G.)
* Correspondence: luis.enrique.mercedes@upc.edu

Abstract: Fabric-reinforced cementitious matrices (FRCMs) are a novel composite material for strengthening structures. Fabric contributes to tying cross-sections under tensile stress. The complexity of the interfaces between the fabric and the matrix does not allow having a simple and accurate model that enables practitioners to perform feasible calculations. This work developed an analytical approach and a numerical simulation based on the reduction of FRCMs' strength capabilities under tensile stress states. The concept of effective strength was estimated for different types of fabrics (basalt, carbon, glass, poly p-phenylene benzobisoxazole (PBO), and steel) from experimental evidence. The proposed models calculate the ultimate bending moment for reinforced concrete (RC) structures strengthened with FRCMs. The numerical models performed simulations that reproduced the moment–deflection curves of the different tested beams. Steel fabric showed the highest contribution to strength (78%), while PBO performed the worst (6%). Basalt and carbon showed irregular contributions.

Keywords: concrete beam; bending; cementitious matrix; FRCM; analytical model; numerical model

1. Introduction

Fabric-reinforced cementitious matrices (FRCMs) have been shown to be one of the most promising retrofitting techniques for reinforced concrete elements [1]. The matrix of this composite material avoids using organic products, which provides FRCMs with a better chemical compatibility with inorganic substrates and diminishes several of the drawbacks of resin-based composite materials [2]. These drawbacks include a lack of vapor permeability, poor fire resistance, non-applicability on wet surfaces or at low temperatures, and chemical hazards for workers who apply strengthening solutions.

Cement-based composites used as a flexural strengthening system for reinforced concrete (RC) structures have been studied experimentally by several authors [3–19]. In all of these studies, FRCMs provided an enhancement of the ultimate flexural capacity of the RC element, demonstrating similar performance to fiber-reinforced polymer (FRP) systems in certain cases. Nevertheless, the use of a cementitious matrix presents new complex failure modes, such as complex slip between the matrix and fibers (see D'Ambrisi and Focacci [3]) and matrix–substrate interactions (see Tommaso et al. [20] and Leone et al. [21]), providing non-optimal development of the ultimate tensile capacity of fibers. Kurtz and Balaguru [4] tested carbon fiber layers, as did Wiberg [12] and Toutanji and Deng [13]. Meanwhile, Barton et al. [14] and Sneed et al. [15] used steel fibers. Brückner et al. [16] included bending and shear. Papanicolaou and Papantoniou [17], Ombres [18], Larrinaga [19], and Babaeidarabad et al. [5] developed particular flexural formulations. Si Larbi et al. [6] analyzed hybrid solutions. Elsanadedy et al. [7] included numerical models to deal with textile reinforcement. Pellegrino and D'Antino [8] studied the reinforcement on prestressed beams, as did Gil et al. [9]. Napoli et al. [10] used steel-reinforced polymer/grout (SRP/SRG) systems, and finally, Ebead et al. [11] developed an effective approach for strengthening beams.

Citation: Mercedes, L.; Escrig, C.; Bernat-Masó, E.; Gil, L. Analytical Approach and Numerical Simulation of Reinforced Concrete Beams Strengthened with Different FRCM Systems. *Materials* **2021**, *14*, 1857. https://doi.org/10.3390/ma14081857

Academic Editors: Sanichiro Yoshida and Giovanni Pappalettera

Received: 24 February 2021
Accepted: 31 March 2021
Published: 8 April 2021

Publisher's Note: MDPI stays neutral with regard to jurisdictional claims in published maps and institutional affiliations.

Copyright: © 2021 by the authors. Licensee MDPI, Basel, Switzerland. This article is an open access article distributed under the terms and conditions of the Creative Commons Attribution (CC BY) license (https://creativecommons.org/licenses/by/4.0/).

Regarding analytical studies, little research has proposed models for predicting the contribution of FRCMs to the ultimate bending moment. D'Ambrisi and Focacci [3] developed an analytical model based on the evaluation of the effective strain of FRCMs, which is related to the debonding fiber strain. Concurrently, Ombres [18] evaluated the effectiveness of FRP models for predicting the flexural capacity provided by poly p-phenylene benzobisoxazole (PBO)-FRCMs. This investigation concluded that FRP models can predict the capacity of FRCMs in cases in which premature failure modes of the RC elements are avoided. However, these predictions were not accurate when debonding failures occurred. Finally, Babaeidarabad et al. [5] and Ebead et al. [11] assessed the capacity for prediction of the model included in the FRCM code published at the present time (ACI 549.4R-13 [22]) with different results: While the first authors showed that the predicted flexural strengths were conservative with respect to the experimental results, the second authors concluded that the values obtained from the theoretical formulation satisfactorily predicted the mechanical behavior of carbon- and PBO-FRCMs.

Focusing on the numerical models for FRCMs, there are studies such as the one presented by Donnini et al. [23] that used a variational model to obtain information about the mechanical behavior of FRCM composites, particularly at the interface between the tuft and the matrix. The numerical models presented in this study were able to reproduce the behavior of the FRCMs during debonding tests in the case of fabrics constituted by uncoated fibers. This model also allowed establishing an effective joint length of approximately 200 mm, since the maximum load did not increase as the bond length between the mesh and the matrix increased further.

Complementarily, a study by Grande and Milani [24] presented a numerical model dedicated to investigating the influence of different variables on the FRCMs' strength. This work analyzed the progressive damage of the upper mortar layer (the layer that is not in contact with the element to be strengthened) that affects both the local mechanism of transfer of the shear stresses between the fibers and the upper mortar layer and the overall response of the strengthening system. For this purpose, the authors proposed a simple but effective spring model, where each component of the FRCM system (mortar, fabric, support, and fabric–mortar interface) is modeled along springs with linear or non-linear behavior.

Additionally, one study presented by Sucharda [25] used a nonlinear analysis for detailed failure modeling based on a 3D model and a fracture plastic material model for concrete. This approach made it possible to describe the overall load-bearing capacity, as well as the mechanism of failure and the collapse of the analyzed beams.

Another study, presented by Valikhani et al. [26], designed a numerical simulation to characterize the interfacial properties of concrete substrates and their effect on the bond strength between them and a ultra-high-performance concrete used as a repair material. From all possible modes of failure, the most desirable is that which drives to the breakage of the strengthening material. In this type of failure, the composite material collaborates at its maximum capacity and respects the integrity of the substrate and the bonding material.

This paper provides an analytical methodology to calculate the FRCM strengthening systems acting as flexural RC retrofitting. The proposed model introduces a reduction of the tensile capacity of the fibers to define an effective strength in order to represent complex inner phenomena, including slipping or partial fiber breakage inside the matrix. This approach aims to simplify the calculation methods to make the design of FRCM strengthening solutions easier. To fulfil this aim, an experimental campaign consisting of flexural tests on RC beams strengthened with different FRCM systems was conducted by authors [27]. Furthermore, in order to verify the utility of this effective strength, a numerical model was developed. This numerical model was useful to reproduce the experimental moment–deflection curves from the strengthened RC beam. Hence, the proposed novel simplified method of effectiveness strength was doubly validated by comparison with experimental and numerical solutions.

2. Experimental Campaign, Materials, and Methods

The authors developed an extensive campaign testing 11 full-scale RC beams (see Figure 1) deficient in steel-reinforced flexure, cast with high-strength concrete and flexurally strengthened with different types of cementitious-matrix composite materials. Details of the materials, test setup (four-point flexural test), failures, and results can be found in Escrig et al. [27]. Nevertheless, it is worth noting that strengthening fabrics cover a variety of materials: Basalt (B), carbon (C), glass (G), PBO (P), and steel (S). Additionally, four different additive-modified mortars were used as FRCMs: Mortars P and X1 were designed to be applied to masonry, whereas mortars R and X2 are usually smeared on concrete. Three different batches of concrete were used. In all cases, the concrete was HA/40/F/12/I and the cement used to produce the concrete was CEM 42.5 according to the Spanish concrete designation [28]. The mechanical properties of the concrete and steel data were obtained according to the specifications included in EN 12390-1 [29], EN 12390-3 [30], and EN ISO 15630-1 [31]. The concrete compression strength (fc) of each batch is shown in Table 1, and the tensile strength, yield stress, and Young's modulus of the steel reinforcement were 634 MPa, 517 MPa, and 198 GPa, respectively.

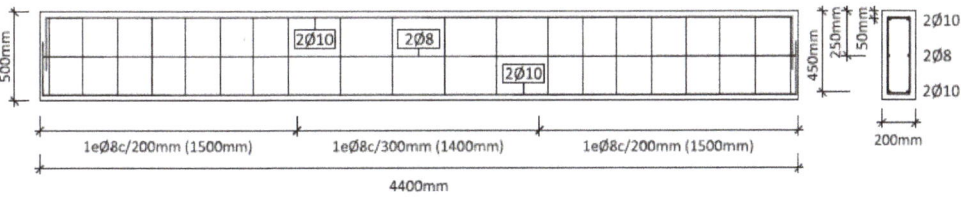

Figure 1. Geometry and steel reinforcement of the beams.

Table 1. Concrete and mortar properties.

Beam State	Concrete Compression Strength (MPa)		Mortar-FRCM Compression Strength (MPa)	
Control beam	42.35	(3%)	-	-
Beam with basalt-FRCM	55.42	(1%)	24.65	(6%)
Beam with carbon-FRCM	42.35	(3%)	24.95	(7%)
Beam with glass-FRCM	46.52	(3%)	35.4	(7%)
Beam with PBO-FRCM	42.35	(3%)	30.02	(7%)
Beam with steel-FRCM	46.52	(3%)	24.65	(6%)

(%) Coefficient of variation. FRCM, Fabric-reinforced cementitious matrix; PBO, poly p-phenylene benzobisoxazole.

The compression strength of the cementitious matrix (experimental test) and the mechanical properties (supplied by the manufacturer) of the fabrics necessary to define the material properties of the analytical and numerical models are summarized in Tables 1 and 2.

Table 2. Fabrics and cementitious matrix properties.

FRCM Properties	Basalt	Carbon	Glass	PBO	Steel
Fabric tensile strength (MPa)	3080	4320	2610	5800	3200
Fabric modulus of elasticity (MPa)	95,000	240,000	90,000	270,000	206,000
Fabric area (mm^2)	10.6	9.4	8.4	9.1	15

The specimens were tested under a four-point flexural test with a free span between supports of 4.00 m in length. The supports were steel cylinders that allowed free rotation in the plane of the beams. The load was applied using a hydraulic actuator of 250 kN, and transferred to the tested specimens through a steel distribution beam with two application points separated 1.40 m. The deflections were obtained using six potentiometers placed symmetrically in pairs on each side of the section. Deflections of the other two analyzed sections, corresponding to the load application points, were controlled by laser position transducers. All data were continuously recorded at a frequency of 50 Hz using a data acquisition system (more information in [27]).

A summary of the experimental results of the bending loading tests are shown in Table 3. This shows the experimental values of the maximum flexural moment ($M_{max,exp}$), the yielding flexural moment ($M_{y,exp}$), and the deflections recorded at the mid-span of the specimens when the maximum flexural moment ($\delta_{max,exp}$) and the yielding flexural moment ($\delta_{y,exp}$) were reached.

Table 3. Experimental results of the bending tests.

Beam	No. of Test	$M_{max,exp}$ (kN-m)	$M_{y,exp}$ (kN-m)	$\delta_{max,exp}$ (mm)	$\delta_{y,exp}$ (mm)
Control	1	67.89	48.66	135.08	10.05
Basalt	2	75.05 (4%)	58.62 (1%)	120.78 (2%)	15.10 (4%)
Carbon	2	71.62 (2%)	60.12 (1%)	120.44 (2%)	16.30 (%)
Glass	2	72.35 (3%)	58.04 (%)	128.18 (14%)	14.54 (9%)
PBO	2	66.26 (3%)	63.56 (1%)	60.35 (16%)	20.39 (8%)
Steel	2	82.10 (2%)	69.64 (1%)	46.47 (0%)	22.82 (8%)

(%) Coefficient of variation.

The results presented in Table 3 show coefficients of variation between 0% and 14%, indicating good accuracy and repeatability of the experiments.

3. Analytical Model

3.1. Considerations

A new approach for a classical analytical method for estimating the ultimate bending moment of RC beams flexurally strengthened with FRCMs is proposed. The increase in the ultimate flexural capacity provided by the strengthening system depends on the capacity of the grids to distribute the stresses uniformly inside the matrix. Papanicolaou and Papantoniou [17] and D'Ambrisi and Focacci [3] listed the multiple aspects that can affect the stress transfer mechanism between the RC beam and the FRCM-strengthening material:

- In the grid: The type of fiber, the arrangement of the fibers in the yarn (dry fiber fabric or coated yarn grid), and the geometric configuration of the grid.
- In the matrix: The chemical composition of the mortars and the size of the fine grain.
- In the RC beam: The type of concrete and the substrate treatment before the application of FRCMs.

It is easy to understand that the complexity of interfaces and the chemical mechanism among yarns, grids, coatings, and mortars that determine the stress distribution in the composite material is a matter that goes far beyond the scope of the present work. It is an extremely complex problem that requires a multi-scale approach with the assessment of chemists and physicists. The following analytical method predicts the flexural ultimate capacity of the strengthened RC beams for each specific strengthening system by reducing the tensile capacity of the FRCM materials using a reduction coefficient (β) applied to the ultimate tensile capacity of the fibers. Hence, it approaches the problem from a simplification, reducing the complexity to a single effective fiber strength capacity limited by β. This parameter is determined for each type of strengthening fabric by means of the adjustment of the analytical predictions of the maximum flexural moment ($M_{max,an}$) with the experimental results ($M_{max,exp}$) of the database composed by the tests carried out by the authors and the evidence found in the literature review. The intention of this work was to provide a useful tool to help practitioners to estimate the complex performance from a feasible formulation.

3.2. Formulation

The failure of strengthened systems happens at a lower capacity of the theoretical maximal strengthening capacity of a composite. The explanation for this combines different complex inner failures that weaken the capacity of the composite, as stated in the previous paragraph, particularly the relative slip between the fibers and matrix in the contact surfaces, as well as the relative slip between the yarns in contact with mortar and the core yarns and the effect of coating on the fibers. These are properties difficult to estimate during the execution of the reinforcement cast-in-place. Another additional mechanism is the partial breakage of yarns in the fabric. The overall problem is obviously very complex because it depends on the surface properties, penetration of mortar, yarn shapes, boundary contacts, and execution.

Hence, the analytical method for determining the ultimate flexural capacity of the strengthened RC beams is based on the following assumptions: (1) Failure of the strengthening composite while the substrate and bonding maintain their capacities, (2) strain compatibility during the loading process, (3) equilibrium of the forces of the load-bearing cross-section, and (4) reduction of the tensile fiber capacities by a mixed failure of fibers and mortar relative slip.

The constitutive behavior of concrete, steel, and fibers is shown in Figure 2 according to Eurocode 2 [32]. In the case of concrete, bilinear simplification was considered (Figure 2a). Regarding the steel, an elastic–plastic diagram was used, considering the strain-hardening phenomenon after yielding (Figure 2b). The fibers were assumed to be linear–elastic until failure (Figure 2b). The tensile strength of the concrete and the mortar-FRCM was not considered.

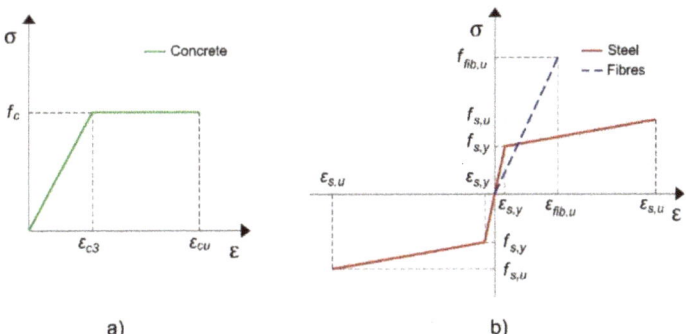

Figure 2. Constitutive behavior of the materials: (a) Concrete and (b) steel and fibres.

Similar to the analytical approach carried out by Wiberg [12] to calculate the maximum flexural moment ($M_{max,an}$), concrete and tensile steel can reach their ultimate capacities in compression and tension according the failure domain. In this case, the ultimate strain of the steel reinforcement ($\varepsilon_{s,u}$) is considered to be 90‰, and the ultimate strain of the concrete in compression is considered to be 3.5‰. The ultimate tensile capacity of the fibers is reduced by the coefficient β to compensate the slipping effect of the yarns inside the matrix. Hence, the contribution of the composite is an effective strength over the total strength of the cross section. Figure 3 shows the internal force equilibrium and the strain distribution of a rectangular RC beam cross-section flexurally strengthened with

FRCMs. In this figure, it can be observed that the model considers the different levels of steel reinforcement separately. Furthermore, the compression concrete block is simplified by the assumption that the block works at its maximum capacity (f_c), and its high equals 0.8 times the neutral axis depth (x) [32].

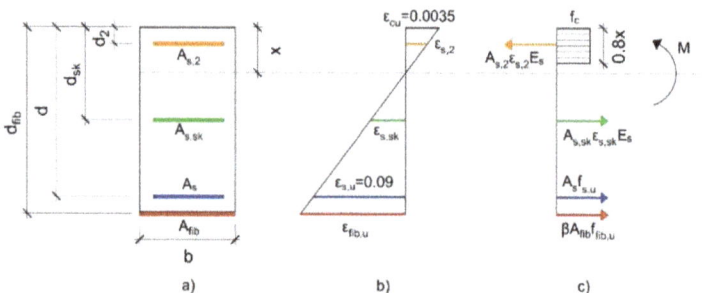

Figure 3. Analysis of the cross-section for the ultimate limit state in bending: (a) Geometry, (b) strain distribution, and (c) force equilibrium.

According to Figure 3, the analytical maximum flexural moment ($M_{max,an}$) is calculated as follows (Equation (1)):

Unstrengthened beam:

$$M_{cmax,an} = M_{c,c} + M_{c,s} + M_{c,s,sk} + M_{c,s,2} \tag{1}$$

Strengthened beam:

There was a discrepancy between the analytical ultimate moment ($M_{cmax,an}$) and the experimental ultimate moment ($M_{cmax,exp}$), probably due to the conservative analysis. The ratio $\frac{M_{cmax,exp}}{M_{cmax,an}}$ is 1.26 for this test. Assuming that all of the other concrete beams have the same performance, the strength contribution of the pure RC beam for each one can be modified with this ratio. In Equation (2), it is assumed that this ratio is the same in all of the plastic stages of the strengthened beam. Then, Equation (2) is only useful for the plastic stage from the strengthened beam.

$$M_{cmax,an} = M_{c,c} + M_{c,s} + M_{c,s,sk} + M_{c,s,2} \tag{2}$$

where

$M_{c,c}$, $M_{s,c}$—the flexural contributions of the concrete (control and strengthened beams);
$M_{c,s}$, $M_{s,s}$—the flexural contributions of tensile steel reinforcement (control and strengthened beams);
$M_{c,s,sk}$, $M_{s,s,sk}$—the flexural contributions of skin steel reinforcement (control and strengthened beams);
$M_{c,s,2}$, $M_{s,s,2}$—the flexural contributions of compressive steel reinforcement (control and strengthened beams);
$M_{sma,an}$—the analytical ultimate maximum strength of the strengthened beam.

The contributions of each withstanding material and neutral axis depth (x) can be determined according to the following equations (Equations (3)–(9)):

Ultimate flexural contributions of the concrete

$$M_c = \frac{0.8 f_c b x^2}{2} \tag{3}$$

Tensile steel reinforcement:

$$f_{s,u} = \varepsilon_s E_s \quad \text{if } \varepsilon_s < \varepsilon_{s,y} \tag{4}$$

$$f_{s,u} = \varepsilon_s \frac{f_{s,u} - f_{s,y}}{\varepsilon_{s,u} - \varepsilon_{s,y}} + f_{s,y} \quad \text{if } \varepsilon_s \geq \varepsilon_{s,y} \tag{5}$$

$$M_s = A_s f_{s,u}(d - x) \tag{6}$$

where $\varepsilon_{s,y}$ is the elastic limit deformation equal to $\varepsilon_{s,y} = \frac{f_{s,y}}{E_s}$.

Compressive steel reinforcement:

In this case, the same criteria from Equation (6) are fulfilled, and the bending moment is calculated from the following equation:

$$M_{s,2} = f_{s,u} A_{s,2}(x - d_2) \tag{7}$$

Skin steel reinforcement:

$$M_{s,sk} = f_{s,u} A_{s,k}(d_{sk} - x) \tag{8}$$

Fibers of the strengthening system:

$$M_{fib} = \beta A_{fib} f_{fib,u}\left(d_{fib} - x\right) \tag{9}$$

where the means of all the variables use standard steel–concrete code notation and are depicted in Figure 3. The effectiveness is defined by the reduction coefficient (β) introduced in Equation (9).

In the case of this study, fibers that conform have yarns uniformly distributed along the width of the beams. Thus, the area of the fibers (A_{fib}) is calculated as follows (Equation (10)):

$$A_{fib} = n b_f t_{tex} \tag{10}$$

where n is the number of grid layers of the FRCMs, b_f is the width of the fabrics, and t_{tex} is the equivalent thickness of the textile. This last variable is a parameter provided by the corresponding textile manufacturer and represents the thickness of the textile for a continuous distribution of the fibers. In this research, $b = b_f$.

Finally, to determine the coefficient β for each type of fiber, the equality between the analytical predictions of the ultimate bending moment for RC beams flexurally strengthened with FRCMs and the corresponding experimental values is imposed. As a result, an equation of a line in which the reduction coefficient β represents its slope is obtained (Equation (11)).

$$a = \beta k \tag{11}$$

where a is the experimental contribution of the FRCMs to the ultimate flexural capacity of the strengthened specimen (Equation (12)), and k represents the theoretical contribution of the fibers to the ultimate bending capacity of the beams (Equation (13)):

$$a = M_{su,exp} - \frac{M_{cu,exp}}{M_{cu,an}}(M_c + M_s + M_{s,sk} + M_{s,2}) \tag{12}$$

$$k = A_{fib} f_{fib,u}\left(d_{fib} - x\right) \tag{13}$$

In the case of the FRCM-strengthened concrete beam, it is known that failure of the meshes happens before than the failure of other materials. The ultimate deformation of mesh ($\varepsilon_{f,u}$) is taken to set the point where the maximum load is reached in this case. From the strain compatibility:

$$\varepsilon_c = \frac{\varepsilon_{f,u} x}{d_f - x} \tag{14}$$

$$\varepsilon_s = \frac{\varepsilon_{f,u}(d - x)}{d_f - x} \tag{15}$$

$$\varepsilon_{s,2} = \frac{\varepsilon_{f,u}(x - d_2)}{d_f - x} \quad (16)$$

$$\varepsilon_{s,2} = \frac{\varepsilon_{f,u}(d_{s,k} - x)}{d_f - x} \quad (17)$$

For the control beam, it is considered that the crushing failure of the concrete occurs before the breakage of the steel. Compressive ultimate deformation of the concrete is taken ($\varepsilon_{c,u} = 0.0035$) to set the point where the maximum tension is reached in this case. Other strains are calculated from this.

$$\varepsilon_s = \frac{\varepsilon_{c,u}(d - x)}{x} \quad (18)$$

$$\varepsilon_{s,2} = \frac{\varepsilon_{c,u}(x - d_2)}{x} \quad (19)$$

Once the ultimate strains of the materials are known, one of the following conditions must be fulfilled to determine the maximum moment of the reinforced beam:
$\varepsilon_c \leq 0.0035$ (Code [28]).
$\varepsilon_s \leq 0.09$ (experimental results [27]).

This analytical model does not consider other failures such as debonding of FRCM strengthening systems. Notice that this case represents the desirable situation for practitioners, in which FRCMs may develop their maximum tensile capacity as a flexural strengthening material.

3.3. Analytical Results

The results obtained from the analytical model are presented in Table 4. This shows the results of the maximum experimental bending moment ($M_{max,exp}$) supported by the beams, and the ultimate contributions of the different materials (ε_c, f_c, $f_{s,u}$, $f_{s,uk}$, and $f_{f,u}$), where $f_{f,u}$ is the tensile stress supported by the fabric after reaching the maximum experimental moment. This was calculated by $f_{f,u} = \beta f_{fib,u}$. Moreover, Table 4 includes the parameters a and k, both of which are necessary to obtain the effectiveness coefficient (β).

Table 4. Analytical model results.

Beam	$M_{u,exp}$ (kN-m)	ε_c (‰)	f_c (MPa)	$f_{s,u}$ (MPa)	$f_{s,uk}$ (MPa)	a (kN-m)	K (kN-m)	β	$f_{f,u}$ (MPa)
Control	67.89	0.0035	42.35	579.634	549.85	14.12	-	-	-
Basalt	75.05	0.00173	55.42	500	520.00	8.43	15.65	0.54	1658.32
Carbon	71.65	0.00241	42.35	538.375	528.25	8.24	19.30	0.43	1844.09
Glass	72.40	0.00169	46.52	551.456	535.21	6.06	10.46	0.58	1512.73
PBO	66.25	0.00146	42.35	542.493	530.35	1.42	24.96	0.06	329.41
Steel	82.10	0.00091	46.52	535.455	526.75	17.92	19.30	0.78	2504.40

Table 4 shows that none of the strengthened beams reached the ultimate concrete strain (0.0035) or steel tensile strain (0.09). This means that the FRCM failures occurred before the concrete crushing failure or the steel tensile failure were reached for the strengthened beams. This meets the experimental observations. In the case of the control beam, failure by crushing of the concrete occurred before the breakage of steel.

Additionally, Table 4 shows that the steel-FRCM strengthening systems presented the most efficient behavior with respect to the tensile capacity of the fibers (effective strength of 78%), where the reduction of the flexural capacity provided by the fibers to adapt the mechanical behavior to FRCM systems was less than that of the other fibers. In the case of glass-FRCM, its better efficiency (second, effective strength of 58%) can be explained by the use of a polymer coating of the roving that improves the bonding interface between yarns and mortar. This coating protects the fibers from the breakage caused by friction with the matrix and provides the capacity to distribute the stresses uniformly to the inner and the outer fibers of the yarn (see Voss et al. [33]). This result was also reported by Papanicolaou and Papantoniou [17] in cases in which polymer-coated textiles were used.

However, the FRCMs that presented the highest loss of the withstanding capacity with regard to the tensile strength of the fibers were those made of PBO, carbon, and basalt fabrics. These phenomena may be caused by the poor impregnation of dry fibers by the cementitious matrix, in which the inner filaments of the rovings are not in contact with the mortar, and stress distributions in the yarn are not homogeneous (see Hegger et al. [34]).

The influence of the matrix on obtaining coefficient β is noteworthy. This indicates that the proposed analytical methodology is highly sensitive to changes in FRCM components and can be used only when the strengthening solution selected to strengthen the RC structural element is similar to the strengthening solution used for obtaining the reduction coefficient β.

4. Numerical Models

In order to verify the utility of the β coefficients specified in the analytical approach, a numerical model to reproduce the experimental results was developed.

The commercial mechanical simulation software Abaqus® 6.14-4 [35] was used to implement numerical simulations. This choice was based on the aim of using a widely available general purpose simulation tool capable of representing complex material models. In addition, many previous studies based on analysis of FRCMs and reinforced concrete successfully used this software (see, for example, [36,37]).

4.1. General Materials' Constitutive Formulations

The concrete plastic damage model [38] was used to simulate the FRCMs and reinforced concrete. This model is characterized by two elastic moduli: One corresponding to the elastic zone, and another depending on the damage coefficient, which is a function of the cracking situation or the plasticization achieved.

Regarding the plastic zone of the cementitious matrix in tension, it was necessary to define the following parameters (also used in [39]):

Dilatation angle: The first value used for this parameter was 13. It was chosen on the basis of existing literature [37], but convergence difficulties justified increasing this value to 30.

Eccentricity: The predetermined eccentricity suggested by Abaqus was 0.1, which implies that the material has almost the same angle of expansion in a significant range of confining pressure values.

Form parameter of the plasticizing surface (K): The default value was equal to 2/3.

Relationship between the maximum uniaxial and biaxial compression stress at the beginning of the loading process ($fb0/fc0$). The default value was equal to 1.16.

Viscoplastic regularization: The values of 5×10^{-6}, 5×10^{-5}, and 5×10^{-4} were tested for an objective choice, proving that the value of 5×10^{-5} achieved results that better fit with the experimental results, as well as better model convergence.

Once these material properties were defined, the matrix's stress–strain curves and the corresponding damage variables were calculated. To calculate these damage variables, the procedure published by [37] was followed.

4.2. Unstrengthened Beam Model

For this analysis, a deformable solid was used to simulate the concrete part defined by a length of 4.4 m, a width of 200 mm, and height of 500 mm (Figure 4a).

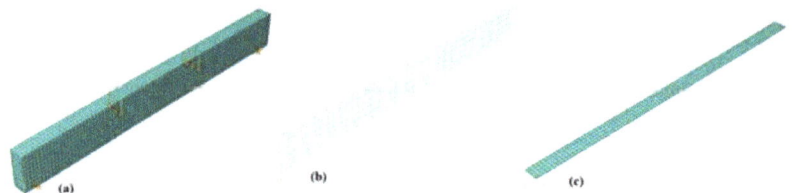

Figure 4. Numerical model of beams: (**a**) Deformable solid beam, (**b**) truss element-reinforced steel, and (**c**) shell element FRCMs.

Truss elements were used to simulate steel reinforcement (longitudinal steel bars and stirrups) (Figure 4b). A simplified steel elastic–plastic curve (Figure 2b) obtained from the experimental results was assigned to these reinforcement elements.

The concrete damage plastic model (previously presented) was used to define the concrete response. The corresponding material properties were obtained by their relationship with the concrete experimental compression strength. These properties refer to the secant modulus of deformation (E_{ci}) and the characteristic tensile strength ($f_{ct,k}$) that were calculated from Equations (20)–(22) [28].

$$E_{ci} = 8500\sqrt[3]{f_c} \tag{20}$$

$$f_{ct,m} = 0.3 f_c^{2/3} \tag{21}$$

$$f_{ct,k} = 0.7 f_{ct,m} \tag{22}$$

where $f_{ct,m}$ is the medium tensile strength.

It was necessary to calibrate this property with the aim of reproducing the experimental stiffness and maximum load of the control beam. With this purpose, the modulus of deformation was reduced to 36%, and the 0.7 coefficient from Equation (22) was reduced to 0.35. These modifications were useful to fit the numerical moment–displacement curve with the experimental curve of the control beam.

Two mesh sizes for concrete discretization were tested for convergence analysis: 0.05 m and 0.025 m. The 0.05 m mesh was chosen because no significant difference between the 0.05 mm and 0.025 m meshes was observed (4.7% variation of the maximum reaction force), and the calculation time was 30 times less using the 0.05 m mesh.

Boundary conditions were set according to the four-point flexural setup adopted in the experimental campaign [27]. The displacement in the "y" and "x" directions in one support and the displacement in the "y" and "z" directions in the other were restrained (see Figure 4a) for providing stability to the numerical model.

The load was directly applied by imposing the vertical displacement that caused the failure of the experimental test of the control beam.

Finally, to identify and check the breaking condition, the steel experimental tensile strength (634 MPa) was considered as the governing criterion. This allowed to reproduce all of the experimental moment–deflection curves.

4.3. FRCM-Strengthened Beam Model

In the case of the beams strengthened with FRCMs, the same unstrengthened beam model was used, with the difference that shell elements were added to simulate the FRCMs (see Figure 4c). In the case of the beams strengthened with basalt, glass, and steel FRCMs were used, the corresponding concrete compression strengths (f_c) are presented in Table 4 (different batch).

Shell elements are intended to model structures with one dimension significantly smaller than the other two dimensions. The stresses in the thickness direction have to be negligible to properly use shells. An elastic–plastic model was used on the shell elements to represent the FRCMs.

To define the mechanical properties of each FRCM (shell elements), it was necessary to implement FRCM models (fabric and mortar) and to determine its tensile behavior with the procedure presented in [40].

This procedure consisted of simulating the FRCMs with a deformable solid (cementitious matrix) and truss elements (fabrics) (see Figure 5), where the fabric was assumed totally bonded (embedded region) to the matrix without allowing sliding in the fabric–matrix interface. To define the material of the cementitious matrix and the fabric, the values presented in Tables 1 and 2 were used.

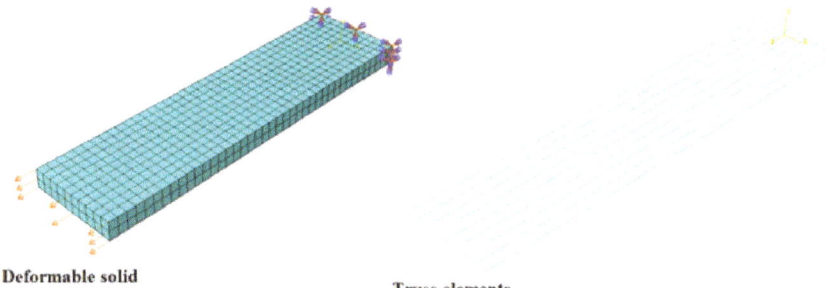

Figure 5. FRCM models.

Once the stress–strain curve (see Figure 6) of each type of FRCM was obtained, the modulus of elasticity was obtained from the initial slope, and the stress and plastic strain data of the FRCMs were introduced as the shell elements properties (plastic model). The ultimate stress from the FRCMs imitated when the fabric reached its analytical ultimate stress presented in Table 4 (the β coefficients were used).

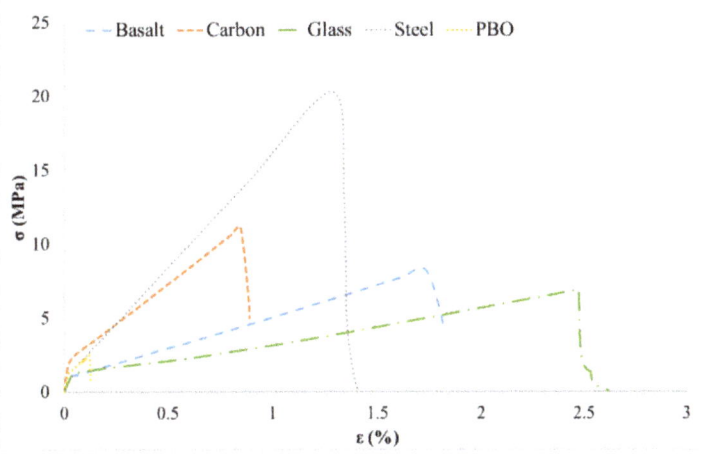

Figure 6. FRCM stress–strain diagrams (numerical model results).

All of the input material properties used in the FRCM simulation are presented in Table 5.

Table 5. Input material properties to the FRCM simulation.

FRCM	Mortar—Young's Modulus (MPa)	Mortar—Compression Strength (MPa)	Mortar—Tension Strength (MPa)	Fabric-Young's Modulus (MPa)	Fabric—Tension Strength × β (MPa)
Basalt	8905.54	24.65	0.89	95,000	1658.32
Carbon	8941.53	24.95	0.90	240,000	2048.99
Glass	10,047.45	35.40	1.13	61,250	1512.73
PBO	9510.24	30.02	1.01	270,000	329.41
Steel	8905.54	24.65	0.89	206,000	2504.40

According to the experimental failures, two types of stress–strain curves were used:

Fabric broke (steel and glass): In these, a discharge slope was defined when the analytical ultimate FRCM stress (f_{fu}; Table 4) was reached. This was calculated with the same procedure used to calculate the discharge slope in the tensile concrete damage.

Fabric sliding failures (basalt, carbon, and PBO): In these, a similar discharge slope was defined, but the tensile stress was assumed constant when the discharge slope was proximal to 45% of the analytical ultimate FRCM stress. This is because of the small contribution of the FRCMs during the sliding process of the fabric.

For the interaction between the shell elements (FRCMs) and deformable solids (the concrete beam), a tie connection was used. This approach was considered in other studies [40,41]. A tie connection is a link restriction that allows merging two regions, even though the meshes created on the surfaces of the regions may be different, so complete beam–FRCM bonding can be assumed.

The same boundary condition to that of the unreinforced beam were imposed.

All of the input parameters used in this numerical simulation are summarized in Table 6.

Table 6. Input materials properties to simulate unstrengthened and strengthened beams.

Concrete Beams	Concrete—Young's Modulus (MPa)	Concrete—Compression Strength (MPa)	Concrete—Tension Strength (MPa)	FRCM—Young's Modulus (First Slope) (MPa)	FRCM—Tension Strength × β (MPa)
Control	10,666.11	42.35	1.28	-	-
Basalt	11,666.58	55.42	1.53	1760.38	7.91
Carbon	10,666.11	42.35	1.28	6187.00	9.63
Glass	11,005.29	46.52	1.36	2438.09	6.35
PBO	10,666.11	42.35	1.28	3518.83	1.50
Steel	11,005.29	46.52	1.36	3887.05	18.60

4.4. Results of the Beams' Numerical Models

Figure 7 shows the stress contour plots of the concrete beam and the reinforcement steel at the state when the reinforcement steel reached the tensile strength taken as the failure criterion.

The principal stress in the concrete beam shown in Figure 7 describes failure modes similar to the experimental results, with the appearance of flexural cracks and their propagation from the tensile side of the specimens to the neutral axis.

Figure 8 shows the stress contour plots of the different FRCMs at the state when the whole structural element reached the maximum reaction force.

Figure 9 shows the complete moment–deflection curves for all of the beams, including the experimental and numerical results. From experimental values of Figure 9 it can be observed that the greatest contribution to the maximum load from the concrete beam was by steel-FRCM, followed by carbon-FRCM, basalt-FRCM, glass-FRCM, and PBO-FRCM. This order corresponds to the tensile capacity presented in Figure 6. For the case of PBO- and glass-FRCM, the FRCMs significantly decreased the tensile contribution at the middle beam, meaning that the FRCMs reached their maximum stress before the whole structure reached the maximum reaction force.

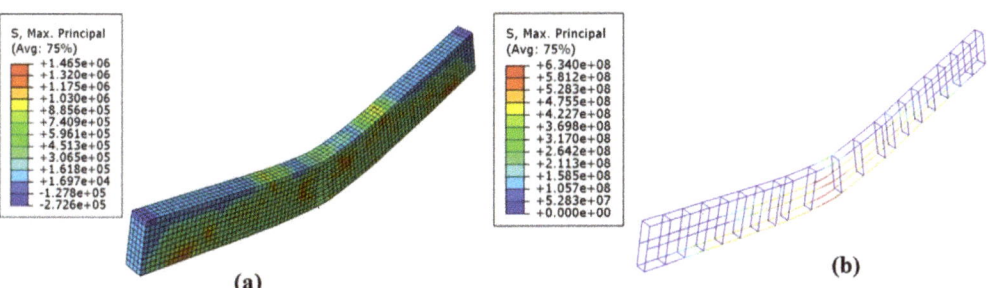

Figure 7. Max. principal stress (N/m²) state of the simulation beam: (**a**) Concrete beam and (**b**) steel reinforcement.

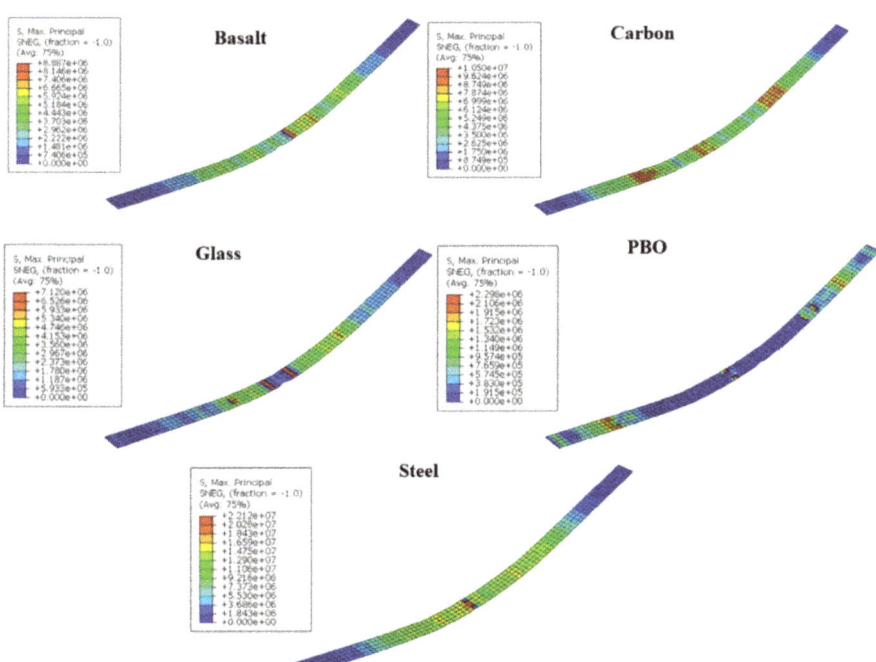

Figure 8. Max. principal stress (N/m²) state of the FRCMs.

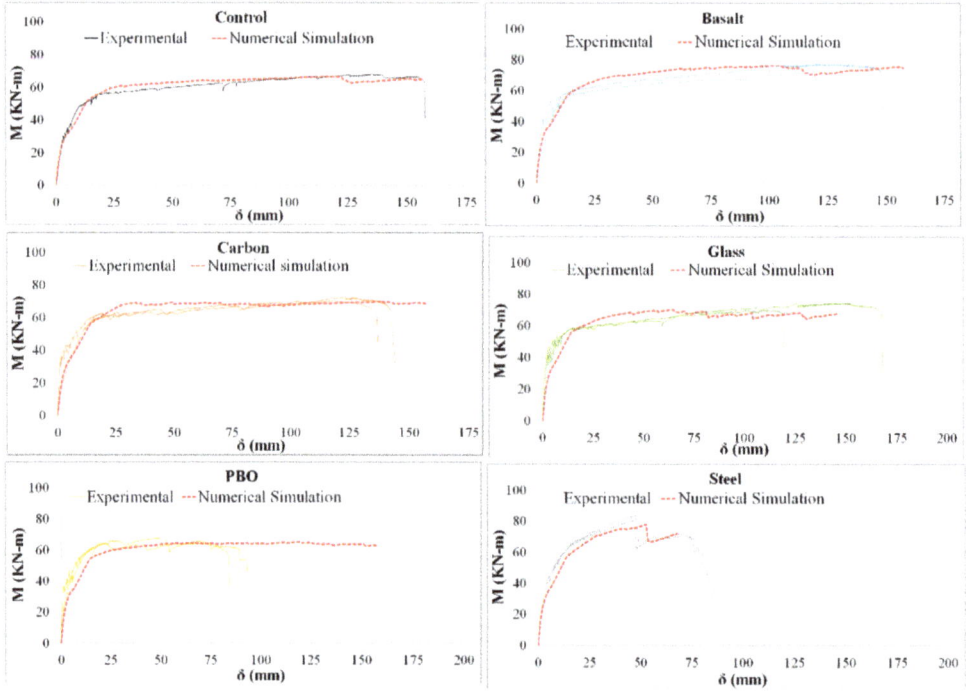

Figure 9. Moment–deflection diagrams.

The results obtained from the numerical model are presented in Table 7. That table shows the results of the maximum flexural moment ($M_{max,num}$), the yielding flexural moment ($M_{y,num}$), and the deflections recorded at the mid-span of the specimens when the maximum flexural moment ($\delta_{max,num}$) and the yielding flexural moment ($\delta_{y,num}$) were reached.

Table 7. Numerical simulation results.

Beam	$M_{max,num}$ (kN-m)	$M_{y,num}$ (kN-m)	$\delta_{max,num}$ (mm)	$\delta_{y,num}$ (mm)
Control	66.60	52.06	120.76	13.99
Δ_{exp}	(−2%)	(7%)	(−11%)	(39%)
Basalt	76.51	57.27	104.01	13.62
Δ_{exp}	(2%)	(−2%)	(−14%)	(−10%)
Carbon	69.78	55.81	138.65	14.18
Δ_{exp}	(−3%)	(−7%)	(15%)	(−13%)
Glass	70.30	54.99	64.71	13.77
Δ_{exp}	(−3%)	(−5%)	(−49%)	(−5%)
PBO	65.30	82.10	119.08	14.02
Δ_{exp}	(−1%)	(29%)	(97%)	(−31%)
Steel	77.93	56.44	53.03	13.62
Δ_{exp}	(−5%)	(−19%)	(14%)	(−40%)

The fitting capabilities of the numerical model were analyzed for all the beams:
- Control beam: The ultimate and yielding moments and the deflection at the ultimate moment properly fit the experimental results with differences ranging between 2%

and 11%. However, the numerically predicted deflection at the yielding moment was much higher than that in the experimental results (39%).
- Beam strengthened with basalt-, carbon-, glass-, and steel-FRCM: The model for the strengthened beam was able to obtain values of the ultimate moment the and deflection at the ultimate moment close to the experimental results with differences ranging from 2% to 19%, except in the case of glass-FRCM, where the numerical model brought a deflection at the maximum moment with a far lower value than that in the experimental tests (49%).
- Beam strengthened with PBO-FRCM: This beam showed the greatest dispersion of the deflection at the ultimate moment, the yielding moment, and the deflection with differences ranging between 29% and 97%. This is because of the premature experimental failure of the beam strengthened with PBO-FRCM.

Figure 9 demonstrates that the proposed numerical models are able to reproduce the experimental response with sufficient approximation.

The results of the numerical simulations validate the effectiveness of using the β coefficients to determine the ultimate effective contribution of the FRCM systems to RC beams, where the interaction between materials is similar to that considered in this study.

5. Conclusions

This work presented an analytical approach and a numerical model for determining the ultimate bending capacity of RC structures flexurally strengthened with FRCMs using an effective strength. These models are based on the lowering of the tensile capacity of the fibers using a reduction parameter, β.

- Comparing the different types of reinforcements, the steel and glass strengthening systems are the FRCMs that developed the lowest reduction of the tensile capacity of the fibers. Hence, they were the most efficient. However, carbon, basalt, and PBO, very promising materials, showed the highest reduction of tensile capacity of the fibers probably due to the incapacity of the corresponding matrices to impregnate fully the dry fibers that compose the fabric.
- The analysis of FRCM systems revealed that the type of matrix used highly affects the reduction parameter of fibers, implying that it is necessary to create a database of experimental results for each combination of matrix and grid used. Thus, the use of combinations of grids and matrices to manufacture FRCMs not guaranteed by the provider should be taken with extreme care.
- The concrete tensile strength and the modulus of deformation were adapted to fit the numerical simulations with the experimental results. The tensile strength coefficient was reduced a 50% (from 0.7 of the normative to a 0.35). The modulus of deformation was reduced to 36%. These values were successfully used for both the control and FRCM-strengthened beams. For all of them, numerical simulations proved that these modified properties were representative of the proposed model.
- The numerical model was effective for carbon-, basalt-, glass-, and steel-FRCM-strengthened beams for predicting the bending moments and displacement (variation between 2% and 15%) due to the coefficient β determined in the analytical approach. However, in the case of PBO-FRMC, it was not possible to reproduce the experimental results, because the failure criterion was reached at a very large deflection bigger (variation of 97%) than its experimental failure.
- In summary, it can be said that although there is limited experimental evidence to determine the coefficient β, it can be successfully used as a representative parameter of the performance capability of different types of FCRM solutions. β can be applied as a reduction of the ultimate tensile strength of fibers and it represents a promising approach that highly simplifies the analytical design of FRCMs when applied for RC flexural strengthening. Practitioners could benefit from using this strategy to calculate feasible strengthening solutions. With the aim of increasing the reliability of the model,

we urge the performance of more experimental tests that allow to expand the results database and fitting the coefficient β more accurately.

Author Contributions: Conceptualization, L.M. and C.E.; methodology, L.M. and E.B.-M.; software, L.M.; validation, L.M., E.B.-M. and L.G.; formal analysis, L.M.; investigation, L.M. and C.E.; resources, L.M.; data curation, L.M.; writing—original draft preparation, L.M. and C.E.; writing—review and editing, L.M., E.B.-M. and L.G. All authors have read and agreed to the published version of the manuscript.

Funding: The authors gratefully acknowledge the financial support from the Ministry of Science, Innovation and Universities of the Spanish Government (MCIU), the State Agency of Research (AEI), as well as that of the ERDF (European Regional Development Fund) through the project SEVERUS (Multilevel evaluation of seismic vulnerability and risk mitigation of masonry buildings in resilient historical urban centers, ref. num. RTI2018-099589-B-I00). The third author is a Serra Húnter Fellow.

Institutional Review Board Statement: Not applicable.

Informed Consent Statement: Not applicable.

Data Availability Statement: Data is contained within the article.

Acknowledgments: The authors wish to acknowledge the support provided by Bernat Almenar Muns during the experimental testing.

Conflicts of Interest: The authors declare no conflict of interest.

Notation
The following symbols were used in this paper:

A_{fib}	Area of fibers
a	Experimental contribution of the FRCMs to the ultimate flexural capacity of the strengthened specimen
β	Coefficient to compensate the slipping effect of the fabric inside the matrix
E_{ci}	Secant modulus of deformation
$\varepsilon_{f,u}$	Fabric ultimate deformation
$\varepsilon_{c,u}$	Deformation of the concrete is taken
$\varepsilon_{s,u}$	Ultimate strain of the steel reinforcement
$\varepsilon_{s,y}$	Elastic limit
$F_{y,exp}$	Experimental yielding flexural force
$fb0/fc0$	Relationship between the maximum uniaxial and biaxial compression stress at the beginning of the loading process
$f_{ct,k}$	Characteristic tensile strength
$f_{ct,m}$	Medium tensile strength
f_c	Concrete compression strength
$f_{s,u}$	Tensile stress supported by tensile steel
$f_{s,uk}$	Tensile stress supported by skin steel
$f_{f,u}$	Tensile stress supported by fabric when the strengthened beam reaches the maximum experimental moment
$K_{y,exp}$	Experimental stiffness coefficient
K	Form parameter of the plasticizing surface
k	Theoretical contribution of fibers to the ultimate bending capacity of concrete beams
$M_{max,exp}$	Experimental maximum flexural moment
$M_{y,exp}$	Experimental yielding flexural moment
$M_{max,an}$	Analytical predictions of the maximum flexural moment
$M_{cmax,an}$	Analytical ultimate moment of control beams
$M_{c,c}, M_{s,c}$	Flexural contributions of concrete (control and strengthened beams)
$M_{c,s}, M_{s,s}$	Flexural contributions of tensile steel reinforcement (control and strengthened beams)
$M_{c,s,sk}, M_{s,s,sk}$	Flexural contributions of skin steel reinforcement (control and strengthened beams)
$M_{c,s,2}, M_{s,s,2}$	Flexural contributions of compressive steel reinforcement (control and strengthened beams)
$M_{sma,an}$	Analytical ultimate maximum strength of strengthened beams
$\delta_{max,exp}$	Experimental deflections recorded at the mid-span of specimens at the maximum flexural moment
$\delta_{y,exp}$	Experimental deflections recorded at the mid-span of specimens when the yielding flexural moment is reached
x	Neutral axis depth

References

1. Awani, O.; El-Maaddawy, T.; Ismail, N. Fabric-reinforced cementitious matrix: A promising strengthening technique for concrete structures. *Constr. Build. Mater.* **2017**, *132*, 94–111. [CrossRef]
2. Triantafillou, T.C.; Papanicolaou, C.G. Textile reinforced mortars (TRM) versus fibre reinforced polymers (FRP) as strengthening materials of concrete structures. In Proceedings of the 7th ACI International Symposium on Fibre-Reinforced (FRP) Polymer Reinforcement for Concrete Structures, Kansas City, MO, USA, 6–9 November 2005; pp. 99–118.
3. D'Ambrisi, A.; Focacci, F. Flexural Strengthening of RC Beams with Cement-Based Composites. *J. Compos. Constr.* **2011**, *15*, 707–720. [CrossRef]
4. Balaguru, P.; Kurtz, S.; Rudolph, J. *GEOPOLYMER for Repair and Rehabitlitation of Reinforced Concrete Beams*; Geopolymer Institute: St. Quentin, France, 1997; pp. 1–5.
5. Babaeidarabad, S.; Loreto, G.; Nanni, A. Flexural Strengthening of RC Beams with an Externally Bonded Fabric-Reinforced Cementitious Matrix. *J. Compos. Constr.* **2014**, *18*, 04014009. [CrossRef]
6. Larbi, A.S.; Contamine, R.; Hamelin, P. TRC and Hybrid Solutions for Repairing and/or Strengthening Reinforced Concrete Beams. *Eng. Struct.* **2012**, *45*, 12–20. [CrossRef]
7. Elsanadedy, H.M.; Almusallam, T.H.; Alsayed, S.H.; Al-Salloum, Y.A. Flexural strengthening of RC beams using textile reinforced mortar—Experimental and numerical study. *Compos. Struct.* **2013**, *97*, 40–55. [CrossRef]
8. Pellegrino, C.; D'Antino, T. Experimental behaviour of existing precast prestressed reinforced concrete elements strengthened with cementitious composites. *Compos. Part B Eng.* **2013**, *55*, 31–40. [CrossRef]
9. Gil, L.; Escrig, C.; Bernat-Maso, E. Bending Performance of Concrete Beams Strengthened with Textile Reinforced Mortar TRM. *Key Eng. Mater.* **2014**, *601*, 203–206. [CrossRef]
10. Napoli, A.; Realfonzo, R. Reinforced concrete beams strengthened with SRP/SRG systems: Experimental investigation. *Constr. Build. Mater.* **2015**, *93*, 654–677. [CrossRef]
11. Ebead, U.; Shrestha, K.C.; Afzal, M.S.; El Refai, A.; Nanni, A. Effectiveness of Fabric-Reinforced Cementitious Matrix in Strengthening Reinforced Concrete Beams. *J. Compos. Constr.* **2017**, *21*, 04016084. [CrossRef]
12. Wiberg, A. *Strengthening of Concrete Beams Using Cementitious Carbon Fibre Composites*; Royal Institute of Technology of Sweden: Stockholm, Sweden, 2003.
13. Toutanji, H.; Deng, Y. Comparison between Organic and Inorganic Matrices for RC Beams Strengthened with Carbon Fiber Sheets. *J. Compos. Constr.* **2007**, *11*, 507–513. [CrossRef]
14. Barton, B.L.; Wobbe, E.N.; Dharani, L.; Silva, P.F.; Birman, V.; Nanni, A.; Alkhrdaji, T.; Thomas, J.; Tunis, G. Characterization of reinforced concrete beams strengthened by steel reinforced polymer and grout (SRP and SRG) composites. *Mater. Sci. Eng. A* **2005**, *412*, 129–136. [CrossRef]
15. Sneed, L.H.; Verre, S.; Carloni, C.; Ombres, L. Flexural behavior of RC beams strengthened with steel-FRCM composite. *Eng. Struct.* **2016**, *127*, 686–699. [CrossRef]
16. Brückner, A.; Ortlepp, R.; Curbach, M. Textile reinforced concrete for strengthening in bending and shear. *Mater. Struct.* **2006**, *39*, 741–748. [CrossRef]
17. Papanicolaou, C.G.; Papantoniou, I.C. Mechanical Behavior of Textile Reinforced Concrete (TRC)/Concrete Composite Elements. *J. Adv. Concr. Technol.* **2010**, *8*, 35–47. [CrossRef]
18. Ombres, L. Flexural analysis of reinforced concrete beams strengthened with a cement based high strength composite material. *Compos. Struct.* **2011**, *94*, 143–155. [CrossRef]
19. Larrinaga, P.; Chastre, C.; San-José, J.T.; Garmendia, L. Non-linear analytical model of composites based on basalt textile reinforced mortar under uniaxial tension. *Compos. Part B Eng.* **2013**, *55*, 518–527. [CrossRef]
20. D'Antino, T.; Sneed, L.H.; Carloni, C.; Pellegrino, C. Influence of the substrate characteristics on the bond behavior of PBO FRCM-concrete joints. *Constr. Build. Mater.* **2015**, *101*, 838–850. [CrossRef]
21. Leone, M.; Aiello, M.A.; Balsamo, A.; Carozzi, F.G.; Ceroni, F.; Corradi, M.; Gams, M.; Garbin, E.; Gattesco, N.; Krajewski, P.; et al. Glass fabric reinforced cementitious matrix: Tensile properties and bond performance on masonry substrate. *Compos. Part B Eng.* **2017**, *127*, 196–214. [CrossRef]
22. ACI Committee 549. *ACI 549.4R-13—Guide to Design and Construction of Externally Bonded Fabric-Reinforced Cementitious Matrix (FRCM) Systems for Repair and Strengthening Concrete and Masonry Structures*; American Concrete Institute: Farmington Hills, MI, USA, 2013.
23. Donnini, J.; Lancioni, G.; Corinaldesi, V. Failure modes in FRCM systems with dry and pre-impregnated carbon yarns: Experiments and modeling. *Compos. Part B Eng.* **2018**, *140*, 57–67. [CrossRef]
24. Grande, E.; Milani, G. Interface modeling approach for the study of the bond behavior of FRCM strengthening systems. *Compos. Part B Eng.* **2018**, *141*, 221–233. [CrossRef]
25. Sucharda, O. Identification of Fracture Mechanic Properties of Concrete and Analysis of Shear Capacity of Reinforced Concrete Beams without Transverse Reinforcement. *Materials* **2020**, *13*, 2788. [CrossRef]
26. Valikhani, A.; Jahromi, A.J.; Mantawy, I.M.; Azizinamini, A. Numerical Modelling of Concrete-to-UHPC Bond Strength. *Materials* **2020**, *13*, 1379. [CrossRef]
27. Escrig, C.; Gil, L.; Bernat-Maso, E. Experimental comparison of reinforced concrete beams strengthened against bending with different types of cementitious-matrix composite materials. *Constr. Build. Mater.* **2017**, *137*, 317–329. [CrossRef]

28. EHE-08. *Normativa de Hormigón Estructural en España*; Gobierno de Espana: Madrid, Spain, 2008; Volume 1, pp. 1689–1699.
29. EN 12390-1. *Testing Hardened Concrete-Part 1: Shape, Dimensions and Other Requirements for Specimens and Moulds*; No. 10; BSI: Milton Keynes, UK, 1993; Volume 38.
30. EN 12390-3. *Testing Hardened Concrete—Part 3: Compressive Strength of Test Specimens*; British Standards Institution: London, UK, 2019.
31. EN ISO 15630-1. *Steel for the Reinforcement and Pre-Stressing of Concrete—Test Methods—Part 1: Reinforcing Bars, Wire Rod and Wire*; International Organization for Standardization: Geneva, Switzerland, 2010; p. 248.
32. Comité Européen de Normalisation. *Eurocode Design of Concrete Structures—Part 1: Common Rules for Building and Civil Engineering Structures*; prEN 1992-1; European Committee for Standardisation, Central Secretariat: Brussels, Belgium, 2004.
33. Voss, S.; Hegger, J.; Hegger, J.; Brameshuber, W.; Will, N. Dimensioning of textile reinforced concrete structures. In Proceedings of the 1st International RILEM Symposium on Textile Reinforced Concrete, RILEM Publication SARL, Paris, France, 6–7 September 2006; pp. 151–160.
34. Hegger, J.; Will, N.; Bruckermann, O.; Voss, S. Load–bearing behaviour and simulation of textile reinforced concrete. *Mater. Struct.* **2006**, *39*, 765–776. [CrossRef]
35. Simulia. Abaqus 6. User´s Manual. 2011.
36. Bertolesi, E.; Carozzi, F.G.; Milani, G.; Poggi, C. Numerical modeling of Fabric Reinforce Cementitious Matrix composites (FRCM) in tension. *Constr. Build. Mater.* **2014**, *70*, 531–548. [CrossRef]
37. Alfarah, B.; López-Almansa, F.; Oller, S. New methodology for calculating damage variables evolution in Plastic Damage Model for RC structures. *Eng. Struct.* **2017**, *132*, 70–86. [CrossRef]
38. Sumer, Y.; Aktaş, M. Defining parameters for concrete damage plasticity model. *Chall. J. Struct. Mech.* **2015**, *1*, 149–155. [CrossRef]
39. Ronanki, V.S.; Burkhalter, D.I.; Aaleti, S.; Song, W.; Richardson, J.A. Experimental and analytical investigation of end zone cracking in BT-78 girders. *Eng. Struct.* **2017**, *151*, 503–517. [CrossRef]
40. Mercedes, L.; Bernat, E.; Gil, L. Numerical model of vegetal fabric reinforced cementitious matrix composites (FRCM) subjected to tensile loads. In Proceedings of the CMMoST 2019—5th International Conference on Mechanical Models in Structural Engineering, Alicante, Spain, 23–25 October 2019.
41. Bertolesi, E.; Milani, G.; Poggi, C. Simple holonomic homogenization model for the non-linear static analysis of in-plane loaded masonry walls strengthened with FRCM composites. *Compos. Struct.* **2016**, *158*, 291–307. [CrossRef]

Article

Modeling of the Achilles Subtendons and Their Interactions in a Framework of the Absolute Nodal Coordinate Formulation

Leonid P. Obrezkov [1,2,*], **Taija Finni** [1] **and Marko K. Matikainen** [2]

1 Faculty of Sport and Health Sciences, University of Jyväskylä, 40014 Jyväskylä, Finland
2 Mechanical Engineering, LUT University, 53850 Lappeenranta, Finland
* Correspondence: Leonid.Obrezkov@lut.fi

Abstract: Experimental results have revealed the sophisticated Achilles tendon (AT) structure, including its material properties and complex geometry. The latter incorporates a twisted design and composite construction consisting of three subtendons. Each of them has a nonstandard cross-section. All these factors make the AT deformation analysis computationally demanding. Generally, 3D finite solid elements are used to develop models for AT because they can discretize almost any shape, providing reliable results. However, they also require dense discretization in all three dimensions, leading to a high computational cost. One way to reduce degrees of freedom is the utilization of finite beam elements, requiring only line discretization over the length of subtendons. However, using the material models known from continuum mechanics is challenging because these elements do not usually have 3D elasticity in their descriptions. Furthermore, the contact is defined at the beam axis instead of using a more general surface-to-surface formulation. This work studies the continuum beam elements based on the absolute nodal coordinate formulation (ANCF) for AT modeling. ANCF beam elements require discretization only in one direction, making the model less computationally expensive. Recent work demonstrates that these elements can describe various cross-sections and materials models, thus allowing the approximation of AT complexity. In this study, the tendon model is reproduced by the ANCF continuum beam elements using the isotropic incompressible model to present material features.

Keywords: biomechanics; Achilles tendon; beam-to-beam contact; arbitrary cross-section; ANCF; elasticity

1. Introduction

The Achilles tendon (AT) is the strongest tendon in the body and serves an important function during locomotion. It can reach loads up to four times the body weight while walking and approximately 10 times while running, with the upper border as much as 9 kN [1]. At the same time, it is vulnerable to traumatic injuries due to chronic or acute overloading [2], with the determinants of good recovery not well understood. The AT possesses a complex structure whereby three subtendons, having subject-specific cross-sections, arise from the soleus and the lateral and medial heads of the gastrocnemius muscles. The three heads twist around each other, counterclockwise in the right AT and clockwise in the left. With the complex structure comes functional consequences. Studies have revealed nonuniform displacements within healthy AT [3], whereas the displacement can be more uniform in an injured tendon [4]. Furthermore, loading from the three different muscles causes nonhomogeneous longitudinal strains, compression, and transverse strains in the AT [5]. Longitudinal and transverse strains have also been reported in human studies [6,7]. These nonhomogeneous strains are likely linked to the architectural structure of the tendon [8]. The AT can endure the large forces transmitted axially, and they are most often studied, whereas less attention is placed on shear forces. Because of its significant role in the human musculoskeletal system, a better understanding of AT can provide valuable information for diagnosis and treatment.

What is the significance of the complex geometry and twist of AT? Previous research has shown that region-specific susceptibility to strain injury changes with the amount of tendon twist [9] and that the changes in AT stress are more sensitive to volumetric tendon shape rather than material properties [10]. The appropriate model can help study stress and strain distributions within the AT and improve understanding of tendon function in health, disease, and rehabilitation.

One of the most popular methods in modern biomechanical research for studying tendons is the finite element method (FEM). Using finite solid elements in a framework of the nonlinear FEM helps to comprehend the tendon's complex geometry and materiality. However, AT modeling with the solid finite elements leads to a significant number of degrees of freedom (DOFs) [11–13], which results in a long computation time to obtain a solution. Solid elements can approximate almost any shape and provide reliable results, but require dense discretization in all three dimensions. Therefore, other types of finite elements are necessary to decrease the computational cost.

This research introduces a new approach to the deformation analysis of the Achilles subtendons and their interactions. The main idea is to use one-dimensional finite element discretization over the subtendon's length (in the longitudinal direction) to decrease computational costs. It can be achieved by considering the tendon as a beam-like structure using ANCF-based continuum beam elements with specific descriptions for geometrically complicated deformable cross-sections. This approach leads to the finite element discretization over the subtendon's length and makes it possible to consider material laws based on the continuum mechanics. Recent studies [14–17] show that this transition is possible without significant losses in the quality of the results within the ANCF framework. For example, the work [16] considers the deformations of the beam-like structures described with the ANCF continuum beam elements and from the various soft material models. In [17], the approximation of the rat Achilles tendon experiment with the ANCF elements is given and verified against experimental results. Ref. [15] provides the approximation way for arbitrary cross-sections, which is suitable for the continuum-based ANCF beams. For example, the provided technique approximates one of the Achilles subtendons. In the case of multibeam construction, the question of mutual interaction between beams arises, i.e., the so-called contact problem. In the work [14], the methods for solving contact problems between beams with arbitrary cross-sections are presented.

In this study, we explored the method given in [15] and modeled the whole AT with the ANCF continuum beam elements. The subtendons' cross-sections are obtained with the integration scheme proposed in [15]. Then, the obtained beams are pretwisted, one around the other. In this study, the neo-Hookean material model describes the soft tissue response [10,18]. There are also possibilities to use others' material models in a way given in [16]. However, Annaidh et al. [19] demonstrated that using anisotropic material models within FEM can have inconsistencies. The possible contact between subtendons can be described via the surface-to-surface procedure, thus, taking into account the complicated border interactions between two bodies.

2. ANCF Beam Element

This section provides the geometrical setup for the continuum-based ANCF beam element. The idea behind this element type is to use the slope vectors for defining the cross-section orientation and deformation. The advantages of these finite elements are discussed in [20–22].

There are various types of ANCF elements, and one can divide them into several groups and subgroups; for more details, the reader is referred to Nachbagauer et al. [23], Obrezkov et al. [24], Patel and Shabana [25]. Here, the higher-order three-nodded element with the second-order interpolation in longitudinal and thickness directions denoted 3363 is used. It does not require any modifications to demonstrate good performance even for complicated loading cases [26] and allows the use of all material laws based on the 3D elasticity.

2.1. Kinematics of the ANCF Continuum Beam Elements

Let $r = r(x,y,z) \in \mathbb{R}^3$ be the position vector field of any particle in the current configuration. The position in the initial configuration is denoted as $\bar{r}$ [15], see Figure 1. The connection between the two vectors is

$$r = \bar{r} + u_h, \quad (1)$$

where u_h is a displacement vector. Hence, the body motion is

$$r(x,y,z,t) = \mathbf{N}_m(x,y,z)q(t), \quad (2)$$

where $\mathbf{N}_m$ is a shape function matrix and q is a vector of nodal coordinates. q contains the position of the nodes as well as their derivatives. Therefore, we accept the following notation for i^{th} node:

$$r^i_{,x} = \frac{\partial r^i}{\partial x}, r^i_{,y} = \frac{\partial r^i}{\partial y}, r^i_{,z} = \frac{\partial r^i}{\partial z},$$

$$r^i_{,yy} = \frac{\partial^2 r^i}{\partial y^2}, r^i_{,yz} = \frac{\partial^2 r^i}{\partial y \partial z}, r^i_{,zz} = \frac{\partial^2 r^i}{\partial z^2}.$$

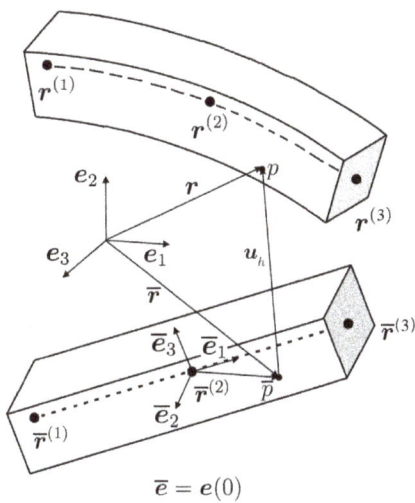

Figure 1. Illustration of a three-nodded beam element with an arbitrary particle p in current and $\bar{p}$ in reference configurations. The three nodes are denoted by $r^{(i)}$ and $\bar{r}^{(i)}$, respectively, $i = 1, 2, 3$ [17].

As mentioned above, in this work, we consider 3363 beam elements [26]. The vectors of nodal coordinates related to this element are presented as follows:

$$q^i = [r^i, r^i_{,y}, r^i_{,z}, r^i_{,yy}, r^i_{,zz}, r^i_{,yz}]. \quad (3)$$

Accordingly, the vector of displacements u_h has the form

$$u_h(x,y,z,t) = \mathbf{N}_m(x,y,z)u(t), \quad (4)$$

where u is a vector of nodal displacements. The element is isoparametric. Here, we introduce a new local coordinate system $\xi = \{\xi, \eta, \zeta\}$ with the range for the local coordinates $[-1,1]$, where $\xi = \frac{2x}{l_x}, \eta = \frac{2y}{l_y}, \zeta = \frac{2z}{l_z}$. Here, l_x, l_y and l_z are the physical dimensions of the element. The substitutions are made to deal with the Gaussian integration procedure [15]. Now, we have

$$\begin{aligned} r(\xi,\eta,\zeta,t) &= \mathbf{N}_m(\xi,\eta,\zeta)q(t), \\ u_h(\xi,\eta,\zeta,t) &= \mathbf{N}_m(\xi,\eta,\zeta)u(t). \end{aligned} \quad (5)$$

Then, the form of the shape function matrix is

$$\mathbf{N}_m(\xi,\eta,\zeta) = [N_1 \mathbf{I} \ N_2 \mathbf{I} \ N_3 \mathbf{I} \ ... \ N_{18} \mathbf{I}], \quad (6)$$

where $\mathbf{I}$ is a 3×3 identity matrix and components of $\mathbf{N}_m$ are

$$N_1 = \frac{1}{2}\xi(\xi-1) \qquad N_2 = \frac{1}{4}l_y\xi\eta(\xi-1) \qquad N_3 = \frac{1}{4}l_z\xi\zeta(\xi-1)$$

$$N_4 = \frac{1}{8}l_zl_y\xi\eta\zeta(\xi-1) \qquad N_5 = \frac{1}{16}l_y^2\xi\eta^2(\xi-1) \qquad N_6 = \frac{1}{16}l_z^2\xi\zeta^2(\xi-1)$$

$$N_7 = 1-\xi^2 \qquad N_8 = \frac{1}{2}l_y\eta(1-\xi^2) \qquad N_9 = \frac{1}{2}l_z\zeta(1-\xi^2)$$

$$N_{10} = \frac{1}{4}l_zl_y\eta\zeta(1-\xi^2) \qquad N_{11} = \frac{1}{8}l_y^2\eta^2(1-\xi^2) \qquad N_{12} = \frac{1}{8}l_z^2\zeta^2(1-\xi^2)$$

$$N_{13} = \frac{1}{2}\xi(\xi+1) \qquad N_{14} = \frac{1}{4}l_y\xi\eta(\xi+1) \qquad N_{15} = \frac{1}{4}l_z\xi\zeta(\xi+1)$$

$$N_{16} = \frac{1}{8}l_zl_y\xi\eta\zeta(\xi+1) \quad N_{17} = \frac{1}{16}l_y^2\xi\eta^2(\xi+1) \quad N_{18} = \frac{1}{16}l_z^2\xi\zeta^2(\xi+1).$$

For further investigation, it is necessary to define the deformation gradient $\mathbf{F}$. From (1) and (2), it can be written as

$$\mathbf{F} = \frac{\partial r}{\partial \bar{r}} = \frac{\partial r}{\partial \xi}\left(\frac{\partial \bar{r}}{\partial \xi}\right)^{-1} = \mathbf{I} + \frac{\partial u_h}{\partial \xi}\left(\frac{\partial \bar{r}}{\partial \xi}\right)^{-1}. \quad (7)$$

The determinant of $\mathbf{F}$ defines the volume ratio of the element, we assume

$$J = \det \mathbf{F} > 0. \quad (8)$$

2.2. Cross-Section Geometry Description

The standard Gaussian quadrature formula for the integration of any function $f(x,y)$ in the general form can be written as follows,

$$\int_\Omega f(x,y)d\Omega = \sum_{i=1}^n \sum_{j=1}^n f(x_i,y_j)w_iw_j, \quad (9)$$

where $2n-1$ is the polynomial exactness degree of function f over one of the axis lines, and w is the weight of the point. For simple cross-sections (circular, etc.), we send our readers to [27], where weights and points in a binormalized coordinate system can be found. Below we present the method for more complicated domains, which can also be found in [15].

Let us consider a closed domain Ω, which has a piecewise border $\partial \Omega$ with points V_i on it:

$$V_i = (\alpha_i, \beta_i), \ i = 1,..,\varphi, \quad (10)$$

$$\partial \Omega = [V_1, V_2] \cup [V_2, V_3] \cup ... \cup [V_\varphi, V_1].$$

The lines $[V_i, V_{i+1}]$ also have several additional "control" points, such as $P_{i1} = V_i, P_{i2}, ..., P_{im_i} = V_{i+1}$, or in the binormalized coordinates as $P_{i1}^{\tilde{z}} = V_i^{\tilde{z}}, ..., P_{im_i}^{\tilde{z}} = V_{i+1}^{\tilde{z}}$. Subsequently, the "cumulative chordal" formula parametrization is recalled:

$$[\alpha_{ij}^{\tilde{z}}, \beta_{ij}^{\tilde{z}}] = \left[0, \sum_{j=1}^{m_i-1} \Delta t_{ij}\right], \ |\Delta t_{ij}| = |P_{ij+1}^{\tilde{z}} - P_{ij}^{\tilde{z}}|, \ j = 1, ..., m_i - 1.$$

Then, each line $[V_i^{\tilde{z}}, V_{i+1}^{\tilde{z}}]$ is tracked by a spline curve $S_i(t) = (S_{i1}(t), S_{i2}(t))$ degree of p_i, where $p_i \leq m_i - 1$, see Figure 2.

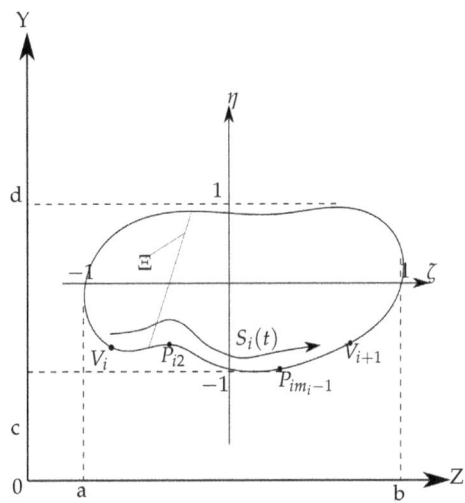

Figure 2. An arbitrary domain in initial and local coordinate systems.

Then the cubature formula with the $2n - 1$ polynomial exactness degree over the Ω domain has the form

$$I_{2n-1} = \sum_{\lambda \in \Lambda_{2n-1}} f(\eta_\lambda, \zeta_\lambda) w_\lambda, \qquad (11)$$

where

$$\Lambda_{2n-1} = \{\lambda = (i, j, k, h) : 1 \leq i \leq \varphi, 1 \leq j \leq m_i - 1,$$
$$1 \leq k \leq n_i, 1 \leq h \leq n\},$$

and w_λ, η_λ and ζ_λ are:

$$\eta_\lambda = \frac{S_{i1}(q_{ijk}) - \Xi}{2} \tau_h^n + \frac{S_{i1}(q_{ijk}) + \Xi}{2},$$

$$\zeta_\lambda = S_{i2}(q_{ijk}),$$

$$w_\lambda = \frac{\Delta t_{ij}}{4} w_k^{n_i} w_h^n (S_{i1}(q_{ijk}) - \Xi) \frac{dS_{i2}(t)}{dt}\bigg|_{t=q_{ijk}},$$

$$q_{ijk} = \frac{\Delta t_{ij}}{2} \tau_k^{n_i} + \frac{t_{ij+1} + t_{ij}}{2}, \ \Delta t_{ij} = t_{ij+1} - t_{ij},$$

$$n_i = \begin{cases} np_i + p_i/2, & p_i \text{ is even,} \\ np_i + (p_i + 1)/2, & p_i \text{ is odd.} \end{cases}$$

Thus, only $\tau_k^{n_i}$, $w_k^{n_i}$ and Ξ need to be defined. Ξ is an arbitrary straight line

$$\Omega \subseteq \mathbb{R}^2 = [a,b] \times [c,d], \Xi(\eta) \in [a,b], \eta \in [c,d].$$

The choice of Ξ does not have any influence. However, it is necessary to obtain the nodes and weights. $\tau_k^{n_i}$, $w_k^{n_i}$ are the nodes and weights, respectively, of the Gauss–Legendre quadrature formula of the exactness degree $2n_i - 1$ on $[-1,1]$.

3. Equilibrium Equation

Our task involves many subroutines, each of them contributing to the energy balance and equilibrium of the whole system. The common approach for calculating is to use the variational formulation. The variations can be grouped as inertia, external, contact, and internal:

$$\delta\Pi_{ext} - \delta\Pi_{int} + \delta\Pi_{inert} - \delta\Pi_{con} = 0. \tag{12}$$

$\delta\Pi_{inert}$ can be written as

$$\delta\Pi_{inert} = \ddot{q}^T \int_V \rho \mathbf{N}^T \mathbf{N} dV \cdot \delta q, \tag{13}$$

where ρ is the mass density, and V is the volume of the element in the reference configuration. The mass matrix is $\mathbf{M} = \int_V \rho \mathbf{N}^T \mathbf{N} dV$. In the case of the static problem, which is the concern of this work, $\delta\Pi_{inert} = 0$. The variation of Π_{int} with respect to the nodal coordinates is [16]

$$\delta\Pi_{int} = \int_V \mathbf{S} : \delta \mathbf{E} dV = \int_V \mathbf{S} : \frac{\partial \mathbf{E}}{\partial q} dV \cdot \delta q. \tag{14}$$

$\mathbf{S}$ is the second Piola–Kirchhoff stress tensor, and its form depends on the material model, which will be presented in Section 4. $\mathbf{E}$ is the Green–Lagrange strain tensor

$$\mathbf{E} = \frac{1}{2}\left(\mathbf{F}^T \cdot \mathbf{F} - \mathbf{I}\right). \tag{15}$$

The last is the variation of the contact force work $\delta\Pi_{con}$, which will be explained.

4. Approximation of the Tendon Tissue

The elastic properties of the Achilles tendon tissue can be presented in different ways. One can find examples in [10,12,13,28,29], where the Helmholtz free energy function Ψ describes elastic features for such material models. In the isotropic case, Ψ depends only on the right Cauchy–Green tensor $\mathbf{C} = \mathbf{F}^T \cdot \mathbf{F}$, therefore, $\Psi = \Psi(\mathbf{C})$. In the case of anisotropy, the additional structural tensor $\mathbf{A}$ can be added to define the preferable deformation direction $\Psi = \Psi(\mathbf{C}, \mathbf{A})$. Models describing AT are usually incompressible. The common approach to deal with it is to split the deformation gradient $\mathbf{F}$ into dilational (volumetric) and distortion (isochoric) parts. Here again, we want to send our readers to the work [19], where the authors point out the possible problems associated with the decomposition of the anisotropic material models. We have

$$\mathbf{F} = J^{\frac{1}{3}}\overline{\mathbf{F}}, \ J = \det \mathbf{F} > 0. \tag{16}$$

This leads to the follow representation of the right Cauchy–Green tensor:

$$\overline{\mathbf{C}} = \overline{\mathbf{F}}^T \cdot \overline{\mathbf{F}}. \tag{17}$$

Thus, after the decomposition, we have

$$\Psi = \Psi_{vol}(J) + \Psi_{iso}(\overline{\mathbf{C}},), \tag{18}$$

where $\Psi_{vol}(J) = k(J-1)^2$, k is a penalty coefficient to guarantee the incompressibility. The part Ψ_{iso} might be reformulated in the terms of the Cauchy–Green deformation tensor invariants,

$$\Psi_{iso} = \Psi_{iso}(\overline{I}_1, \overline{I}_2), \tag{19}$$

where $\overline{I}_1$ and $\overline{I}_2$ have the forms

$$\begin{aligned}\overline{I}_1 &= \operatorname{tr}\overline{\mathbf{C}},\\ \overline{I}_2 &= \tfrac{1}{2}\left(\operatorname{tr}\overline{\mathbf{C}}^2 + \operatorname{tr}^2\overline{\mathbf{C}}\right).\end{aligned} \tag{20}$$

The second Piola–Kirchhoff stress from (14) is formulated as follows:

$$\mathbf{S} = 2\frac{\partial \Psi}{\partial \mathbf{C}} = 2\frac{\partial \Psi}{\partial \overline{\mathbf{C}}} : \frac{\partial \overline{\mathbf{C}}}{\partial \mathbf{C}}. \tag{21}$$

Using (18), it can be expressed as

$$\mathbf{S} = 2\frac{\partial \Psi}{\partial \overline{\mathbf{C}}}\frac{\partial \overline{\mathbf{C}}}{\partial \mathbf{C}} + 2\frac{\partial \Psi_{vol}}{\partial J}\frac{\partial J}{\partial \mathbf{C}} = 2\left(\sum_k \frac{\partial \Psi}{\partial \overline{I}_k}\frac{\partial \overline{I}_k}{\partial \overline{\mathbf{C}}}\right)\frac{\partial \overline{\mathbf{C}}}{\partial \mathbf{C}} + \frac{\partial \Psi_{vol}}{\partial J}J\mathbf{C}^{-1}, \tag{22}$$

$$\frac{\partial J}{\partial \mathbf{C}} = \frac{1}{2}J\mathbf{C}^{-1}.$$

The corresponding volumetric part has the form

$$\mathbf{S}_{vol} = d(J-1)J\mathbf{C}^{-1}. \tag{23}$$

In this study, we consider one type of material model: the neo-Hookean model. The isochoric part of the neo-Hookean model is

$$\overline{\Psi} = c_{10}(\overline{I}_1 - 3), \tag{24}$$

with the expression for the second Piola–Kirchhoff stress tensor:

$$\overline{\mathbf{S}} = 2c_{10}J^{-\frac{2}{3}}\left[\mathbf{I} - \frac{1}{3}\overline{I}_1\overline{\mathbf{C}}^{-1}\right]. \tag{25}$$

5. Contact Formulation

Working with an assembled structure consisting of two or more bodies, the question of interaction between the substructures appears. That problem requires the solution of a contact task. In this study, we are concerned with the description of the bodies of nonstandard forms, such as only surface-to-surface contact formulation, which can describe this contact [14].

Let us describe the task of two contacting beams (denoted as A and B) in the terms of the distances between the two closest position vector fields r^A and r^B. Then, assuming that along the contact surface there is no penetration, the minimum distance problem in the most general case can be formulated as follows:

$$d = \|r^A - r^B\|. \tag{26}$$

The nonpenetration condition is defined via the so-called gap function, which in this work is given as follows,

$$\begin{aligned}g(\xi^A, \xi^B, \eta^A, \eta^B, \zeta^A, \zeta^B) &= \|r^A(\xi^A)_{\eta,\zeta=0} - r^B(\xi^B_c)_{\eta,\zeta=0}\|\\ &\quad - (\|r^A(\xi^A, \eta^A, \zeta^A) - r^A(\xi^A)_{\eta,\zeta=0}\|\\ &\quad + \|r^B_c(\xi^B_c, \eta^B_c, \zeta^B_c) - r^B(\xi^B_c)_{\eta,\zeta=0}\|),\end{aligned} \tag{27}$$

where $g(\xi^A, \xi^B, \eta^A, \eta^B, \zeta^A, \zeta^B) \geq 0$. The subscript $_c$ denotes the orthogonal projection of the point on beam A on the beam B obtained from (27), $r^B(\xi_c^B)_{\eta,\zeta=0}$ is the point projection $r_c^B(\xi_c^B, \eta_c^B, \zeta_c^B)$ on the beam centerline. In the model, we assume that two bodies are closely placed to each other, and only sliding is allowed. Therefore, the nonpenetration condition is $g = 0$. Then the variation reads as follows,

$$\delta \Pi_{con} = p_n \int_{\Omega_c} g \delta g \, d\Omega, \quad (28)$$

where Ω_c is the contacting surface between A and B beams, and p_n is the penalty parameter. The weak form of contact energy (28) presented in Section 3 can be expressed in the discrete form as follows,

$$\delta \Pi_{con} = -\delta u_A^T p_n \sum_{i=1}^{n^i} \sum_{j=1}^{n^j} \sum_{k=1}^{n^k} g(\xi_i^A, \eta_j^A, \zeta_k^A) N_A^T n_{ijk} w_i w_j w_k$$

$$+ \delta u_B^T p_n \sum_{i=1}^{n^i} \sum_{j=1}^{n^j} \sum_{k=1}^{n^k} g(\xi_c^B, \eta_c^B, \zeta_c^B) N_B^T n_{ijk} w_i w_j w_k, \quad (29)$$

where

$$n_{ijk} = n(\xi_c^B(\xi_i^A), \eta_c^B(\xi_i^A, \eta_j^A, \zeta_k^A), \eta_c^B(\xi_i^A, \eta_j^A, \zeta_k^A)),$$

$$N_A^T = N^T(\xi_i^A, \eta_j^A, \zeta_k^A)$$

$$N_B^T = N^T(\xi_c^B(\xi_i^A), \eta_c^B(\xi_i^A, \eta_j^A, \zeta_k^A), \zeta_c^B(\xi_i^A, \eta_j^A, \zeta_k^A)).$$

In (29), n^i is the amount of Gauss points in the A beam, along the ξ direction, w_i are their corresponding weight, η_k and ζ_k are the Gauss points coordinates along the η and ζ directions parameters. ξ_c^B, η_c^B and ζ_c^B are the parameters of the closest projected point $r(\xi_j^A, \eta_k^A, \zeta_k^A)$ on B, n is a normal vector from the B to A beam elements' surfaces.

6. Numerical Examples

Previous studies found that the Achilles tendon consists of three subtendons with each having a complicated cross-section shape [30,31]. Additionally, there are three common types of AT with varying subtendon regions and torsion [30]. In this work, we consider the AT of Type III due to its relatively simple cross-section form. We extracted the geometrical description of subtendons from [30]. Although the exact geometrical data are not presented, we use CAD software to obtain the positions of the points, as in [15]. Here, we also considered the pretwist of the tendon about the centroidal axis (line, where all three subtendons are connected) from $0°$ at $x = 0$ to ψ degrees at $x = L$. The centroidal axis of the beam remains straight. See Figure 3. The representations of the Gauss points for all three subtendons are given in Figure 4a–c.

The length of the tendon was set at $L = 0.07$ m [29]. The geometrical results based on the approximation are 16.31 mm^2 for the soleus subtendon, 15.98 mm^2 for the medial and 19.57 mm^2 for lateral subtendons, with total area equaling to 51.86 mm^2. That slightly exceeds the average female tendon cross-section 51.2 mm^2 and is smaller than the average male cross-section 62.1 mm^2 [29].

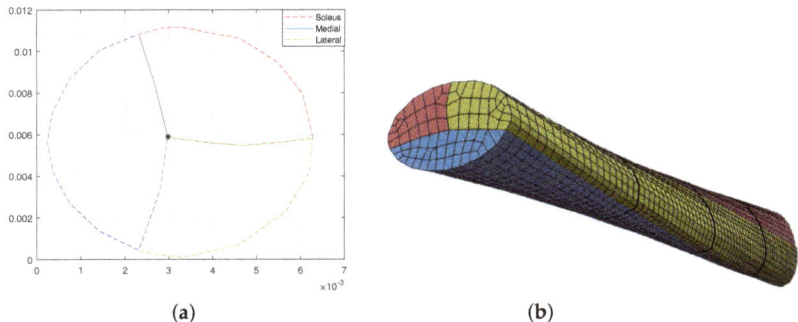

Figure 3. The Type III tendon representation. (a) The Achilles sub-tendons' cross sections. (b) The pretwisted underformed Achilles tendon discretized by four ANCF-based continuum beam elements at each subtendon with $\psi = 45°$.

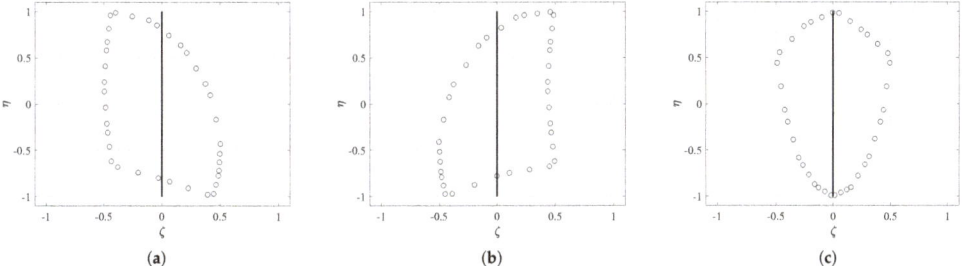

Figure 4. Integration approximations of the three subtendons by the Gauss–Green cubature formula. (a) Soleus. (b) Medial gastrocnemius. (c) Lateral gastrocnemius.

We used the neo-Hookean material model with three shear modulus equal to $c_{10} = 103.1$ MPa for soleus, $c_{10} = 143.2$ MPa, and $c_{10} = 226.7$ MPa for medial and lateral subtendons, respectively [29]. We considered three different pretwisted designs: $\psi = 0$, $\psi = 15$, and $\psi = 45$. The choice is based on the work [28], where the optimal value of twisted is found between 15 and 45 degrees. Then, the soleus subtendon was subjected to forces along the longest direction and applied at the last node, the maximum applied tensile load is 400 N. The applied force exceeds four times the loading conditions given in [15,29], allowing the demonstration of the nonlinear deformations, about 10% of the initial length. On the other edge, $r = 0$ from (3) is fixed at the first node, and this condition forbids the displacement, but allows the cross-sectional contraction.

The results presented in Table 1 are consistent with the ones given in [15], where the pretwisted subtendons show higher elongations under the same load in comparison to straight subtendons. The elongations for other subtendons are near zero, which indicates that there is sliding between the subtendons as in Section 5 holds. Table 2 presents the converge tests, wherein the elongation results for a number of mesh refinements for the straight and pretwisted soleus subtendon of Type III from the neo-Hookean material model subjected to $N = 400$ N tensile force are given.

Table 1. The elongation test results in [mm] for the straight and pretwisted soleus subtendon of Type III tendons from the neo-Hookean material model.

Applied Load [N]	Elongation [mm] of the Soleus Sub-Tendon		
	Variation of $\psi°$		
	$\psi = 0$	$\psi = 15$	$\psi = 45$
10	0.143	0.146	0.168
20	0.286	0.289	0.313
30	0.430	0.433	0.457
40	0.574	0.578	0.602
45	0.647	0.650	0.675
60	0.864	0.868	0.893
80	1.157	1.160	1.186
90	1.304	1.307	1.333
100	1.451	1.454	1.481
150	2.197	2.201	2.228
200	2.958	2.961	2.989
300	4.524	4.527	4.557
400	6.151	6.155	6.187

Table 2. Elongation results in [mm] for several mesh refinements for the straight and pretwisted soleus sub-tendon of Type III from the neo-Hookean material model under $N = 400$ N tensile force.

Element Number per Sub-Tendons $n_{Sol} \times n_{MG} \times n_{LG}$	Elongation [mm] of the Soleus Sub-Tendon		
	Variation of ψ		
	$\psi = 0°$	$\psi = 15°$	$\psi = 45°$
$1 \times 1 \times 1$	6.051	6.055	6.123
$2 \times 2 \times 2$	6.089	6.093	6.123
$4 \times 4 \times 4$	6.151	6.155	6.187

The deformed shapes for straight pretwisted $\psi = 15$ the tendons are given in Figure 5a, where the shapes are discretized by four ANCF-based continuum beam elements at each subtendon.

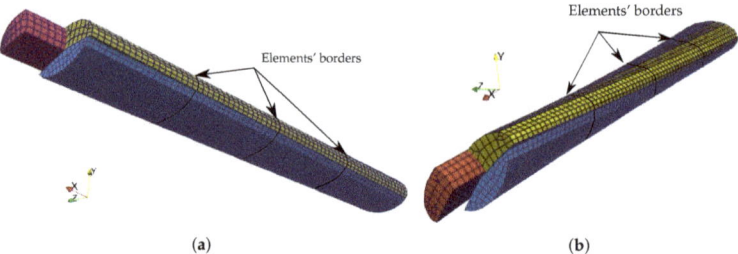

Figure 5. The deformed shapes of the pre-twisted Achilles tendons when the soleus is loaded and discretized by four ANCF-based continuum beam elements at each subtendon. (a) $\psi = 15°$ (b) $\psi = 45°$.

7. Conclusions

This work uses the continuum-based ANCF beam element to describe the human Achilles tendon's deformation due to elongation. In the study, the AT is presented as a combination of three substructures, pretwisted and sliding one around the others. The contact between them is described with the segment-to-segment algorithm. The Gauss–Green cubature integration formula captures the sophisticated cross-section form of each subtendon. The neo-Hookean isotropic material model describes the pure elastic response. The results show that the model is feasible, but more careful verification is necessary. That can include models built with conventional 3D solid elements and the comparison with experimental data.

Additionally, the work possesses certain limitations. For example, the cross-sectional area is taken to be the same for all subtendons along their longitudinal axes. That is a substantial simplification, but there is no available geometrical data to approximate such variation.

Author Contributions: Conceptualization, M.K.M. and T.F.; methodology, M.K.M. and L.P.O.; software, Matlab; validation, M.K.M. and T.F., and L.P.O.; formal analysis, L.P.O.; investigation, L.P.O.; writing—original draft preparation, L.P.O.; writing—review and editing, M.K.M. and T.F. and L.P.O.; visualization, L.P.O.; supervision, M.K.M. and T.F.; project administration, M.K.M. and T.F.; funding acquisition, T.F. All authors have read and agreed to the published version of the manuscript.

Funding: This research was funded by The Academy of Finland (Decisions No. 299033 and 323168).

Data Availability Statement: Not applicable.

Acknowledgments: We would like to thank the Academy of Finland (Decisions No. 299033 and 323168) for funding.

Conflicts of Interest: The authors declare no conflict of interest.

Abbreviations

The following abbreviations are used in this manuscript:

AT	Achilles tendon
FEM	Finite Element Model
ANCF	Absolute Nodal Coordinate Formulation
DOF	Degrees of Freedom

References

1. Komi, P.V.; Fukashiro, S.; Järvinen, M. Biomechanical loading of Achilles tendon during normal locomotion. *Clin. J. Sport Med.* **1992**, *11*, 521–531. [CrossRef]
2. Järvinen, T.A.; Kannus, P.; Maffulli, N.; Khan, K.M. Achilles Tendon Disorders: Etiology and Epidemiology. *Foot Ankle Clin.* **2005**, *10*, 255–266. [CrossRef] [PubMed]
3. Slane, L.C.; Thelen, D.G. Non-uniform displacements within the Achilles tendon observed during passive and eccentric loading. *J. Biomech.* **2014**, *47*, 2831–2835. [CrossRef]
4. Khair, R.M.; Stenroth, L.; Péter, A.; Cronin, N.J.; Reito, A.; Paloneva, J.; Finni, T. Non-uniform displacement within ruptured Achilles tendon during isometric contraction. *Scand. J. Med. Sci. Sport.* **2021**, *31*, 1069–1077. [CrossRef] [PubMed]
5. Obuchowicz, R.; Ekiert, M.; Kohut, P.; Holak, K.; Ambrozinski, L.; Tomaszewski, K.; Uhl, T.; Mlyniec, A. Interfascicular matrix-mediated transverse deformation and sliding of discontinuous tendon subcomponents control the viscoelasticity and failure of tendons. *J. Mech. Behav. Biomed. Mater.* **2019**, *97*, 238–246. [CrossRef] [PubMed]
6. Farris, D.J.; Trewartha, G.; McGuigan, M.P.; Lichtwark, G.A. Differential strain patterns of the human Achilles tendon determined in vivo with freehand three-dimensional ultrasound imaging. *J. Exp. Biol.* **2013**, *216*, 594–600. [CrossRef] [PubMed]
7. Obst, S.J.; Renault, J.B.; Newsham-West, R.; Barrett, R.S. Three-dimensional deformation and transverse rotation of the human free Achilles tendon in vivo during isometric plantarflexion contraction. *J. Appl. Physiol.* **2014**, *116*, 376–384. [CrossRef] [PubMed]
8. Bojsen-Møller, J.; Magnusson, S.S. Heterogeneous Loading of the Human Achilles Tendon In Vivo. *Exerc. Sport Sci. Rev.* **2015**, *43*, 190–197. [CrossRef]
9. Shim, V.; Handsfield, G.; Fernandez, J.; Lloyd, D.; Besier, T. Combining in silico and in vitro experiments to characterize the role of fascicle twist in the Achilles tendon. *Sci. Rep.* **2018**, *8*, 13856. [CrossRef]
10. Hansen, W.; Shim, V.B.; Obst, S.; Lloyd, D.G.; Newsham-West, R.; Barrett, R.S. Achilles tendon stress is more sensitive to subject-specific geometry than subject-specific material properties: A finite element analysis. *J. Biomech.* **2017**, *56*, 26–31. [CrossRef]
11. Kinugasa, R.; Yamamura, N.; Sinha, S.; Takagi, S. Influence of intramuscular fiber orientation on the Achilles tendon curvature using three-dimensional finite element modeling of contracting skeletal muscle. *J. Biomech.* **2016**, *49*, 3592–3595. [CrossRef] [PubMed]
12. Morales-Orcajo, E.; Souza, T.R.; Bayod, J.; Barbosa de Las Casas, E. Non-linear finite element model to assess the effect of tendon forces on the foot-ankle complex. *Med. Eng. Phys.* **2017**, *49*, 71–78. [CrossRef] [PubMed]
13. taş, R.A.; Lucaciu, D.O. Finite Element Analysis of the Achilles Tendon While Running. *Acta Medica Marisiensis* **2013**, *59*, 8–11. [CrossRef]

14. Bozorgmehri, B.; Obrezkov, L.P.; Harish, A.B.; Matikainen, M.K.; Mikkola, A. A contact description for continuum beams with deformable arbitrary cross-section. *Finite Elem. Anal. Des.* **2023**, *214*, 103863. [CrossRef]
15. Obrezkov, L.; Bozorgmehri, B.; Finni, T.; Matikainen, M.K. Approximation of pre-twisted Achilles sub-tendons with continuum-based beam elements. *Appl. Math. Model.* **2022**, *112*, 669–689. [CrossRef]
16. Obrezkov, L.P.; Matikainen, M.K.; Harish, A.B. A finite element for soft tissue deformation based on the absolute nodal coordinate formulation. *Acta Mech.* **2020**, *231*, 1519–1538. [CrossRef]
17. Obrezkov, L.P.; Eliasson, P.; Harish, A.B.; Matikainen, M.K. Usability of finite elements based on the absolute nodal coordinate formulation for the Achilles tendon modelling. *Int. J. Non-Linear Mech.* **2021**, *129*, 103662. [CrossRef]
18. Weiss, J.A.; Maker, B.N.; Govindjee, S. Finite element implementation of incompressible, transversely isotropic hyperelasticity. *Comput. Methods Appl. Mech. Eng.* **1996**, *135*, 107–128. [CrossRef]
19. Annaidh, A.N.; Destrade, M.; Gilchrist, M.D.; Murphy, J.G. Deficiencies in numerical models of anisotropic nonlinearly elastic materials. *Biomech. Model. Mechanobiol.* **2013**, *12*, 781–791. [CrossRef]
20. Escalona, J.L.; Hussien, H.A.; Shabana, A.A. Application of absolute nodal co-ordinate formulation to multibody system dynamics. *J. Sound Vib.* **1998**, *214*, 833–851. [CrossRef]
21. Maqueda, L.G.; Bauchau, O.A.; Shabana, A.A. Effect of the centrifugal forces on the finite element eigenvalue solution of a rotating blade: a comparative study. *Multibody Syst. Dyn.* **2008**, *19*, 281–302. [CrossRef]
22. Shen, Z.; Li, P.; Liu, C.; Hu, G. A finite element beam model including cross-section distortion in the absolute nodal coordinate formulation. *Nonlinear Dyn.* **2014**, *77*, 1019–1033. [CrossRef]
23. Nachbagauer, K.; Gruber, P.; Gerstmayr, J. A 3D Shear Deformable finite element based on the absolute nodal coordinate formulation. *Multibody Dyn.* **2013**, *28*, 77–96. [CrossRef]
24. Obrezkov, L.P.; Mikkola, A.; Matikainen, M.K. Performance review of locking alleviation methods for continuum ANCF beam elements. *Nonlinear Dyn.* **2022**, *109*, 531–546. [CrossRef]
25. Patel, M.; Shabana, A.A. Locking alleviation in the large displacement analysis of beam elements: the strain split method. *Acta Mech.* **2018**, *229*, 2923–2946. [CrossRef]
26. Ebel, H.; Matikainen, M.K.; Hurskainen, V.V.; Mikkola, A. Higher-order beam elements based on the absolut nodal coordinate formulation for three-dimensional elasticity. *Nonlinear Dyn.* **2017**, *88*, 1075–1091. [CrossRef]
27. Abramowitz, M.; Stegun, I.A.; Romer, R.H. Handbook of Mathematical Functions With Formulas, Graphs and Mathematical Tables. *Am. J. Phys.* **1988**, *58*, 958–958. [CrossRef]
28. Handsfield, G.G.; Greiner, J.; Madl, J.; Rog-Zielinska, E.A.; Hollville, E.; Vanwanseele, B.; Shim, V. Achilles Subtendon Structure and Behavior as Evidenced From Tendon Imaging and Computation Modeling. *Front. Sport. Act. Living* **2020**, *2*, 70. [CrossRef]
29. Yin, N.Y.; Fromme, P.; McCarthy, I.; Birch, H. Individual variation in Achilles tendon morphology and geometry changes susceptibility to injury. *eLife* **2021**, *10*, e63204. [CrossRef]
30. Edama, M.; Kubo, M.; Onishi, H.; Takabayashi, T.; Inai, T.; Yokoyama, E.; Hiroshi, W.; Satoshi, N.; Kageyama, I. The twisted structure of the human Achilles tendon. *Scand. J. Med. Sci. Sport.* **2015**, *25*, e497–e503. [CrossRef]
31. Finni, T.; Bernabei, M.; Baan, G.C.; Noort, W.; Tijs, C.; Maas, H. Non-uniform displacement and strain between the soleus and gastrocnemius subtendons of rat Achilles tendon. *Scand. J. Med. Sci. Sport.* **2018**, *28*, 1009-1017. [CrossRef] [PubMed]

Article

On the Stability of Complex Concentrated (CC)/High Entropy (HE) Solid Solutions and the Contamination with Oxygen of Solid Solutions in Refractory Metal Intermetallic Composites (RM(Nb)ICs) and Refractory Complex Concentrated Alloys (RCCAs)

Panos Tsakiropoulos

Department of Materials Science and Engineering, Sir Robert Hadfield Building, The University of Sheffield, Mappin Street, Sheffield S1 3JD, UK; p.tsakiropoulos@sheffield.ac.uk

Citation: Tsakiropoulos, P. On the Stability of Complex Concentrated (CC)/High Entropy (HE) Solid Solutions and the Contamination with Oxygen of Solid Solutions in Refractory Metal Intermetallic Composites (RM(Nb)ICs) and Refractory Complex Concentrated Alloys (RCCAs). *Materials* 2022, 15, 8479. https://doi.org/10.3390/ma15238479

Academic Editors: Giovanni Pappalettera and Sanichiro Yoshida

Received: 18 October 2022
Accepted: 22 November 2022
Published: 28 November 2022

Publisher's Note: MDPI stays neutral with regard to jurisdictional claims in published maps and institutional affiliations.

Copyright: © 2022 by the author. Licensee MDPI, Basel, Switzerland. This article is an open access article distributed under the terms and conditions of the Creative Commons Attribution (CC BY) license (https://creativecommons.org/licenses/by/4.0/).

Abstract: In as-cast (AC) or heat-treated (HT) metallic ultra-high temperature materials often "conventional" and complex-concentrated (CC) or high-entropy (HE) solid solutions (sss) are observed. Refractory metal containing bcc sss also are contaminated with oxygen. This paper studied the stability of CC/HE Nb_{ss} and the contamination with oxygen of Nb_{ss} in RM(INb)ICs, RM(Nb)ICs/RCCAs and RM(Nb)ICs/RHEAs. "Conventional" and CC/HE Nb_{ss} were compared. "Conventional" Nb_{ss} can be Ti-rich only in AC alloys. Ti-rich Nb_{ss} is not observed in HT alloys. In B containing alloys the Ti-rich Nb_{ss} is usually CC/HE. The CC/HE Nb_{ss} is stable in HT alloys with simultaneous addition of Mo, W with Hf, Ge+Sn. The implications for alloy design of correlations between the parameter δ of "conventional" and CC/HE Nb_{ss} with the B or the Ge+Sn concentration in the Nb_{ss} and of relationships of other solutes with the B or Ge+Sn content are discussed. The CC/HE Nb_{ss} has low $\Delta\chi$, VEC and Ω and high ΔS_{mix}, $|\Delta H_{mix}|$ and δ parameters, and is formed in alloys that have high entropy of mixing. These parameters are compared with those of single-phase bcc ss HEAs and differences in ΔH_{mix}, δ, $\Delta\chi$ and Ω, and similarities in ΔS_{mix} and VEC are discussed. Relationships between the parameters of alloy and "conventional" Nb_{ss} also apply for CC/HE Nb_{ss}. The parameters δ_{ss} and Ω_{ss}, and VEC_{ss} and VEC_{alloy} can differentiate between types of alloying additions and their concentrations and are key regarding the formation or not of CC/HE Nb_{ss}. After isothermal oxidation at a pest temperature (800 °C/100 h) the contaminated with oxygen Nb_{ss} in the diffusion zone is CC/HE Nb_{ss}, whereas the Nb_{ss} in the bulk can be "conventional" Nb_{ss} or CC/HE Nb_{ss}. The parameters of "uncontaminated" and contaminated with oxygen sss are linked with linear relationships. There are correlations between the oxygen concentration in contaminated sss in the diffusion zone and the bulk of alloys with the parameters $\Delta\chi_{Nbss}$, δ_{Nbss} and VEC_{Nbss}, the values of which increase with increasing oxygen concentration in the ss. The effects of contamination with oxygen of the near surface areas of a HT RM(Nb)IC with Al, Cr, Hf, Si, Sn, Ti and V additions and a high vol.% Nb_{ss} on the hardness and Young's modulus of the Nb_{ss}, and contributions to the hardness of the Nb_{ss} in B free or B containing alloys are discussed. The hardness and Young's modulus of the bcc ss increased linearly with its oxygen concentration and the change in hardness and Young's modulus due to contamination increased linearly with $[O]^{2/3}$.

Keywords: high entropy alloys; complex concentrated alloys; refractory metal intermetallic composites; high entropy phases; complex concentrated phases; Nb silicide-based alloys; alloy design

1. Introduction

The interdepended targets for performance and environmental impact of future aero engines could be met with materials that would allow high pressure turbines to operate at significantly higher than current temperatures. In other words, ultra-high temperature materials (UHTMs) with capabilities beyond those of Ni-based superalloys are needed [1]. UHTMs must meet property goals for fracture toughness, oxidation resistance and creep [2]. The fracture toughness property goal necessitates the new materials to show some degree of metallic behaviour to distinguish them from ceramic UHTMs [3]. Research and development work is in progress to find metallic UHTMs that can be used in structural engineering applications [2–8].

Metallic UHTMs depend on refractory metal (RM) additions and include RM intermetallic composites (ICs), i.e., RMICs, RM high entropy alloys (HEAs), i.e., RHEAs and RM complex concentrated alloys (CCAs), i.e., RCCAs. This classification is logically and pragmatically exhaustive. Not all RHEAs or RCCAs are RMICs, but some are. Moreover, not all RMICs are RHEAs or RCCAs, but some are. RMICs based on the Nb-Si system, i.e., RM(Nb)ICs or the Mo-Si system, i.e., RM(Mo)ICs are under development [3,8]. Some of the former are also high-entropy or complex concentrated alloys, i.e., RM(Nb)IC/RHEA or RM(Nb)IC/RCCA [3,9]. In this paper ceramic UHTMs and RM(Mo)ICs are not considered.

The RM(Nb)ICs, RM(Nb)ICs/RCCAs and RM(Nb)ICs/RHEAs are multiphase alloys with phases such as bcc solid solution(s), silicide(s), C14 Laves and A15 compounds, and other intermetallics [10–13]. These phases can be "conventional" phases or high entropy (HE) phases or complex concentrated (CC) (compositionally complex) phases [3,10,14]. HE or CC eutectics and/or HE or CC lamellar microstructures also can form in their microstructures [13,15,16]. The "conventional" phases can co-exist with the CC/HE phases in the as cast (AC) and/or heat treated (HT) conditions or after oxidation [13,14,17,18]. Phase transformations of CC intermetallics can generate unusual microstructures in RM(Nb)ICs [15]. HEAs and HE phases are those where the maximum and minimum concentrations of elements are not above or below, respectively, 35 and 5 at.%, whereas RCCAs and CC phases are those where the maximum and minimum concentrations of elements are above 35 at.% (up to about 40 at.%) and below 5 at.% [3,9,19].

The microstructures of RHEAs and RCCAs can be single phase or multiphase, namely solid solution(s) with/without intermetallics, for example M_5Si_3 silicides owing to Si addition (M = transition metal (TM) and/or RM) or Laves phases [19]. RMICs, RHEAs and RCCAs share the same alloying elements [3,9,14]. In the pairings RMIC-RHEA and RMIC-RCCA the two terms are mutually complementary (the same is the case for the pairings HEA-CCA, RHEA-RCCA, RM(Nb)IC-RM(Mo)IC). The development of RM(Nb)ICs is linked with the study of intermetallics and the development of intermetallic-based alloys (e.g., [1,20,21]), in contrast with the development of RHEAs and RCCAs that resulted from research on HEAs [19]. For these three categories of metallic UHTMs there is a significant volume of research [3,19]. Methods of preparation of metallic UHTMs are discussed in [2,3,19].

1.1. Alloy Design and the Alloy Design Methodology NICE

Groups of alloys (e.g., Ni-based superalloys for blade or disc applications in gas turbine engines) exhibit striking regularities [22]. Metallurgists who develop new alloys can have data that might not be directly intelligible as they stand and with relationships that are not immediately apparent [9]. Time and again, enthalpy and entropy of mixing, electronegativity, atomic size, electron-to-atom ratio and relationships based on these parameters provide an intermediate step to link the data, to weave them into a framework of understanding that is subtle and mathematical. Parameters based on the aforementioned thermo-physical and structure properties can reflect, albeit imperfectly, actually existing properties of alloys that help us uncover new things about alloys and their phases, sometimes things we never suspected, to uncover regularities and linkages and to establish relationships between different properties [9]. This has been demonstrated for rapidly solidified crystalline and amorphous alloys, bulk metallic glasses, HEAs and RM(Nb)ICs, for example [3,9,10,14,23–27]. Relationships between parameters of alloys and their phases, between the same parameters and properties of alloys and their phases have shown that there is an elegant simplicity that is underpinned by definite mathematical relationships that interweave each other to form via their interrelatedness and interdependent influences a subtle and harmonious methodology to process alloy design/selection through progressive goal-oriented approach [9,10,14,28–31]. This design methodology is known as NICE [10]. It was founded on data for RM(Nb)ICs [10] and has been expanded to cover RHEAs and RCCAs with Nb and Si addition [3,9,15,32–35]. The papers [28–31] dealt closely with questions that pertain to the alloying behaviour and properties of key phases in RM(Nb)ICs, RM(Nb)ICs/RCCAs and RM(Nb)ICs/RHEAs. In [9], a succinct account was given of the approach and aspirations upon which the said papers and [26] "converged" and were "unified" in NICE [10]. One could visualise this research as a "fruit producing tree". The study in [26] forms the trunk, the "sprinkle of water" that "feeds its growth" is from [9] and new research, [28–31] are its "branches" and [10], i.e., NICE, is its "fruit". Manifestations of the "juiciness" of this "fruit" are [13,14,16–18,32–38] and this paper.

As it will be demonstrated in this paper, NICE helps the alloy developer to find unexpected new relationships as the range of investigation of metallic UHTMs is expanded. NICE depends on high quality chemical analysis data for the calculation of parameters based on aforementioned properties, namely the parameters ΔH_{mix}, ΔS_{mix}, δ, $\Delta \chi$, VEC and Ω, which are the same parameters used to study HEAs and CCAs [3,10,26–28,39–41]. With NICE, a material system suitable for application in high pressure turbine and comprising a metallic UHTM substrate plus metallic bond coat of an environmental coating of the bond coat/thermally grown oxide/ceramic top coat type can be designed [14].

Although metallic UHTMs can be complex, they are clearly not random. We observe regularities and patterns, and organise these into relationships which are used in NICE and give it predictive power [9,14,26,28]. For example, the boron containing RM(Nb)ICs and RM(Nb)ICs/RCCAs occupy a specific corner in the $\Delta \chi$ versus δ map or a specific area in the ΔH_{mix} versus $\Delta \chi$ map [3,14,26], oxidation resistant RM(Nb)ICs and RM(Nb)ICs/RCCAs have low VEC and high δ values [10,18,32,34–36,42]. In these metallic UHTMs the "behaviour" of one element is inextricably entangled with those of the others via the aforementioned parameters and the relationships that have been found between them, for example see Figures 12 and 16 in [10], Figures 12–18 in [15], Figures 1 and 2 in [29], Figures 1–6 in [30], Figures 1–11 in [31], Figures 12–14 in [33], Figures 9, 12 and 13 in [34], Figures 7–15 in [35], Figures 4 and 5 in [36], Figure 13 in [37], Figures 10 and 11 in [43] and Figures 8 and 9 in [44]. The available data give a realistic (workable, effective, consistent) account of how the alloying behaviour and properties of alloys and their phases are "determined (controlled)" by different groups of elements working in synergy in a metallic UHTM [3,9,10,14].

One way of expressing this "quality" of metallic UHTMs, meaning the regularities that they show, is to say that these materials have organised complexity. This organisation is captured by NICE, which focuses on the amount of information needed and its quality and value. Regularities are systematised into relationships [9,10,29–31]. Given a property goal, these relationships are used in NICE to calculate the chemical composition of an alloy, properties of which also can be computed [3,9,10,14,33,34,37,38,45]. Underlying the complexity of metallic UHTMs is the apparent simplicity of relationships that enable organised complexity to emerge. The organizational properties of these complex alloys are attributed to the relationships of parameters that reflect the specific nature of the alloys concerned. Regularities possess contingent features, meaning they depend upon something beyond themselves, for example, contamination by interstitials (see below in this section and Section 3.1) owing to interaction with the environment, and thus parameter values and relationships change (see Section 3 below).

The design/selection of new alloys is possible using NICE [10,14]. Design constraints pertaining to an alloy of interest can be traced to the wider alloying environment, for example see [33,34,37,38,45]. One of the main features of NICE is that the "affairs" of alloys cannot be separated from the "affairs" of phases and the parameters that describe alloying behaviour and properties of alloys and phases. It is a linkage that has profound implication for the design of metallic UHTMs [14].

NICE shows that metallic UHTMs must be understood holistically and that the properties of a metallic UHTM are comprehended by studying the alloying behaviour and properties of its constituent phases. In other words, NICE proposes two complementary ways of studying alloy development using both reductionist and holistic approaches. Akin to all alloys, a RM(Nb)IC, RHEA or RCCA is a physical system with a collection of atoms of different elements with similar or different concentrations and different levels of structure (meaning the different or similar structures of elements and of the phases such as solid solution(s) and intermetallic(s) that make up the alloy microstructure with a particular "architecture" (e.g., co-continuous solid solution(s)-intermetallic(s)), influenced by internal processes (e.g., solute partitioning) or the environment (e.g., contamination with interstitials) in which the alloy is produced and/or operates. For example, partitioning of solutes can result (i) to change in crystal structure (the case of Ti partitioning to Nb_5Si_3 and substituting Nb, thus causing a change in structure from tetragonal to hexagonal [46]) or (ii) formation of sub-grains in Nb_5Si_3 [47], while change in structure also can occur with contamination with interstitials (for example, the case of hexagonal instead of tetragonal Nb_5Si_3 stabilised in Nb-Si alloys with C contamination [48]).

In RM(Nb)ICs, RM(Nb)ICs/RCCAs or RM(Nb)ICs/RHEAs and single phase or multi-phase RCCAs or RHEAs the solid solution(s) will be contaminated with oxygen, and the severity of contamination will differ, depending on alloying additions and their concentrations, and exposure conditions [13]. Alloying strategies might be able to counterbalance effects of interstitial contamination on properties. For example, grain-boundary segregation of oxygen caused room-temperature brittleness of the as cast (AC) single phase solid solution NbMoTaW RHEA. Alloying with B from 400 ppm (0.04 at.%) to 8000 ppm (0.8 at.%) offset this effect of O and improved the mechanical properties at room temperature. Both strength and plasticity were improved and reached maximum values at around 5000 ppm (0.5 at.%) B addition. Specifically, the plasticity increased from <2% to >10% and the fracture strength increased from 1211 MPa to 1780 MPa, respectively, for the base RHEA and the RHEA alloyed with 5000 ppm B. However, the plasticity of the said RHEA decreased with further increase in the B concentration [49]. Contamination with oxygen can have a strong effect of the near surface properties of phases and alloy [3]. This paper will show how NICE helps the alloy developer to understand the effect of contamination with oxygen on the properties of the bcc solid solution.

1.2. Aim of This Work

HE or CC phases can co-exist with "conventional" phases and can be stable in RM(Nb)ICs, RM(Nb)ICs/RCCAs or RM(Nb)ICs/RHEAs [10,14]. Phase transformations of CC silicides give new simple and/or complex microstructures the importance of which for the properties of alloys has not been studied or considered in modelling research, e.g., modelling of creep [50]. Are the CC or HE bcc solid solutions stable? Is their stability dependent on alloying additions, alloy condition (meaning AC or heat treated (HT)) and contamination with oxygen? Boron or Ge and Sn have a distinctive effect on the alloying behaviour and properties of the aforementioned materials [26,32–34,45] and the Nb_5Si_3 silicide [9,14,29]. Is the stability of CC/HE Nb_{ss} dependent on the presence of B or Ge and Sn in the alloy? Are there similarities regarding the dependence of other solute addition concentrations on the B or Ge+Sn content of solid solutions? How does the contamination of bcc Nb_{ss} with oxygen or alloying with boron affect its properties? The motivation for this paper was to provide answers to these questions.

I shall consider some of the possible permutations of available data between aforementioned parameters and between parameters and solutes under two major headings, namely "complex concentrated bcc solid solution" and "contamination of the bcc solid solution with oxygen". There is a logic behind this approach in this paper, as I shall aim to show. The four solutes Ge, Sn, B and O will be a focus, and the latter two will be the point of reference when I shall discuss the hardness of the bcc solid solution. All four solutes are remarkably untypical in RHEAs and RCCAs studied to date (e.g., see [19]) even though they are essential additions in RM(Nb)ICs, RM(Nb)ICs/RCCAs and RM(Nb)ICs/RHEAs for balance of properties. The first three, Ge and Sn together and B on its own or in synergy with Ge or Sn can assist the alloy developer to obtain metallic UHTMs with a balance of properties by making use of the synergies of these three elements with Al, Cr, Hf, Si and Ti, as suggested by research on RM(Nb)ICs/RCCAs and RM(Nb)ICs/RHEAs, e.g., see [9,14,32–36]. Oxygen is a solute the presence of which cannot be avoided in UHTMs with RM additions owing to the sensitivity of RMs to interstitial contamination (e.g., see [3] and Section 3.1 below). The contamination with oxygen has profound implications for properties of phases (this will be demonstrated for the solid solution in this paper) that (a) should not be ignored in studies of processing-microstructure-property relationships in UHTMs, as discussed in [3,19], and (b) can be used to design specific microstructures to improve properties, for example see the "design and selection of Nb-Al-Si-Hf-Ti alloys" in [37,38].

The paper is consciously selective. It does not deal with CC/HE silicides, C14 Laves and A15 compounds, eutectics and lamellar microstructures and their contamination with oxygen. It is intended to open further questions about bcc solid solutions in metallic UHTMs and to suggest future research. It is not a system of polarities (opposite characteristics) (meaning "conventional"–CC/HE, contaminated –"uncontaminated" phase) that we have to deal with but an overlapping set of interrelationships (see below) and transformations [13,16], which are viewed in the context of metallic UHTM development and provide a useful route and compass for exploring the microstructures of these materials.

Given that the analysis of data for bcc solid solutions will be based on aforementioned parameters, the calculation of which requires high quality chemical analysis data [9,10,14], this paper concentrates only on the bcc solid solutions in RM(Nb)ICs, RM(Nb)ICs/RCCAs and RM(Nb)ICs/RHEAs for which such data are available, and cannot include RCCAs or RHEAs, for example, like those included in the review in [19], owing to lack of data for the latter metallic UHTMs.

2. Complex Concentrated Bcc Solid Solution

The bcc Nb_{ss} in RM(Nb)ICs and RM(Nb)ICs/RCCAs with nominal Si concentration 18 at.% and alloying addition of Al, B, Cr, Ge, Hf, Mo, Nb, Sn, Ta, Ti or W can be (i) "conventional" Nb_{ss}, (ii) CC/HE Nb_{ss}, or (iii) Nb_{ss} with no Si and (iv) not stable after heat treatment. These types of bcc solid solution are shown schematically in Figure 1 where the colours for (i) to (iv) are dark blue, red, light purple and light blue, respectively. Note that Figure 1 has data for bcc Nb_{ss} in AC and HT alloys. "Conventional" Nb_{ss} can be Ti rich only in AC alloys, meaning Ti rich Nb_{ss} is not observed in HT alloys. In B containing alloy the Ti rich solid solution is usually CC/HE Nb_{ss}.

For presentation purposes, in Figure 1 the numbers 15, 10 and 5 have been assigned, respectively, to "conventional" Nb_{ss}, CC/HE Nb_{ss} and not stable Nb_{ss}. The nominal compositions of the alloys are shown in the Appendix A. For most of the alloys in Figure 1 the CC/HE Nb_{ss} that was formed in the AC alloy was not stable after heat treatment.

The data in Figure 1 show (a) that CC Nb_{ss} was stable after heat treatment only in alloys where Mo and W simultaneously were in synergy with Hf and with the simultaneous addition of Ge and Sn in the alloy (compare the alloys JZ3+, JZ4, JZ5 and the OHS1), whereas (b) when Mo was substituted with Ta a higher concertation of Sn was required to stabilise the CC Nb_{ss} in the heat treated alloy (compare the alloys JZ3+ and JZ3). The (a) is also supported by the data for the alloy JN2, which in the AC condition had "conventional" Nb_{ss} plus two CC solid solutions and only "conventional" Nb_{ss} in the HT condition [51]. Furthermore, (c) in the alloy JZ3+ the CC Nb_{ss} was formed in the AC and HT conditions, whereas (d) in the alloys JZ4 and JZ5 Nb_{ss} was not formed in the AC condition and the CC Nb_{ss} with no Si formed after heat treatment, while (e) the opposite was the case in the alloy OHS1, where CC Nb_{ss} was formed in the AC condition and the Nb_{ss} was not stable after heat treatment. In the B containing TT4, TT5, TT7 and TT8 alloys and the Ta containing alloy KZ6 "conventional" and CC/HE Nb_{ss} formed in the AC condition and only "conventional" Nb_{ss} after heat treatment, but in the Sn containing alloys EZ8, JG6 and OHS1 the CC Nb_{ss} was formed in the AC condition and the Nb_{ss} was not stable after heat treatment. In the alloys EZ5 and TT6, "conventional" Nb_{ss} formed in the AC condition and the Nb_{ss} was not stable after heat treatment. Note that both alloys contain Sn, whereas B was present only in the alloy TT6.

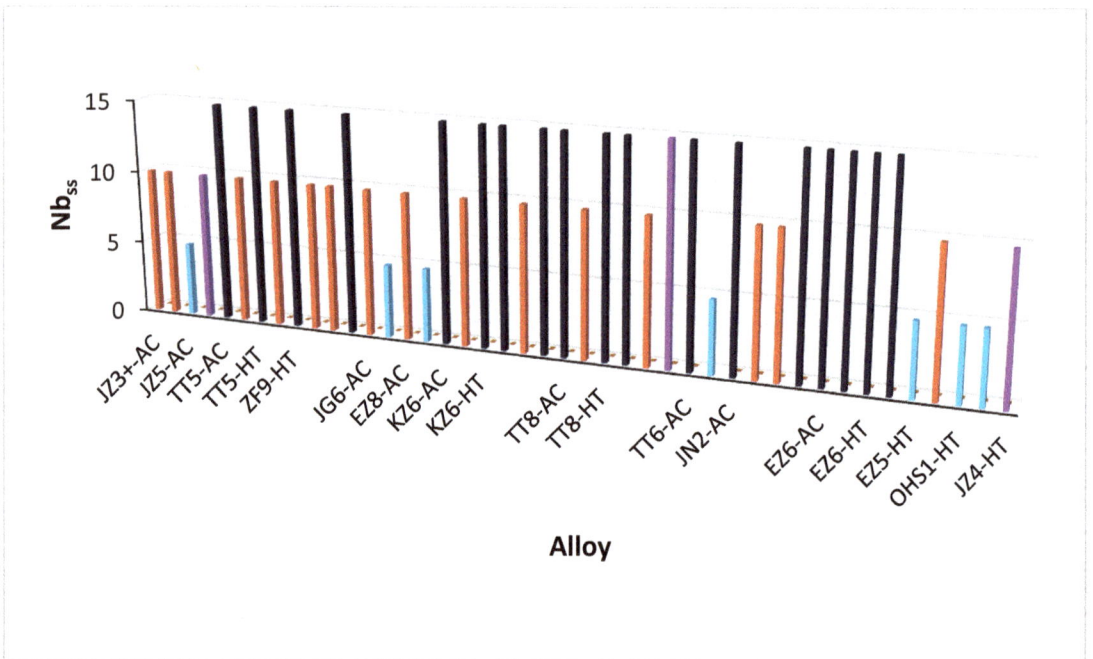

Figure 1. Nb_{ss} in RM(Nb)ICs and RM(Nb)ICs/RCCAs with nominal Si content 18 at.% and alloying elements Al, B, Cr, Ge, Hf, Mo, Nb, Sn, Ta, Ti, W: "conventional" Nb_{ss} (dark blue), CC/HE Nb_{ss} (red), (iii) Nb_{ss} with no Si (light purple) (iv) not stable Nb_{ss} (light blue). For presentation purposes the numbers 15, 10 and 5 have been assigned, respectively, to "conventional" Nb_{ss}, CC/HE Nb_{ss} and not stable Nb_{ss}. AC = as cast, HT = heat treated. For nominal alloy compositions and references see the Appendix A. RM(Nb)ICs/RCCAs the alloys JZ3+, JZ5, TT5, ZF9, JG6, EZ8, TT7, OHS1, JZ4. HE Nb_{ss} in TT4-AC.

In other words, considering the three elements B, Ge and Sn, which in synergy with Al, Cr, Hf and Ti, are key for improving the oxidation resistance and obtaining a balance of properties in RM(Nb)ICs, RM(Nb)ICs/RCCAs and RM(Nb)ICs/RHEAs [3,9,10,14,17,18,33–36,42,52–54] and (I would suggest) in RCCAs and RHEAs, it is advised that alloying with B plus Hf or Mo or Ta is unlikely to stabilise CC/HE Nb_{ss} compared with the simultaneous addition of Ge and Sn with Hf, Mo and W in the aforementioned metallic UHTMs.

The relationship between the entropies of mixing of alloys and their bcc solid solutions is shown in Figure 2a. Data for solid solutions and alloys can be found, respectively, in the Table 1 in [28] and the Table 1 in [26] and the nominal alloy compositions are given in the Appendix A. In Figure 2a the linear fit of all the data is good (R^2 = 0.9019) and shows that the CC/HE Nb_{ss} has high entropy of mixing (see below), and is formed in RM(Nb)ICs, and RM(Nb)ICs/RCCAs or RM(Nb)IC/RHEAs that also have high entropy of mixing (12.4 < ΔS_{mix}^{alloy} < 13.65 Jmol^{-1}K^{-1}). Relationships of the solid solution parameter Ω with the solid solution enthalpy of mixing, and the parameters δ and $\Delta\chi$ are shown in Figure 2b–d. The CC/HE Nb_{ss} has low Ω (<2.4), high $|\Delta H_{mix}|$, high and low δ and $\Delta\chi$ (Pauling electronegativity) parameters (>5.7 and <0.18, respectively) and low VEC (figure not shown). In the plots of Ω_{ss} with δ_{ss}, $\Delta\chi_{ss}$ and VEC_{ss} (figure not shown) only δ_{ss} can show the effect of specific alloying additions. Indeed, in Figure 2c the blue data are for the alloying additions Al, B, Cr, Hf, Mo, Nb, Si, Sn, Ti and W, the brown data for Al, B, Cr, Ge, Hf, Mo, Nb, Si, Ta, Ti, W and the red data for Al, B, Cr, Ge, Hf, Mo, Nb, Si, Sn, Ta, Ti and W. Note that the blue and brown lines in Figure 2c are essentially parallel, and that the red line is for alloys with 24 at.% Ti and 18 at.% Si (nominal). In other words, (i) the addition of Ta and the replacement of Sn with Ge reduces both the δ_{ss} and Ω_{ss} parameters (shift from blue to brown line), whereas the simultaneous addition of the said elements "bridges the gap" with further decrease in Ω_{ss} and formation of CC/HE Nb$_{ss}$ (red data) and (ii) the parameters δ_{ss} and Ω_{ss} are key in the alloy design stage for designing alloys with "conventional" and CC/HE Nb$_{ss}$.

How do the values of the parameters for CC/HE Nb$_{ss}$ of the alloys in Figure 2 compare with those of single-phase bcc solid solution HEAs? Whereas there are similarities for the entropy of mixing ($10.8 < \Delta S_{mix}^{CC/HE\ Nbss} < 12.8$ Jmol^{-1}K^{-1}, compared, for example, with 11.47, 11.53 and 13.38 Jmol^{-1}K^{-1}, respectively, for the HE$_{ss}$ Hf$_{21}$Mo$_{20}$Nb$_{21}$Ti$_{17}$Zr$_{21}$, WNbMoTa and WNbMoTaV) and VEC$_{CC/HE\ Nbss}$ ($4.44 < VEC_{CC/HE\ Nbss} < 4.74$, compared, for example, with 4.4, 4.6, 4.7, 5.5 and 5.4, respectively, for the high entropy solid solution (HE$_{ss}$) HfNbTaTiZr, HfMoTaTiZr, HfMoNbTaTiZr, WNbMoTa and WNbMoTaV) there are significant differences for the other parameters. Indeed, the $\delta_{CC/HE\ Nbss}$ values are higher ($5.7 < \delta_{CC/HE\ Nbss} < 9.7$, compared, for example, with 2.31, 3.15, 5.51 and 6.3, respectively, for the HE$_{ss}$ WNbMoTa, WNbMoTaV, HfNbTaTiZr and HfMoNbTaTiZr), the $\Omega_{CC/HE\ Nbss}$ values are lower ($1.9 < \Omega_{CC/HE\ Nbss} < 2.44$, compared, for example, with 12.37, 17.8, 24.9 and 43.3, respectively, for the HE$_{ss}$ HfNbTaTiZr, HfMoTaTiZr, HfMoNbTaTi and HfMoNbTaTiZr), the enthalpy of mixing is more negative ($-15.04 < \Delta H_{mix}^{CC/HE\ Nbss} < -8.32$ KJ mol^{-1}, compared, for example, with -0.9, -1.9, -4.64 and -6.5 KJ mol^{-1}, respectively, for the HE$_{ss}$ HfMoNbTaTiZr, HfMoTaTiZr, WNbMoTaV and WNbMoTa) and $\Delta\chi_{CC/HE\ Nbss}$ values are smaller ($0.067 < \Delta\chi_{CC/HE\ Nbss} < 0.179$, compared, for example, with 0.34 and 0.36, respectively, for the HE$_{ss}$ WNbMoTaV and WNbMoTa).

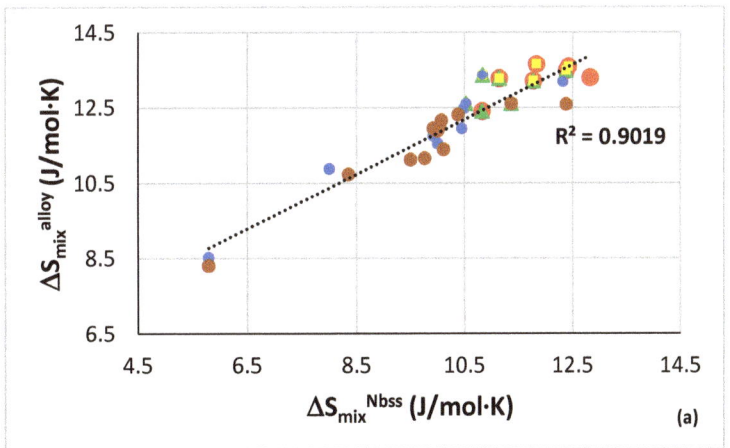

Figure 2. Cont.

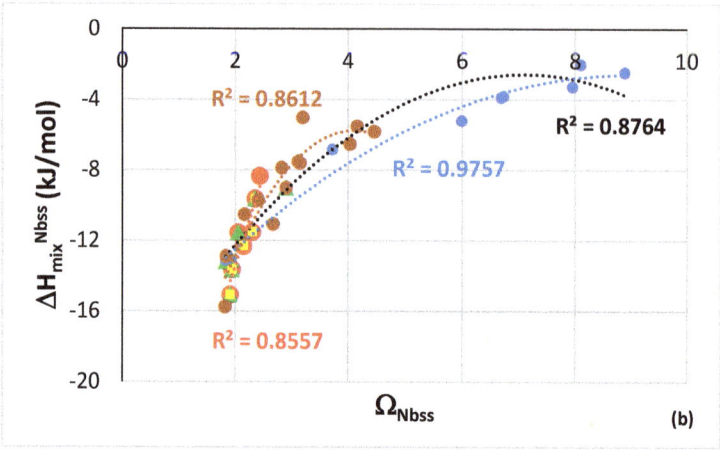
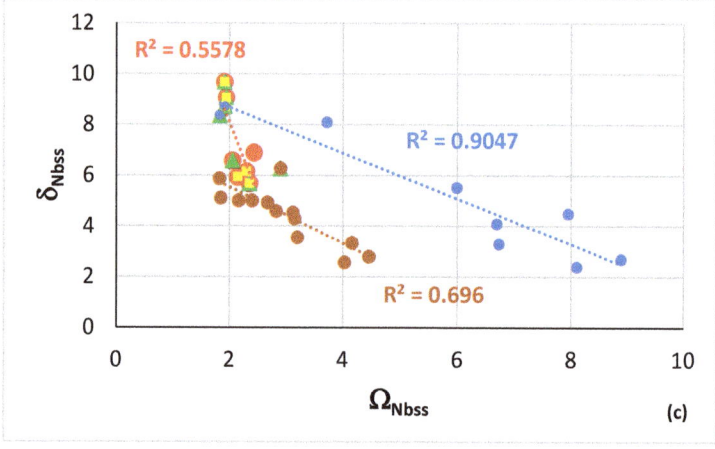
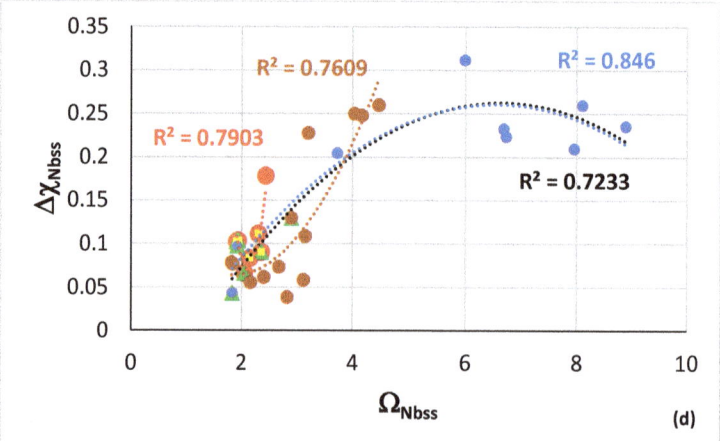

Figure 2. (**a**) Alloy entropy of mixing versus solid solution entropy of mixing, and (**b**–**d**) relationships of the solid solution parameter Ω with the solid solution (**b**) enthalpy of mixing, (**c**) parameter δ

and (**d**) parameter $\Delta\chi$. Red data CC/HE Nbss alloys JN2-AC, TT4-AC, TT7-AC, EZ8-AC, TT6-AC, ZF9-AC, TT5-HT, blue and brown data "conventional" Nbss, blue data AC and HT alloys JN3 and JN4, and HT alloys YG8, YG10, TT4, TT7, brown data AC alloys YG8, YG10, AC and HT alloys YG11, KZ5, JN1 and ZF6, and HT alloys KZ6, JG3 and TT8. Green triangles for B containing alloys, yellow squares for RCCAs. For nominal alloy compositions and references see the Appendix A. In (**a**) for all data $R^2 = 0.9019$, blue and brown data linear fit with $R^2 = 0.9127$, brown data linear fit with $R^2 = 0.8854$, in (**b**) $R^2 = 0.8557$ is for linear fit and $R^2 = 0.9757$, $R^2 = 0.8612$ and $R^2 = 0.8764$ are for parabolic fit, the latter value is for all the data, in (**c**) all the R^2 values are for linear fit of data, in (**d**) all the R^2 values are for parabolic fit, and $R^2 = 0.7233$ is for all the data. HE Nb$_{ss}$ in TT4-AC.

The higher values of $\delta_{CC/HE\ Nbss}$ are attributed to the alloying with B, the more negative $\Delta H_{mix}^{CC/HE\ Nbss}$, and the low $\Omega_{CC/HE\ Nbss}$ and $\Delta\chi_{CC/HE\ Nbss}$ values are attributed to the alloying with B, Ge or Sn. The aforementioned alloying elements have not been used in studies of single-phase bcc solid solution HEAs.

Relationships between the parameters $\Delta\chi$ and VEC of alloys and Nb$_{ss}$ are shown in Figure 3, where the CC/HE Nb$_{ss}$ is indicated with the green data points. Note that this type of solid solution was mostly observed in AC alloys (Figure 1). The data in Figure 3 are for the same alloys as in Figure 2. In Figure 3a the $R^2 = 0.8061$ is for the linear fit of all the data and $R^2 = 0.8867$ is for the data of the CC/HE Nb$_{ss}$. Notice (i) the gap (green double arrow) in $\Delta\chi_{Nbss}$ values, in agreement with [28], which means that the CC/HE Nb$_{ss}$ follows the same rules as the "conventional" Nb$_{ss}$ [10], and (ii) that CC/HE Nb$_{ss}$ is found on either side of this gap. In Figure 3b all the data have $R^2 = 0.628$, the brown data points give $R^2 = 0.8322$, and the green data points give $R^2 = 0.6978$.

Even though the same alloying additions were in the alloys and their solid solutions represented by the green and blue data points in Figure 3, the alloys and their solid solutions indicated with the brown data points did not contain Ge and their Ti content was not fixed at 24 at.% nominal, as is the case for the alloys represented with the green data points. Instead, they were either Ti free (alloy YG8) or their Ti concentration was lower (alloys YG10, YG11). In other words, the parameter VEC (Figure 3b) shows that not only the alloying additions but also their concentrations in an alloy are key regarding the formation or not of CC/HE Nb$_{ss}$. Furthermore, only with the parameter VEC we can differentiate the data for CC/HE Nb$_{ss}$ and "conventional" Nb$_{ss}$, as indicated with the brown and green lines compared with the blue line in Figure 3b. Thus, the co-existence of CC/HE Nb$_{ss}$ with "conventional" Nb$_{ss}$ in most alloys [14] is supported by the data in Figures 2 and 3. Additionally, Figure 3 confirms (iii) that the relationships between the alloy and solid solution parameters $\Delta\chi$ and VEC, which are fundamental relationships in NICE [10], apply also for CC/HE Nb$_{ss}$ and (iv) that the parameters VEC$_{alloy}$ and VEC$_{ss}$ are key in the alloy design stage for designing alloys with "conventional" and CC/HE Nb$_{ss}$. To summarise, the design of alloys with "conventional" and CC/HE Nb$_{ss}$ must make use of the relationships of the parameter $\Delta\chi_{alloy}$ with the concentrations of solute additions in NICE [10] and the relationships between the parameters VEC$_{alloy}$ and VEC$_{ss}$ and δ_{ss} and Ω_{ss}.

The parameters VEC, δ and $\Delta\chi$ of alloys in which CC/HE Nb$_{ss}$ was observed are shown in Figure 4, and the parameters VEC, δ and $\Delta\chi$ of the CC/HE Nb$_{ss}$ in the same alloys are shown in Figure 5. Figure 4a shows significantly higher values of δ_{alloy} for B containing alloys (range 12.57 to 13.35, compared with 8.55 to 9.66 for B free alloys) and essentially similar VEC$_{alloy}$ values (4.403 to 4.584). Figure 4b shows small range of $\Delta\chi_{alloy}$ values (0.131 to 0.21) (also see Table 1 in [26]), and wider range and higher values of VEC$_{CC/HE\ Nbss}$ (4.51 to 5.38, Figure 5a). Significantly wider range of $\Delta\chi_{CC/HE\ Nbss}$ values (0.067 to 0.369) is shown in Figure 5b with strikingly lower values for B containing CC Nb$_{ss}$ (Figure 6a), noticeably higher values of $\delta_{CC/HE\ Nbss}$ for RM(Nb)ICs/RCCAs where B was simultaneously present with Hf (alloy 11) or Ta (alloy 12) (Figure 5a) and overall markedly lower values of $\delta_{CC/HE\ Nbss}$ (4.239 to 9.69) compared with δ_{alloy} (also see Table 1 in [28]). The parameter $\Delta\chi_{CC/HE\ Nbss}$ increases with increasing $\Delta\chi_{alloy}$ (Figure 6a), $\Delta\chi_{"conventional"\ Nbss}$ (Figure 6b) and VEC$_{CC/HE\ Nbss}$ (Figure 6c). Remarkably, there is a strong correlation between the δ parameters of "conventional" and CC/HE Nb$_{ss}$ with the B concentration of the solid solution, as shown in Figure 7a.

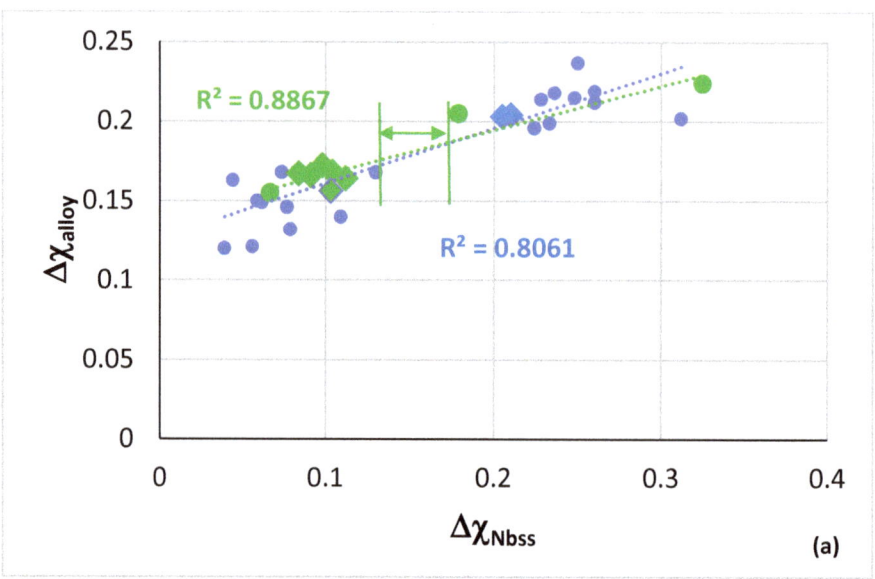

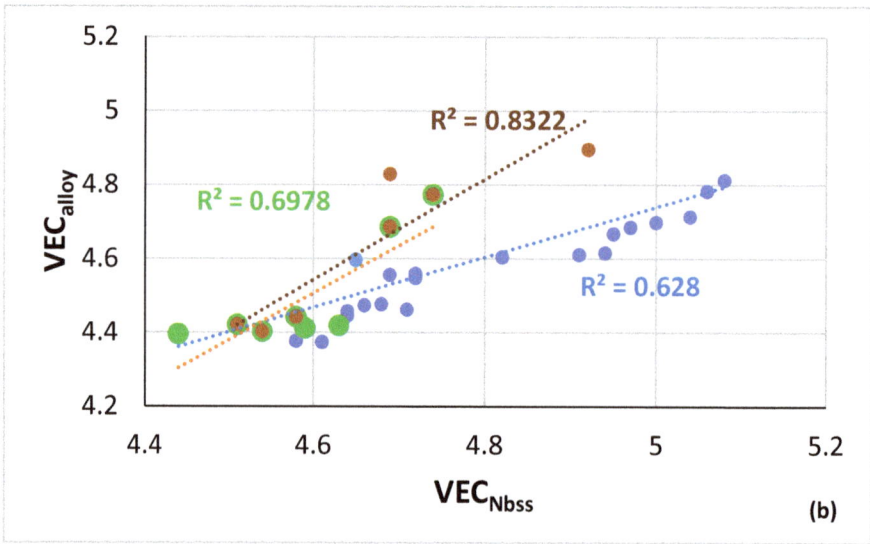

Figure 3. (a) Δχ$_{alloy}$ versus Δχ$_{Nbss}$ and (b) VEC$_{alloy}$ versus VEC$_{Nbss}$. The data are for the same alloys as in Figure 2. In (a) green data points R^2 = 0.8867, all data points R^2 = 0.8061. In (b) the brown data points (R^2 = 0.8322) are for the alloys JN2-AC, YG8-HT, YG11-HT, TT7-AC, EZ8-AC, TT5-HT, the green data points (R^2 = 0.6978) are for the alloys JN2-AC, TT4-AC, TT7-AC, EZ8-AC, TT6-AC, ZF9-AC, TT5-AC, and the blue data points (R^2 = 0.628) are for the alloys JN2-HT, JN3, JN4, YG8-AC, YG10, YG11-AC, KZ5, KZ6-HT, JN1, TT4-HT, TT7-HT, ZF6, JG3-HT, TT8-HT (see the Appendix A for nominal alloy compositions and references). In (a) the diamonds indicate RM(Nb)IC/RCCA. Diamonds not shown in (b) for clarity of presenting the different groups. HE Nb$_{ss}$ in TT4-AC.

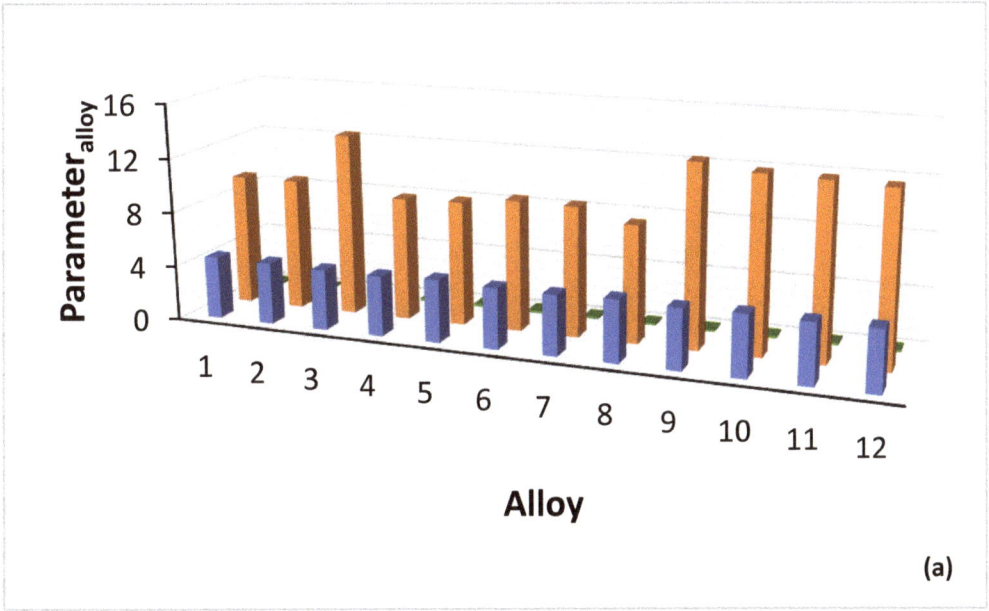

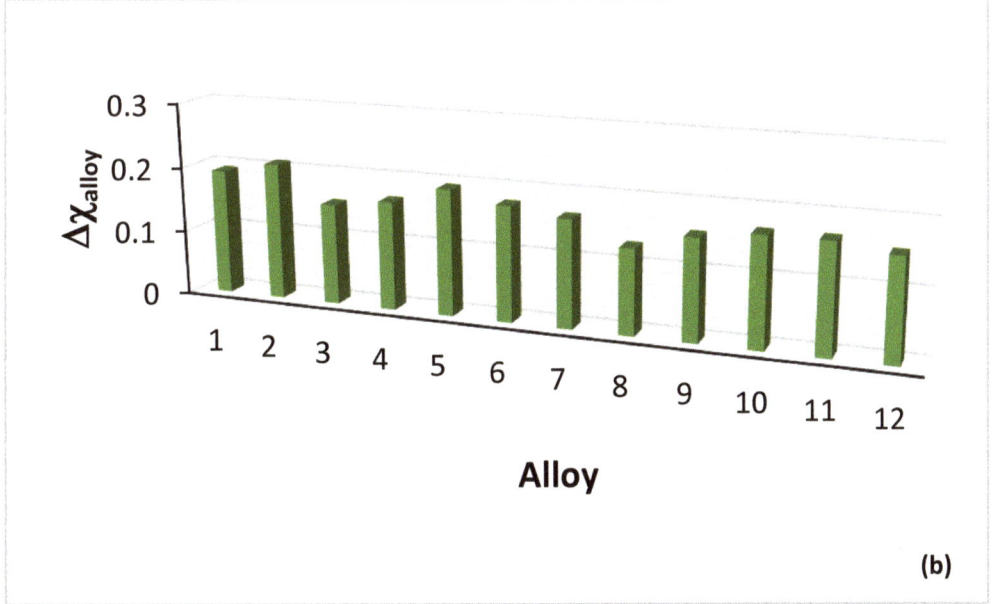

Figure 4. (**a**) Parameters VEC, δ and Δχ of alloys in which CC/HE Nb$_{ss}$ was observed and (**b**) details of Δχ$_{alloy}$. Colours: blue VEC, brown δ, green Δχ. Alloys 1 to 12 contain Al, Cr, Nb, Si, Ti plus in (1) Ge, Hf, Sn, Ta, W, in (2) Ge, Hf, Mo, Sn, W, in (3) B, Ta, in (4) Ge, Hf, in (5) Ge, Hf, Sn, Ta, W, in (6) Hf, Mo, Sn, in (7) Hf, Sn, in (8) Ta, in (9) B, in (10), B, Mo, in (11) B, Hf, in (12) B, Ta. 1 = JZ3+-AC, 2 = JZ5-HT, 3 = TT5-AC, 4 = ZF9-AC, 5 = JZ3-AC, 6 = JG6-AC, 7 = EZ8-AC, 8 = KZ6-AC, 9 = TT4-AC, 10 = TT8-AC, 11 = TT7-AC, 12 = TT5-HT. For nominal alloy compositions and references see Appendix A. RM(Nb)ICs (5, 8–10) and RM(Nb)ICs/RCCAs (1–4,6,7,11,12).

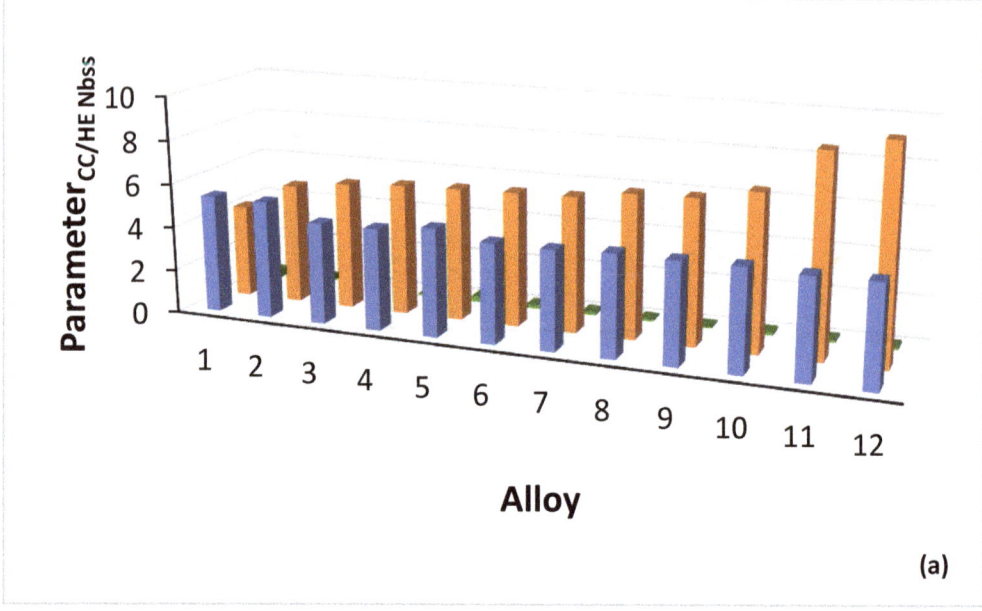

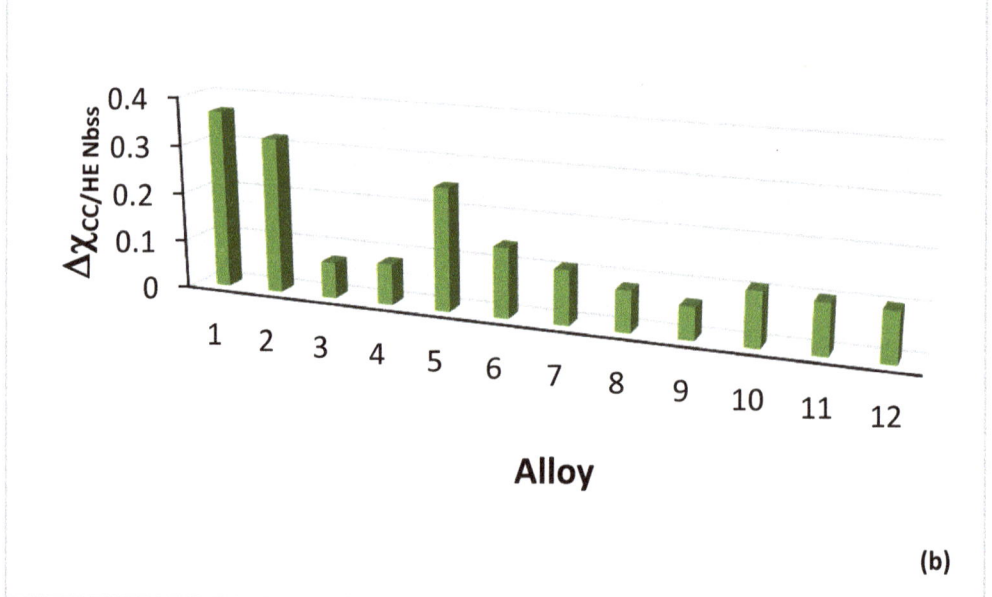

Figure 5. (a) Parameters VEC, δ and $\Delta\chi$ of CC/HE Nb$_{ss}$ in alloys where this type of solid solution was observed and (b) details of $\Delta\chi_{CC/HE\ Nbss}$. Colours: blue VEC, brown δ, green $\Delta\chi$. Alloys 1 to 12 the same as in Figure 4. HE Nb$_{ss}$ in TT4-AC. For nominal alloy compositions and references see Appendix A. RM(Nb)ICs (5, 8–10) and RM(Nb)ICs/RCCAs (1–4,6,7,11,12).

The co-existence of CC/HE Nb$_{ss}$ with "conventional" Nb$_{ss}$ in most alloys [14] is further supported by the data in Figure 6 that also confirm that the relationships between the alloy and solid solution parameters Δχ and VEC, which are fundamental relationships in NICE [10], apply also for CC/HE Nb$_{ss}$.

Boron, Ge and Sn are key elements for obtaining a balance of properties in metallic UHTMs but their roles regarding the stability of CC/HE Nb$_{ss}$ differ, see above. The CC/HE Nb$_{ss}$ was stable after heat treatment in alloys with simultaneous addition of Mo, W with Hf, Ge and Sn (Figure 1). Figure 7 shows relationships of the solid solution parameter δ versus the B or Ge+Sn concentration in the solid solution. In both cases the parameter δ increases with increasing B or Ge+Sn concentration in the solid solution.

Note that in Figure 7a, the data are for "conventional" and CC/HE Nb$_{ss}$, whereas in Figure 7b the data are only for CC/HE Nb$_{ss}$. The co-existence of CC/HE Nb$_{ss}$ with "conventional" Nb$_{ss}$ in boron containing alloys is further supported by the data in Figure 7a. Lowest B concentration in the solid solution and thus lowest δ parameter was found when B was simultaneously present with Sn or Ta in the alloy [35]. Correlations of boron concentration in Nb$_{ss}$ with the parameters VEC and Δχ are not strong (figures not shown).

Relationships between B or Ge+Sn concentration and other solute additions in CC/HE Nb$_{ss}$ in RM(Nb)ICs and RM(Nb)ICs/RCCAs are shown in Figures 8 and 9. Figure 8a–d shows relationships of B concentration with "reactive" solutes in CC/HE Nb$_{ss}$ in boron containing RM(Nb)ICs and RM(Nb)ICs/RCCAs and Figure 8e,f shows correlations with the Nb/Ti ratio in the CC/HE Nb$_{ss}$. The correlation between the B and Si concentrations in the solid solution is shown in Figure 8g. In Figure 8 the solid solution in RM(Nb)ICs/RCCAs is indicated with diamonds. The same correlations for "conventional" Nb$_{ss}$ are not strong (figures not shown).

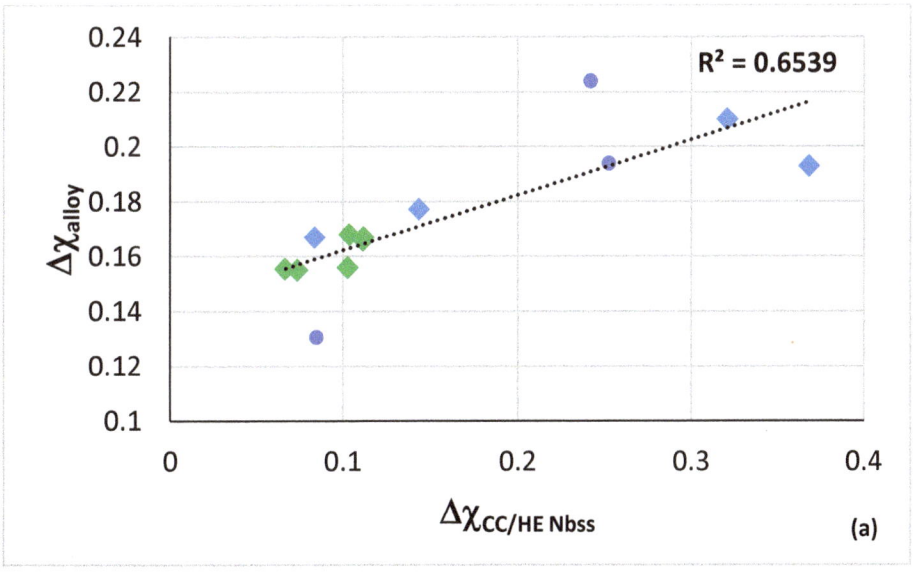

Figure 6. Cont.

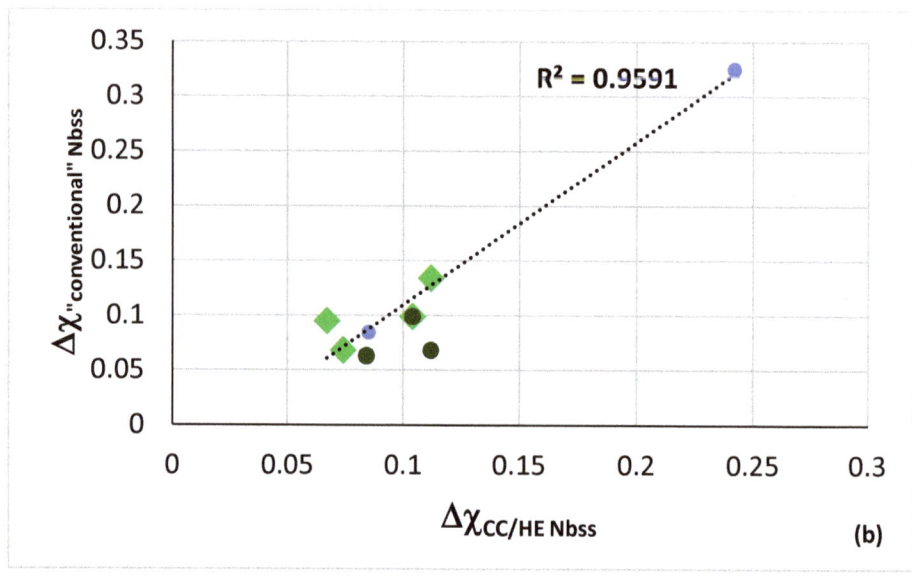

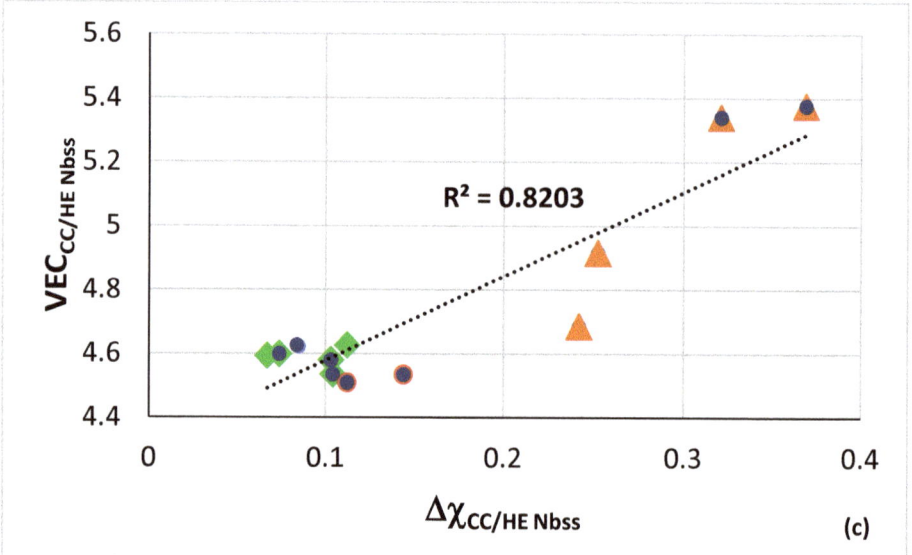

Figure 6. (a) $\Delta\chi_{alloy}$ versus $\Delta\chi_{CC/HE\ Nbss}$, where green colour indicates B containing alloys and diamonds are for RM(Nb)ICs/RCCAs. (b) $\Delta\chi_{"conventional"\ Nbss}$ versus $\Delta\chi_{CC/HE\ Nbss}$, where green diamonds are solid solutions in B containing alloys and green circles for solid solutions in RM(Nb)ICs/RCCAs, (c) $VEC_{CC/HE\ Nbss}$ versus $\Delta\chi_{CC/HE\ Nbss}$, where green diamonds are for solid solutions in B containing alloys, brown triangles are for solid solutions in alloys with simultaneous addition of Ge and Sn, red circles are for solid solutions with Sn, and blue circles are for solid solutions in RM(Nb)ICs/RCCAs. In each part the R^2 value is for the linear fit of all the data. (**a**,**c**) data for the AC alloys EZ8, JG6, JZ3, JZ3+, KZ6, TT4, TT5, TT7, TT8, ZF9 and the HT alloys JZ5 and TT5, (**b**) data for the AC alloys KZ6, TT4, TT5, TT7, TT8, ZF9. HE Nb_{ss} in TT4-AC. See the Appendix A for nominal alloy compositions and references.

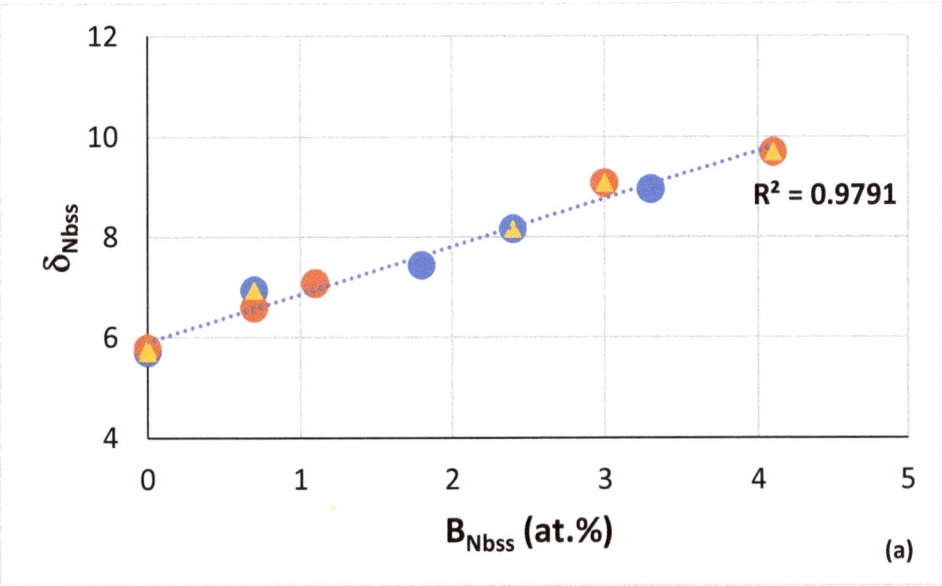

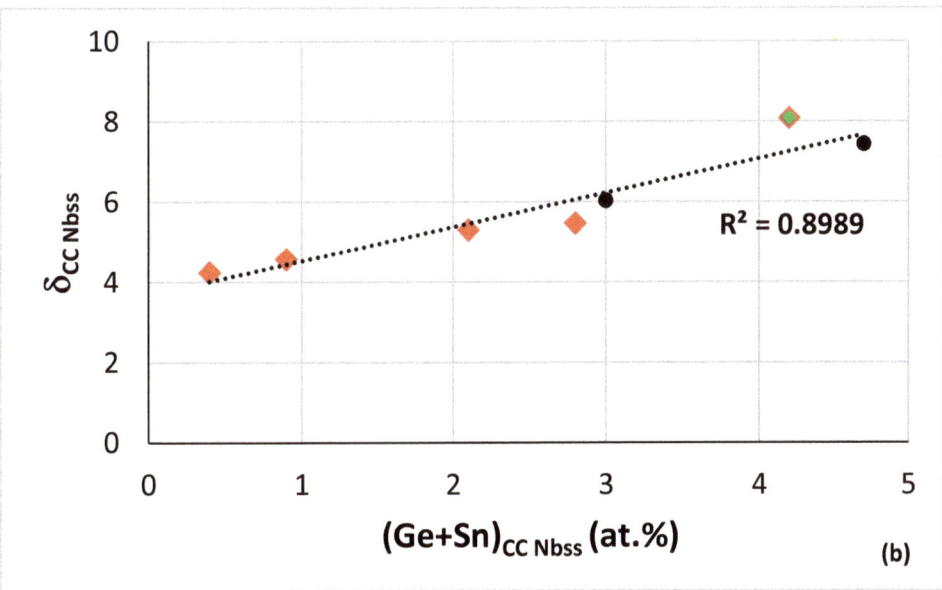

Figure 7. (a) Parameter δ of solid solution versus its B concentration. Red data points for CC/HE Nb_{ss}, blue data points for "conventional" Nb_{ss}. Yellow triangles indicate solid solution was formed in RM(Nb)IC/RCCA. All data R^2 = 0.9791, data for CC/HE Nb_{ss} has R^2 = 0.9884 and data for "conventional" Nb_{ss} has R^2 = 0.9656. Data are as follows: "conventional Nb_{ss} in AC alloys TT4, TT5, TT6, TT7, TT8, CC/HE Nb_{ss} for the AC alloys TT4, TT5, TT7, TT8 and the HT alloy TT5. HE Nb_{ss} in TT4-AC. (b) Parameter δ versus (Ge+Sn) content of CC Nb_{ss}. Data for the AC alloys JZ3, JZ3+ and OHS1 and the HT alloys JZ3+, JZ4 and JZ5. All data R^2 = 0.8989. Diamonds for solid solutions in RM(Nb)ICs/RCCAs. Green colour for the solid solution in the alloy OHS1. For nominal alloy compositions and references see Appendix A.

The Al, Cr, Ti, Al+Cr and Si concentrations in the CC/HE Nb$_{ss}$ decrease as its boron concentration increases. The Nb/Ti ratio of the CC/HE Nb$_{ss}$ increases with its boron concentration and decreases with its Al+Cr content. The parabolic fit of data in Figure 8e give R^2 = 0.9981 with maximum for Nb/Ti = 0.82 and (Al+Cr) = 23.22 at.%. For Nb/Ti = 0.82 Figure 8f gives B = 0.16 at.%. Using this B concentration, from Figure 8a we obtain Ti = 39.98 at.%, from Figure 8b Cr = 16.26 at.%, from Figure 8c Al = 7.18 at.%, from Figure 8d Al+Cr = 23.44 at.%, from Figure 8g Si = 1.37 at.%. Finally, for Ti = 39.98 at.% and the ratio Nb/Ti = 0.82 we obtain Nb = 32.78 at.%, in other words we calculate the chemical composition of CC Nb$_{ss}$ as 32.78Nb-39.98Ti-16.26Cr-7.18Al-1.37Si-0.16B or 32.8Nb-40Ti-16.3Cr-7.2Al-1.4Si-0.2B.

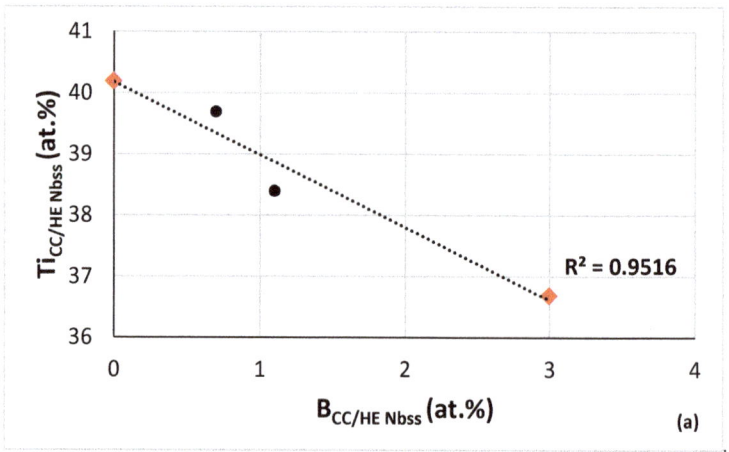

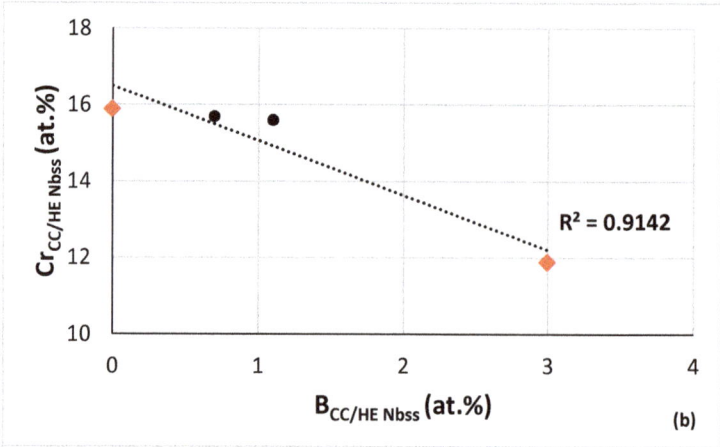

Figure 8. Cont.

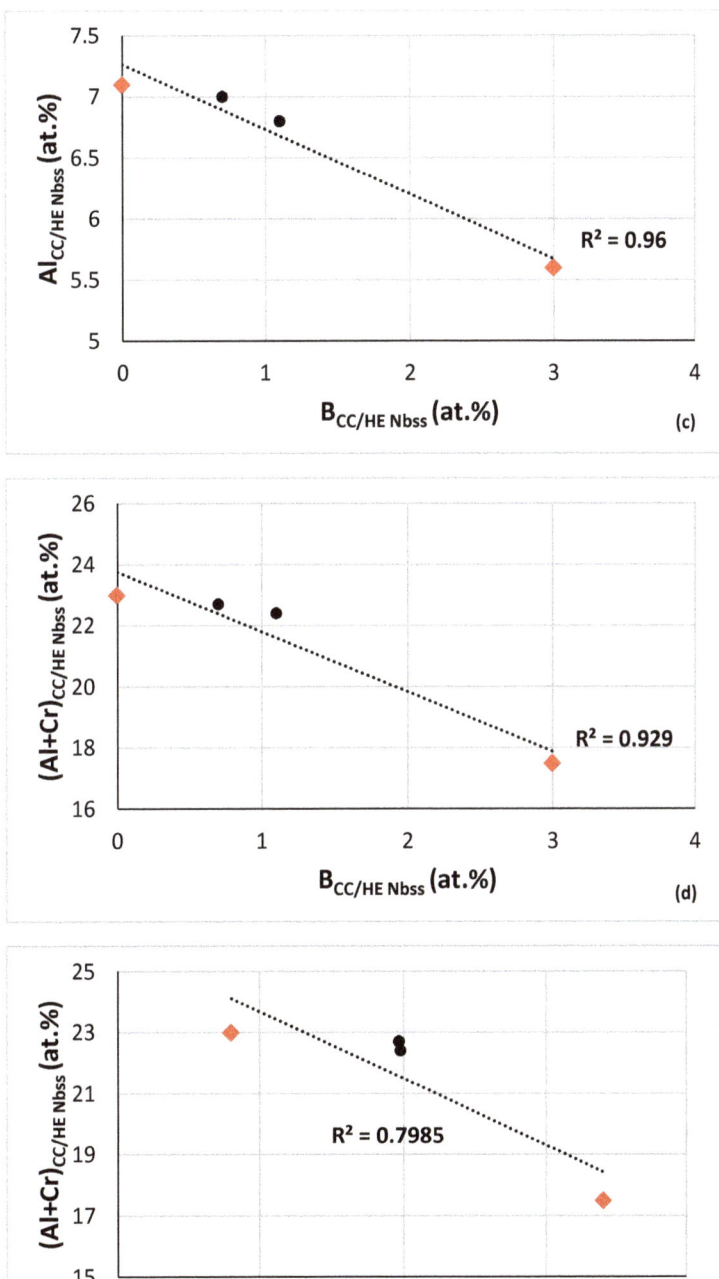

Figure 8. *Cont.*

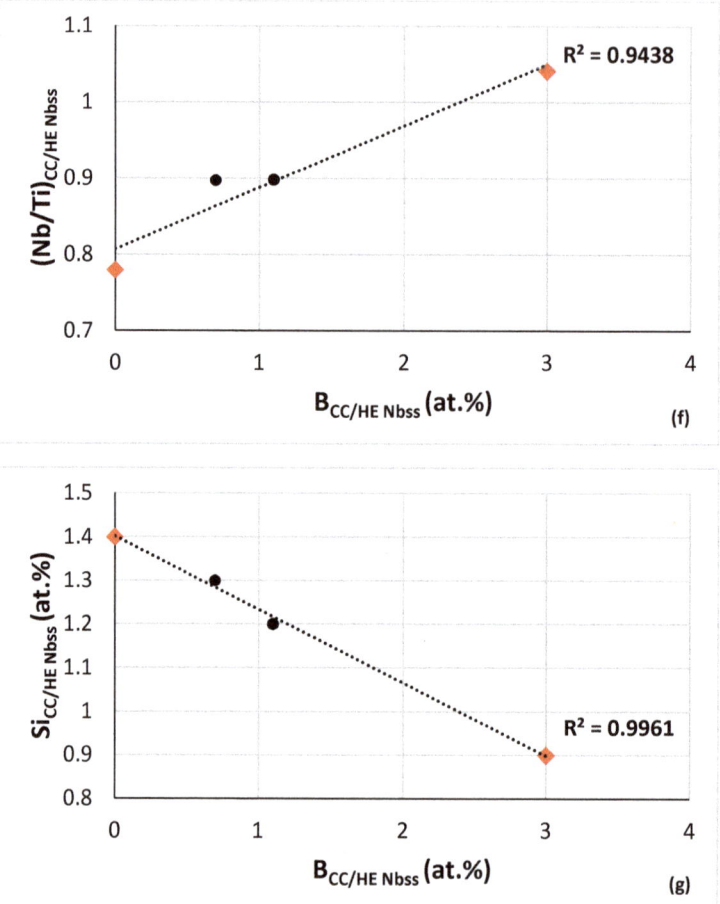

Figure 8. (a–g) data for the as cast B containing alloys TT4, TT5, TT7, TT8. Diamonds for RM(Nb)ICs/RCCAs. Concentration of B versus (**a**) Ti, (**b**) Cr, (**c**) Al, (**d**) Al+Cr and (**g**) Si in CC/HE Nb$_{ss}$. (**e**) Al+Cr concentration versus Nb/Ti ratio and (**f**) Nb/Ti ratio versus B concentration in CC/HE Nb$_{ss}$. R^2 values are for the linear fit of all data in each part. Parabolic fit of data in (**e**) gives R^2 = 0.9981 with maximum at Nb/Ti = 0.82 and (Al+Cr) = 23.22 at.%. HE Nb$_{ss}$ in TT4-AC. For nominal alloy compositions and references see Appendix A.

Whereas the Al, Cr, Ti and Al+Cr concentrations in the CC/HE Nb$_{ss}$ in B containing alloys decrease as the B concentration in the solid solution increases (Figure 8), the opposite is the case when the concentrations of the same solute additions are plotted versus the Ge+Sn concentration of the CC Nb$_{ss}$ in B free alloys (Figure 9). Note that there is no correlation between the Si and Ge+Sn concentrations in CC Nb$_{ss}$. Similarly with the B containing alloys, the Nb/Ti ratio of the CC Nb$_{ss}$ increases with decreasing Al+Cr content (Figures 8e and 9e), but unlike the B containing Nb$_{ss}$, the Nb/Ti ratio decreases with increasing Ge+Sn concentration (Figures 8f and 9f). Furthermore, there is a good correlation between the total RM concentration in CC Nb$_{ss}$ and its Ge+Sn content that shows the former decreasing as the latter content increases (RM = Nb + Mo + Ta + W). Note that also there are good correlations between the W and Ti content, the Ti concentration with the W/RM ratio and the Al+Cr sum with the Sn/Ge ratio of the Nb$_{ss}$ of B free RM(Nb)ICs and

RM(Nb)ICs/RCCAs with Ge, Sn, and RM additions (see Figure 12 in [33] and Figure 12 in [34]).

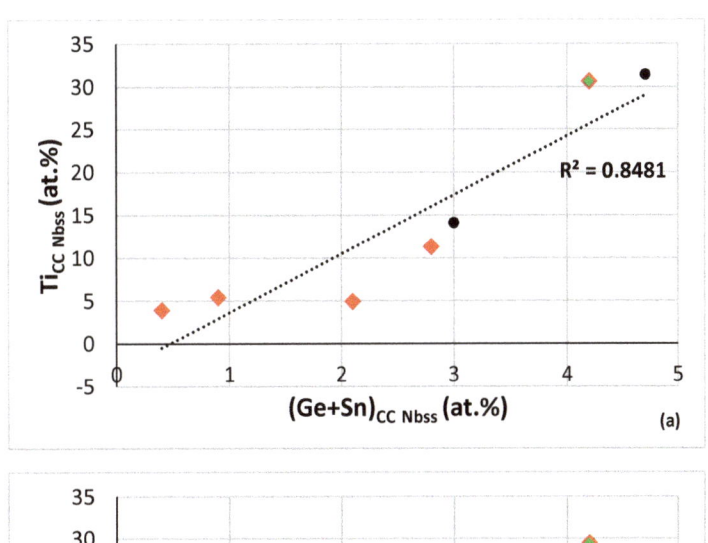

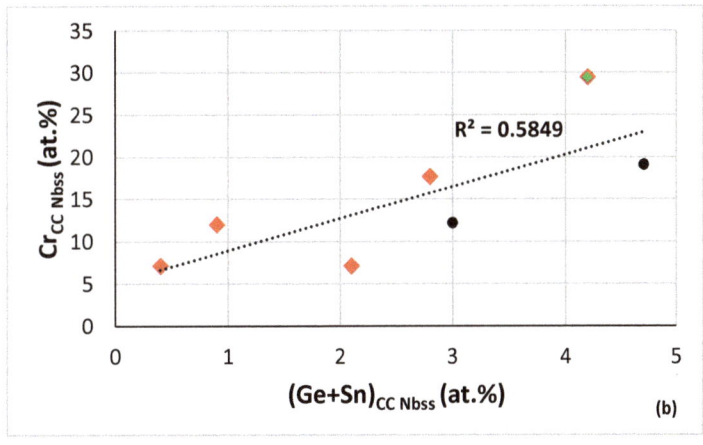

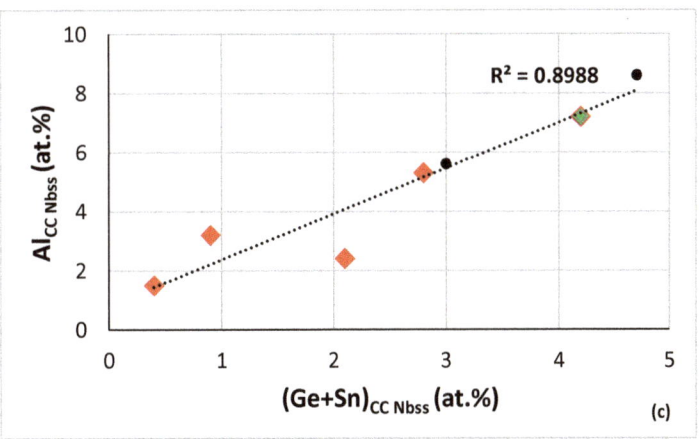

Figure 9. Cont.

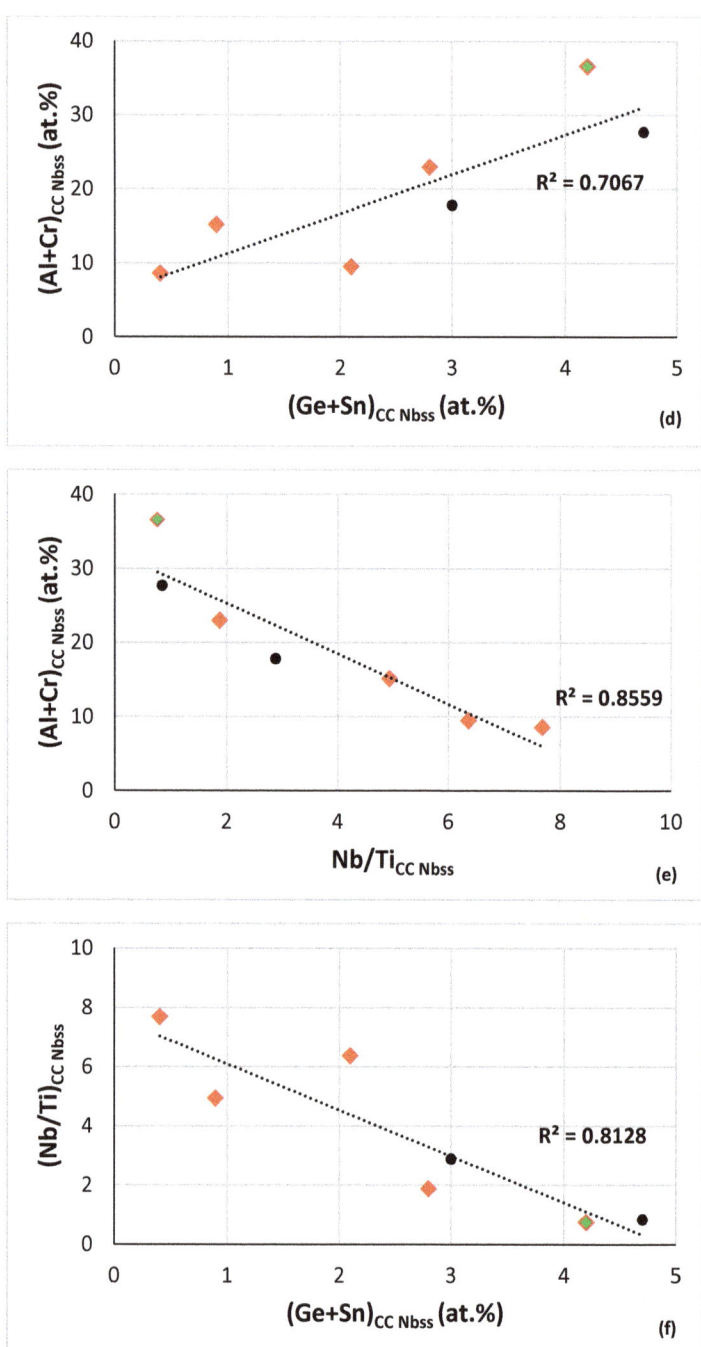

Figure 9. *Cont.*

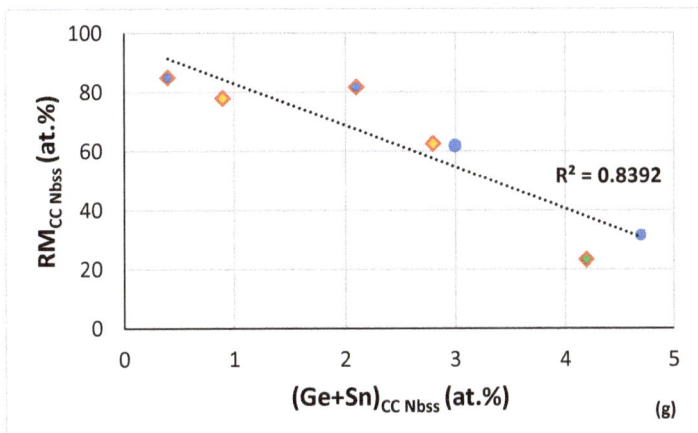

(g)

Figure 9. (a–g) data for the AC alloys JZ3, JZ3+, OHS1 and the HT alloys JZ3+, JZ4, JZ5 with simultaneous addition of Ge+Sn. Concentration of (Ge+Sn) versus (**a**) Ti, (**b**) Cr, (**c**) Al, (**d**) Al+Cr in CC Nb$_{ss}$. (**e**) Al+Cr concentration versus Nb/Ti ratio, (**f**) Nb/Ti ratio versus (Ge+Sn) concentration and (**g**) total RM (=Nb + Mo + Ta + W) concentration versus (Ge+Sn) concentration in CC Nb$_{ss}$. R^2 values are for the linear fit of all data in each part. Solid solution in RM(Nb)IC/RCCA alloy indicated with diamond and the green colour is for the alloy OHS1. In (g) blue colour for alloys where RM = Nb + Ta + W, yellow for RM = Nb + Mo + W and green for RM = Nb, i.e., for the alloy OHS1. See Appendix A for nominal alloy compositions and references.

In B containing RM(Nb)ICs and RM(Nb)ICs/RCCAs, the parameter δ_{Nbss} increases with B_{Nbss} (Figure 7a) and the hardness of the solid solution decreases with increasing δ_{Nbss} (see the descending part (green data) of the HV$_{ss}$ versus δ_{ss} data in Figure 7 in [3]). From the two linear relationships the dependence of HV$_{ss}$ on B_{Nbss} can be derived. The hardness of Nb$_{ss}$ in B free or B containing alloys is discussed in the next section. Note that the alloying with B has the opposite effect on the hardness of tetragonal Nb$_5$Si$_3$ compared with the effect of Ge or Sn, meaning the hardness increases upon alloying with B (see Figure 14 in [9]).

The ductile behaviour and yield strength of bcc Nb-rich solid solution alloys with Al, Cr and Ti additions (i.e., (Nb,Ti,Cr,Al)$_{ss}$) has been studied for different Nb/Ti ratios and Al+Cr sums [55]. At low Nb/Ti ratios, brittle behaviour was observed at higher Al+Cr content compared with high Nb/Ti ratios. For example, for Nb/Ti ≈ 0.8 brittle behaviour was observed for Al+Cr higher than about 22 at.%, and for Nb/Ti ≈ 1 or 2 ductile behaviour was observed for Al+Cr less than about 20 at.% and 18 at.%, respectively,. The room temperature yield strength decreased with decreasing Nb/Ti ratio. For example, for Nb/Ti ≈ 1 and Al+Cr ≈ 20 at.% the yield strength was about 980 MPa, whereas for Nb/Ti ≈ 0.8 it was about 825 MPa for Al+Cr ≈ 22 at.%. For Nb/Ti ≈ 1 increasing the Al content gave strengthening at room temperature and weakening at high temperatures, the Cr addition gave significant strengthening at all temperatures, approximately doubling the strength at 1200 °C. Reduced Ti concentration improved the high temperature strength. Note that for B containing RM(Nb)ICs and RM(Nb)ICs/RCCAs the Nb/Ti ratio of the solid solution increases with increasing B content (Figure 8f). It is suggested that it would be possible to "ductilize" the "conventional" or CC/HE Nb$_{ss}$ with B addition and "fine tuning" of the Nb/Ti ratio, and the Al+Cr sum of the Nb$_{ss}$ in multiphase RM(Nb)ICs, RM(Nb)ICs/RCCAs and RM(Nb)ICs/RHEAs (Figure 8).

Unlike the B containing RM(Nb)ICs and RM(Nb)ICs/RCCAs, currently there are no hardness data for the solid solutions in B free alloys with Ge, Sn and RM additions. Like the B-containing alloys, the latter alloys (i) exhibit exceptional oxidation resistance at pest and high temperatures with no scale spallation [33–36] and (ii) are expected to have good creep properties [14,34]. A material system suitable for high pressure turbine comprising a RCCA substrate of the Nb-Al-Cr-Ge-Hf-Mo-Si-Sn-Ti-W alloy system and a HEA bond coat of the Nb-Al-Hf-Si-Ti alloy system has been proposed using NICE [14,34].

3. Contamination of the Bcc Solid Solution with Oxygen

3.1. Contamination of Nb with Interstitials

The contamination of Nb with carbon, hydrogen, nitrogen and oxygen has been reported in the literature. Pionke and Davis found out that in the temperature range 200 to 600 °C, both carbon and nitrogen had very limited solubility in Nb (<0.1 at.% (0.014 wt.%)), oxygen had slightly more (<0.6 at.%, (0.1 wt.%)) while hydrogen had very large solubility, about 10 at.% (0.1 wt.%). Unlike the other interstitial elements, the solubility of hydrogen in Nb decreased with increasing temperature. The equilibrium concentration of hydrogen was affected by pressure [56].

The use of reactive alloying elements (Hf, Ti, and Zr) in Nb tends to lower the oxygen solubility. The addition of Zr is of particular interest because Zr is an effective strengthener of Nb. Zirconium additions to Nb have the effect of lowering the apparent oxygen solubility limit but increasing the Nb solubility limit. This increase is roughly a factor of 4 for a given temperature and pressure [57].

There are conflicting reports about the solution hardening of Nb with oxygen and nitrogen. For example, Harris [58] reported that oxygen was three times more effective in solution hardening than nitrogen or carbon, whereas Seigle [59] found the latter two elements to be twice as effective as oxygen and Szkopiak ([60] and references within) reported that nitrogen was twice as effective as oxygen.

Oxygen contents as high as 0.41 wt.% increased the room temperature tensile strength of Nb from 276 MPa to 896 PMa and reduced the elongation from 30 to 10%. Contamination of Nb with oxygen increased its hardness [61] and caused embrittlement [62,63]. The latter has been attributed to screw dislocations moving through a repulsive field imposed by oxygen atoms, forming cross kinks and emitting excess vacancies in Nb which bind with oxygen and hinder dislocation motion [64].

Oxygen also affected the elevated temperature properties [62]. Tensile tests conducted on Nb with varying oxygen concentrations (10, 200 and 4300 wppm) revealed brittle failures below 400 °C for oxygen concentration of 0.43% [65]. At higher temperatures ductile failures were produced. The amount of ductility exhibited by Nb−O alloys at elevated temperatures was sensitive to strain-rate. For example, Nb containing 0.15% oxygen exhibited a decrease in the reduction in area from 90% to 30% at 467 °C due to a change in strain rate of 5×10^{-5} to 2×10^{-1} s^{-1} [66].

The DBTT of Nb depends on solute additions and increases with oxygen concentration as does the yield strength [67]. Interstitial elements have a significant effect on the DBTT, in that it can be raised as these impurities are increased. This trend in the data indicates that the interstitials progressively cause embrittlement and that the relative order of embrittlement is hydrogen (which is most potent), followed by oxygen and carbon (which is least potent). The effect of nitrogen is difficult to separate primarily because of uncertainties as to whether the solubility limit has been exceeded; however, based on very limited data, it appears to be more embrittling than either carbon or oxygen [68].

The sensitivity of the group V bcc metals to contamination with oxygen is greater compared with the group VI bcc metals Mo, W [69]. Contamination in air-reacted niobium, was similar to that in oxygen-reacted niobium, suggesting that oxygen is the primary diffusing contaminant [70]. Alloying Nb with Mo reduced the oxygen solubility, whereas alloying with Ti or Zr, respectively, increased and decreased it [71]. Contamination with hydrogen affected the shear moduli $C' = (C_{11} - C_{12})/2$ and C_{44} and the bulk (K) and Young's (E) moduli of V, Nb and Ta (group V bcc metals), of which the C' decreased, the C_{44} increased, the K remained nearly constant, whereas the E of polycrystalline V or Ta with random orientation decreased and that of Nb increased with increasing hydrogen contamination. The effect on the C' of V was about four times the effect in Nb and Ta, whereas the change in C_{44} with hydrogen was greatest for Nb and weakest for Ta [72]. Contamination with oxygen resulted in a small increase in C_{44} and K for V, but in the case of Nb, the C' did not change with O $\leq$ 0.6 at.%, compared with the significant change in the C_{44} with O $\leq$ 0.7 at.%, and both C' and C_{44} increased, respectively, by 1% and 7% with O $\leq$10 at.%, which could be associated with precipitation of Nb oxide. Furthermore, the change in the C_{44} of Nb was similar to that caused by the hydrogen contamination [72]. Hydrogen contamination increased the Young's modulus of all three group V bcc metals [73] and the increase in E_{110} was very significant for Nb [74]. Contamination of Nb with oxygen (about 0.35 at.%) did not cause noticeable changes in the C' and C_{44} shear moduli [75].

Regarding solid solutions of Nb with other bcc metals, contamination with oxygen affects mechanical properties. For example, for Nb-V alloys the addition of 500-ppm nitrogen and 1500-ppm (by weight) oxygen to Nb–2V and Nb–4V (wt.%) alloys caused pronounced increases in the DBTT of (Nb,V,I)$_{ss}$ where I = O or N. Nitrogen was found to be more potent than oxygen as a strengthener. The influence of both nitrogen and oxygen on the mechanical properties increased with increasing V content [76]. The affinity of Al, Cr, Hf, Ti and Zr for oxygen is high (for example Hf or Zr is used to scavenge oxygen in RM alloys [56]).

The bcc solid solution in RM(Nb)ICs, RM(Nb)ICs/RCCAs, RM(Nb)ICs/RHEAs, RHEAs and RCCAs is contaminated with oxygen, owing to the sensitivity of RMs on interstitial contamination and the presence of reactive elements in solution [3,10,13,16–19,42,77]. Contamination can be severe depending on alloying elements (e.g., see Figure 17 in [13]). There are limited data for contaminated Nb$_{ss}$ and such data are available only for RM(Nb)ICs. These data show remarkable correlations between the parameters δ, $\Delta\chi$ and VEC. The data in Figure 10 are for "conventional" Nb$_{ss}$ and Ti rich Nb$_{ss}$ in AC alloys and for the diffusion zone (DZ) and bulk of alloys after isothermal oxidation at 800 °C for 100 h.

Figure 10 shows that δ_{Nbss} increases with increasing $\Delta\chi_{Nbss}$ or VEC$_{Nbss}$ and that VEC$_{Nbss}$ increases with increasing $\Delta\chi_{Nbss}$. The contaminated Nb$_{ss}$ in the DZ is CC/HE Nb$_{ss}$, whereas that in the bulk of oxidised alloys can be "conventional" Nb$_{ss}$ or CC/HE Nb$_{ss}$. The parameters of the solid solution in the AC alloys have the lowest values. There are also linear relationships between the parameters $\Delta\chi$ (Figure 10d), VEC and δ (figures not shown) of the contaminated Nb$_{ss}$ in the bulk of alloy after isothermal oxidation at 800 °C versus the same parameter of "uncontaminated" Nb$_{ss}$ in AC alloy that show the same trend as in Figure 10d, meaning the parameter of the former is higher the higher the parameter of the latter.

Remarkably, strong correlations also exist for the oxygen concentration in contaminated solid solutions in the diffusion zone and the bulk of isothermally oxidised alloys at 800 °C with the parameters $\Delta\chi_{Nbss}$ (Figure 11a), δ_{Nbss} (Figure 11b) and VEC$_{Nbss}$ (Figure 11c), the values of which increase with increasing oxygen concentration in the solid solution. Note (i) that the chemical analysis data have been obtained using electron probe microanalysis [13,16–18] and (b) the strong correlation with the parameter $\Delta\chi_{Nbss}$. Moreover note that there are relationships between the concentrations of solutes in alloy and solid solution and the parameters $\Delta\chi_{alloy}$ and $\Delta\chi_{ss}$, respectively, which are key in alloy design using NICE [10].

The co-existence of CC/HE Nb$_{ss}$ with "conventional" Nb$_{ss}$ in most alloys [14] is further supported by the data in Figure 10, which also show that such relationships between the parameters δ, $\Delta\chi$ and VEC can be used in NICE to predict whether the microstructure of a designed alloys will consist of "conventional" Nb$_{ss}$ and CC/HE Nb$_{ss}$ and what the chemical compositions of such solid solutions would be. Furthermore, Figure 11 shows that contamination with oxygen affects all three parameters, which are related with atomic size, electronegativity and electron concentration in the valence band [10,28] and can account for changes in mechanical properties (creep, strength) and oxidation [9,10,14,17,18,32–34,42].

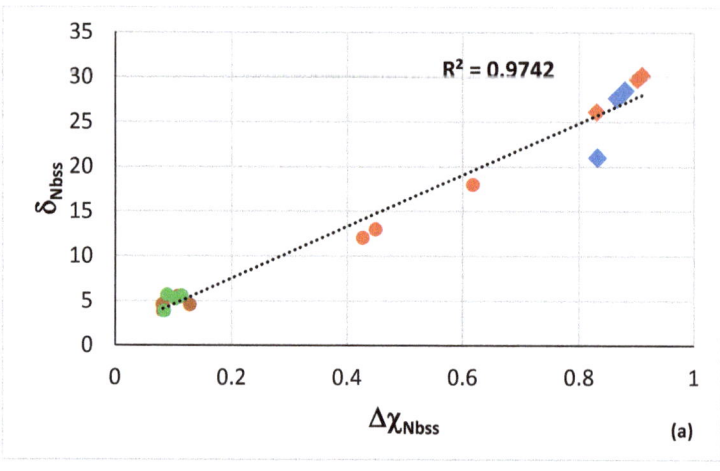
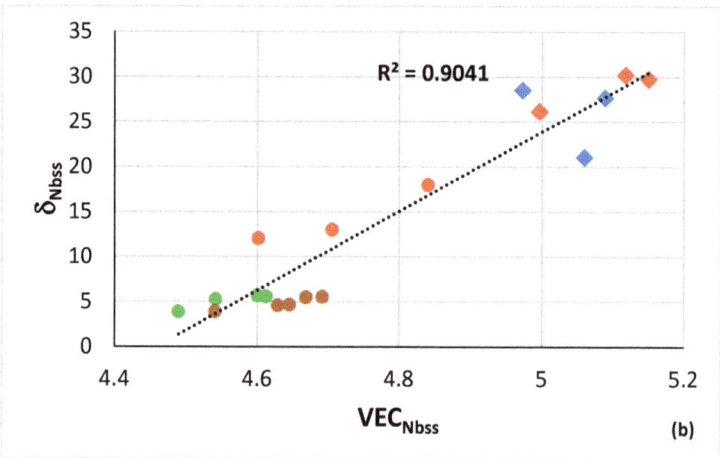
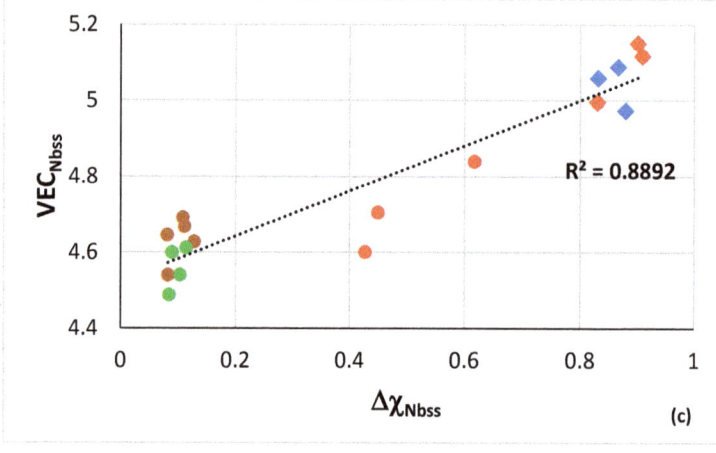

Figure 10. Cont.

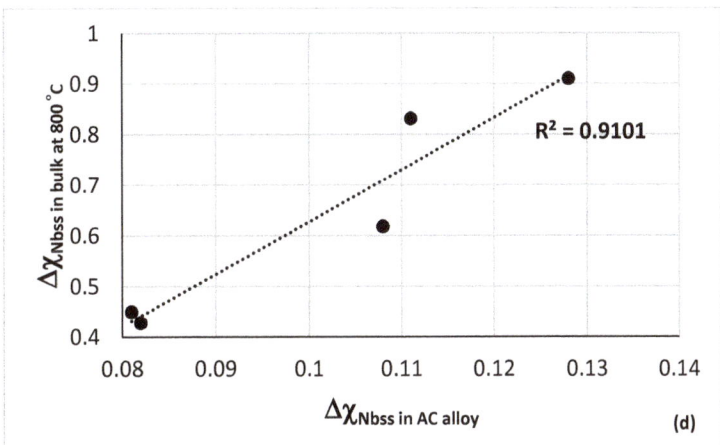

Figure 10. (a) δ_{Nbss} versus $\Delta\chi_{Nbss}$ (b) δ_{Nbss} versus VEC_{Nbss}, (c) VEC_{Nbss} versus $\Delta\chi_{Nbss}$ and (d) $\Delta\chi$ of contaminated Nb_{ss} in the bulk of alloy after isothermal oxidation at 800 °C versus $\Delta\chi$ of "uncontaminated" Nb_{ss} in AC alloy. (a–c) colours: brown for Nb_{ss} in AC alloy, green for Ti rich Nb_{ss} in AC alloy, blue for Nb_{ss} in diffusion zone (DZ) formed at 800 °C, red for Nb_{ss} in bulk of alloy isothermally oxidised at 800 °C. Diamonds for CC/HE Nb_{ss}. In each part the R^2 value is for the linear fit of all the data. Data for the alloys NV1, NV2, NV5, ZX5 and ZX7. See Appendix A for nominal alloy compositions and references.

3.2. Effect of Contamination with Oxygen on Properties of the Solid Solution

3.2.1. Hardness

Contamination of Nb with oxygen increases the Vickers hardness and yield strength of $(Nb,O)_{ss}$ and also increases its DBTT (Section 3.1). Contamination of the Nb solid solution in RM(Nb)ICs, RM(Nb)ICs/RCCAs or RM(Nb)ICs/RHEAs would affect its mechanical properties, in particular its hardness/yield strength and Young's modulus [3]. In each of these types of alloys and other RCCAs and RHEAs, for example those included in the review in [19], the contamination of the bcc solid solution will be different as it depends on the chemical composition of the solid solution, alloy condition (AC, HT) and environment of operation. For example, the contamination of the solid solution of the alloy NV1 was very sever, compared with other RM(Nb)ICs, see Figure 17 in [13]. I shall demonstrate the effects of contamination of bcc solid solution with oxygen on properties using data for the Nb_{ss} in the RM(Nb)IC alloy NV1.

Why the alloy NV1? The high vol.% Nb_{ss} (about 81%) in this alloy made feasible the measurement of the nanohardness of the Nb_{ss} using nanoindentation, as discussed in [16]. Furthermore, the solute additions included key solute elements in metallic UHTMs, namely Al, Cr, Hf, Nb, Ti and V as well as Si and Sn.

The alloy NV1 was heat treated at 1500 °C for 100 h in a Ti-gettered argon atmosphere [16]. Contamination of the alloy could not be avoided even under these HT conditions. The nanohardness (nanoH, GPa) and the reduced elastic modulus E_r (GPa) of the Nb_{ss} of NV1-HT was measured from the surface of the heat-treated specimen to 2000 μm below the surface. A Hysitron TriboScope nano-mechanical testing system was used [16], with 8000 μN indenter load. A 4 × 4 testing array was created over a 50 μm × 50 μm area, 16 indents per area with 10 μm spacing. Data were collected from the surface and areas below it every 40 μm to a depth of 400 μm, then at 470 μm, then every 100 μm to 770 μm depth, and then at 940, 1070, 1220, 1390, 1590 and 2000 μm [78]. The microstructure of NV1-HT was shown in Figure 2 in [16].

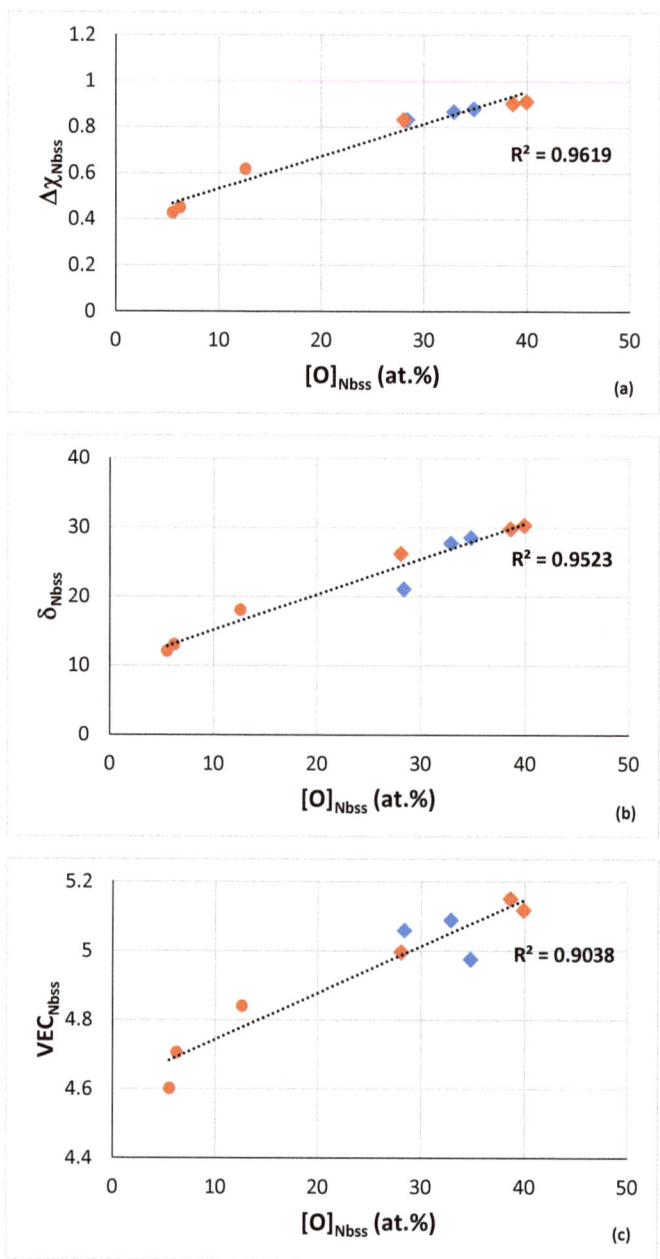

Figure 11. Relationships of the oxygen concentration in contaminated solid solutions in the diffusion zone and the bulk of alloys that were oxidised isothermally at 800 °C. Oxygen concentration (a) versus $\Delta\chi_{Nbss}$ (b) versus δ_{Nbss} and (c) versus VEC_{Nbss}. Colours: blue for Nb_{ss} in diffusion zone (DZ) formed at 800 °C and red for Nb_{ss} in bulk of alloy isothermally oxidised at 800 °C. Diamonds for CC/HE Nb_{ss}. In each part the R^2 value is for the linear fit of all the data. Data for the alloys NV1, NV2, NV5, ZX5, ZX7. See Appendix A for nominal alloy compositions and references.

The data in Figure 12a show that the nanoH$_{ss}$ increased to a maximum value in the area that was 570 µm below the surface, and then decreased to the "bulk" value of the HT specimen (blue data point). In Figure 12a, all the data fit to the 4th order polynomial nanoH$_{ss}$ = $-3 \times 10^{-12}d^4 + 2 \times 10^{-08}d^3 - 3 \times 10^{-05}d^2 + 0.0175d + 7.4127$ with $R^2 = 0.921$. First, there was a rapid increase in nanohardness (red data points, R^2 for linear fit of data) to about 120 µm, then the change in nanohardness with distance decreased (green data, R^2 for linear fit of data) and the nanohardness reached its maximum value in the area 570 µm below the surface, then the nanohardness decreased with distance from 570 µm to about 1220 µm (brown data points, R^2 for linear fit of data) followed with minor changes for distances greater than 1590 µm below the surface. A similar hardness profile to that shown in Figure 12a was reported in [70] for contamination of Nb with oxygen (i.e., for (Nb,O)$_{ss}$) after 1.62 h at 1000 °C, where the depth of contamination was at least 760 µm.

In Figure 12a the surface nanohardness is 7.29 GPa or 743.3 HV and corresponds to microhardness (microH) 548.2 HV based on the relationship microH$_{Nbss}$ = 0.7357 × nanoH$_{Nbss}$ (see [16]), whereas the maximum nanohardness of 10.77 GPa or 1098 HV at 570 µm below the surface corresponds to microhardness 807.9 HV. In the area 2000 µm below the surface the nanohardness was 6.1 GPa or 622 HV and corresponds to microH$_{Nbss}$ = 457.6 HV. This is lower than the hardness of the solid solution (523 HV) reported in [16], where the area of hardness measurement below the surface was not recorded. The surface hardness and the maximum hardness of the alloyed and contaminated with oxygen Nb$_{ss}$ in NV1-HT below the surface, respectively, was more than 7 and 10 times that of "uncontaminated" Nb. Up to 15 times increase in hardness has been reported for Nb contaminated with 16 at.% C, i.e., for (Nb,C)$_{ss}$ [79].

The hardness of the contaminated with oxygen Nb$_{ss}$ in NV1-HT at the surface (548 HV), and 570 µm below the surface (808 HV) was higher than the hardness of the (un-contaminated?) single bcc solid solution phase RHEAs HfMoTaTiZr (542 HV), MoNbTaVW (535 HV), HfMoNbTaTiZr (505 HV), MoNbTaV (504 HV), NbTaVW (493 HV), MoNbTaW (454 HV), NbTaTiVW (447 HV), MoNbTaTiV (443 HV), TaNbHfZrTi (409 HV), HfNbTaTiZr (390 HV), NbTiVZr (335), NbTaTiV (298 HV) [80].

The EPMA analyses of Nb$_{ss}$ grains in areas about 600 µm below the surface gave the average composition of the contaminated Nb$_{ss}$ as 54.1(±4, 49.9–58.7)Nb–17.6(±3.2, 13.3–21.8)Ti–0.6(±0.3, 0–0.9)Si–5(±0.2, 4.8–5.5)Al–2.7(±0.5, 2.1–3.5)Cr–5.6(±0.7, 4.9–6.6)V–2.2(±0.5, 1.2–2.7)Sn–0.2(±0.1, 0–0.4)Hf–12.1(±2, 8.8–15.2)O, where in parenthesis is given the standard deviation and the minimum and maxim analysis value. There was no second phase precipitation in the solid solution. The oxygen concentration of 12.1 at.% and Figure 11b give δ_{ss} = 16.24. The ascending part of the HV$_{ss}$ versus δ_{ss} data (brown data) in Figure 7 in [3] gives microH$_{ss}$$^{600\ µm}$ = 797 HV that corresponds to nanoH$_{ss}$$^{600\ µm}$ = 1083 HV or nanoH$_{ss}$$^{600\ µm}$ = 10.62 GPa. The highest measured nanohardness of the Nb$_{ss}$, which was for the area 570 µm below the surface (see above), and Figure 7 in [3] give δ_{ss} = 16.66 and from Figure 11b we obtain the oxygen content of 12.96 at.%. Both oxygen concentrations are higher than the maximum solubility of oxygen in Nb (9 at.% at 1915 °C) according to the Nb-O binary phase diagram [69] and would suggest that the hardness increased with distance below the surface to the area where the Nb$_{ss}$ most likely became saturated with oxygen. Note (i) the 9 at.% O solubility is for the (Nb,O)$_{ss}$, (ii) that the maximum solubility of oxygen in the Nb$_{ss}$ of NV1 is not known, (iii) that the Nb$_{ss}$ in NV1 was heavily alloyed and its contamination was more severe compared with the Nb$_{ss}$ in other RM(Nb)ICs (see Figure 17 in [13]) and (iv) that no precipitation of a second phase in the Nb$_{ss}$ was observed in the areas below the surface where nanoindentation was performed.

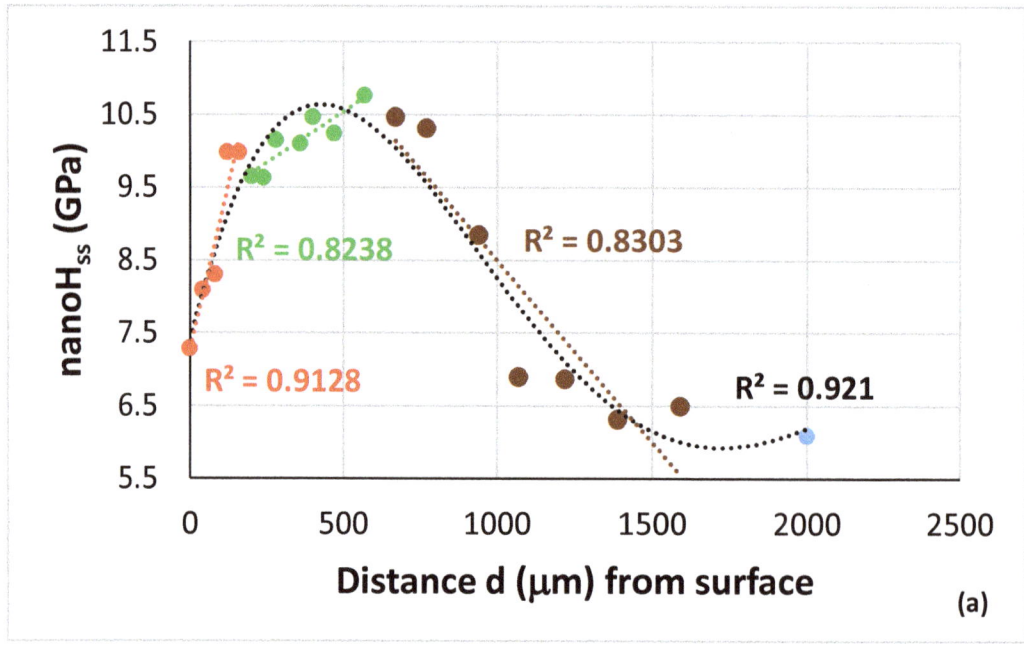

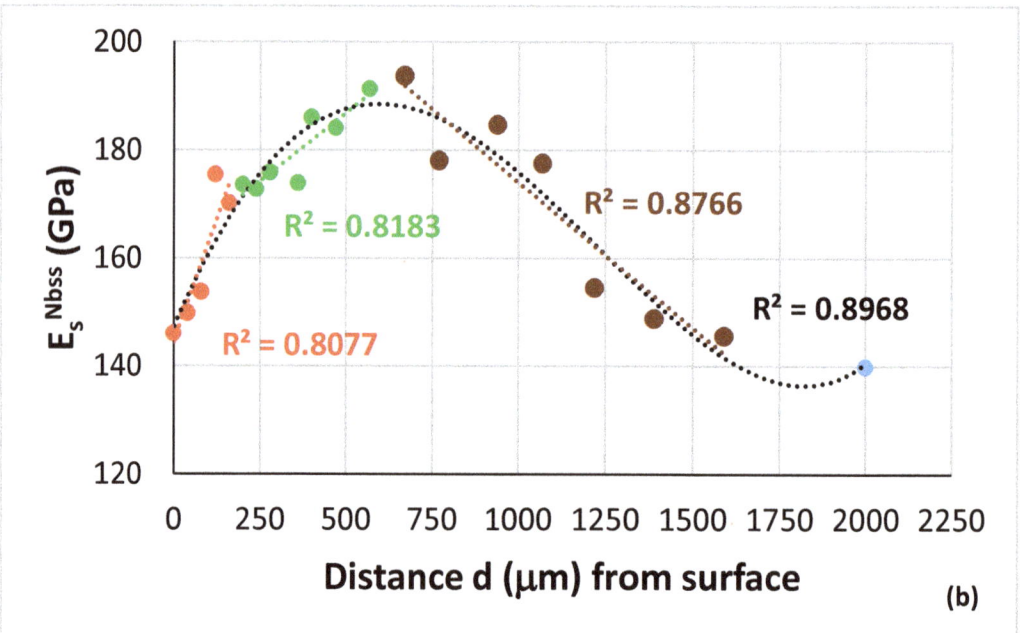

Figure 12. Cont.

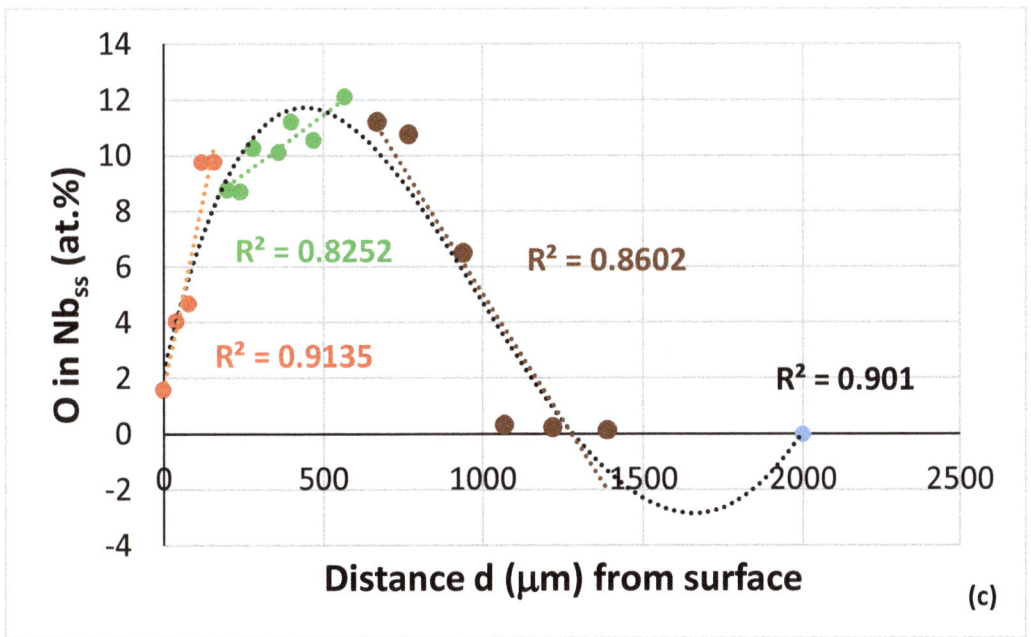
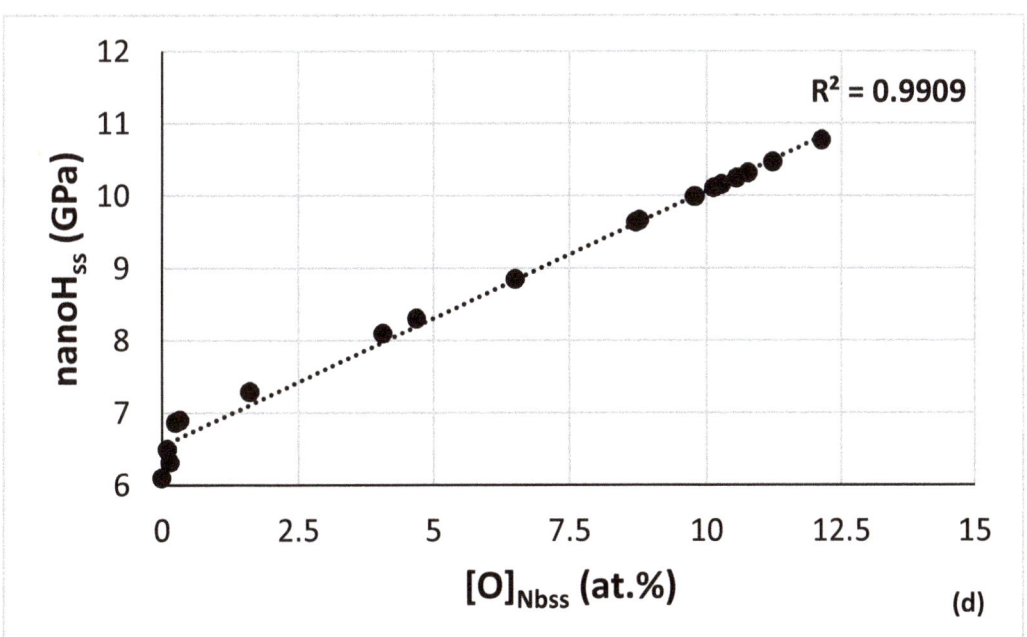

Figure 12. (a) Average nanohardness, (b) average Young's modulus and (c) average oxygen concentration of the Nb_{ss} in NV1-HT (1500 °C/100 h) as a function of distance from the surface of heat treated specimen. In each part, all the data fit to a 4th order polynomial (see text) with R^2 values 0.921, 0.8968 and 0.9016, respectively, for (a–c). (d) Nanohardness of Nb_{ss} versus oxygen content.

NICE has demonstrated how the relationships between parameters of alloys and their phases and between parameters and properties of alloys and their phases can assist the alloy designer to design/select new alloys worthy of R&D work [9,10,14]. Below it will be shown that it is possible to use such relationships to understand/predict how contamination with oxygen or alloying with boron affect properties of the solid solution.

Owing to contamination of Nb with oxygen the hardness and the lattice parameter of the $(Nb,O)_{ss}$ increase. This is well documented in the literature, for example see [60] and references within, and [61]. The effect of oxygen contamination on the hardness of Nb is given with linear relationships of the form $HV_{(Nb,O)ss} = A[O] + HV°_{Nb}$ where [O] is concentration of oxygen and $HV°_{Nb}$ is the hardness of "uncontaminated" "pure" Nb. In the literature the values of the constant A and $HV°_{Nb}$ differ because they depend on the purity of the starting "uncontaminated" Nb, the method of preparation of the $(Nb,O)_{ss}$ and the analysis method used. For example, when the main impurities of the Nb were Ta (860 ppm) and W (460 ppm) Kotch et. al. gave $HV_{(Nb,O)ss} = 90.903[O] + 77.566$ with $R^2 = 0.9762$. The lattice parameter of the contaminated Nb was given by the same researchers as $\alpha_o^{(Nb,O)ss}$ (Å) $= 0.0039[O] + 3.3$ with $R^2 = 0.9622$ [81]. Furthermore, they reported that the contamination of Nb with oxygen decreased the density of electronic states at the Fermi level N(0) and the "band structure" density of states $N_{bs}(0)$ [81], both of which correlate well with the parameter VEC of the $(Nb,O)_{ss}$ (Figure 13a,b). The importance of the parameter VEC for the properties (oxidation, creep) of RM(Nb)ICs was discussed in [10].

Boron Free RM(Nb)ICs

For boron free KZ series alloys (KZ series alloys are RM(Nb)ICs based on Nb-24Ti-18Si (at.%, nominal) with addition of Al, Cr individually or simultaneously, for example the alloys KZ4, KZ5 and KZ7, or with simultaneous addition of Al, Cr and Ta, for example the alloy KZ6, see Appendix A for nominal compositions) the hardness of the Nb_{ss} depends on δ with a linear relationship of the form $HV_{ss} = a\delta + b$ of which the constants a and b are both positive (for example, see the ascending data (brown data points) in Figure 7 in [3]). The values of these constants change when Sn or Ge is present in the alloy with/without Hf but they are still positive. The constant b is the hardness of Nb_{ss} for which the type of solute elements and their concentrations give $\delta = 0$ (solute additions and contamination with oxygen will change the lattice parameter). The parameter δ of the Nb_{ss} depends on oxygen concentration with a linear relationship of the form $\delta = c[O] + d$ (Figure 11b) where both the constants c and d are positive and [O] is the concentration of oxygen in the Nb_{ss}. The constant d is the value of the parameter δ of the "uncontaminated" Nb_{ss}.

For a specific alloy 1, $HV_{ss1} = a_1\delta_1 + b_1$ and $\delta_1 = c_1[O] + d_1$. Thus $HV_{ss1} = a_1(c_1[O] + d_1) + b_1 = a_1c_1[O] + a_1d_1 + b_1$ or $HV_{ss1} = A_1[O] + B_1$, where $A_1 = a_1c_1$ and $B_1 = a_1d_1 + b_1$. Both A_1 and B_1 are positive. The value of A_1 will be deferent from the value of A for $(Nb,O)_{ss}$ (see previous section), and will depend on the solute elements in Nb_{ss}, which sequentially affect the severity of contamination of the solid solution (see Figure 17 in [13]). In other words, the value of A_1 will depend on the specific RM(Nb)IC, RM(Nb)IC/RCCA or RM(Nb)IC/RHEA being considered. For the specific alloy 1 the hardness of its Nb_{ss} for zero [O], i.e., the value of B_1, is made of two parts, one (the constant b_1) is the hardness of a Nb_{ss} with the same solute elements and concentrations that give $\delta = 0$ and the other part depends (i) on how changes in atomic size, owing to alloying additions and their concentrations (excluding oxygen contamination) affect hardness (the constant a_1) and (ii) on how oxygen contamination affects atomic size (the constant d_1).

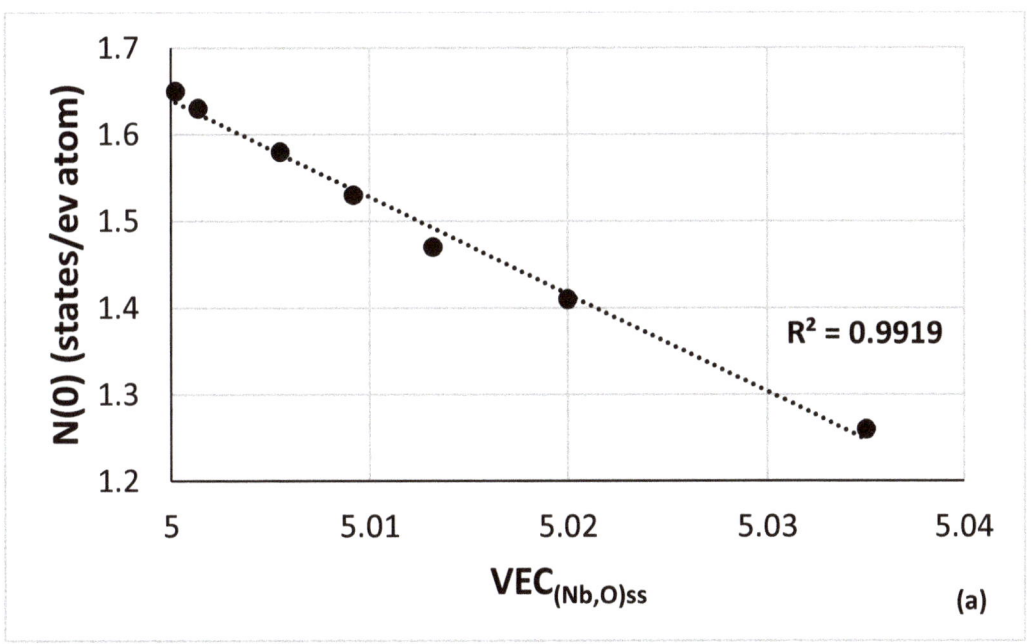

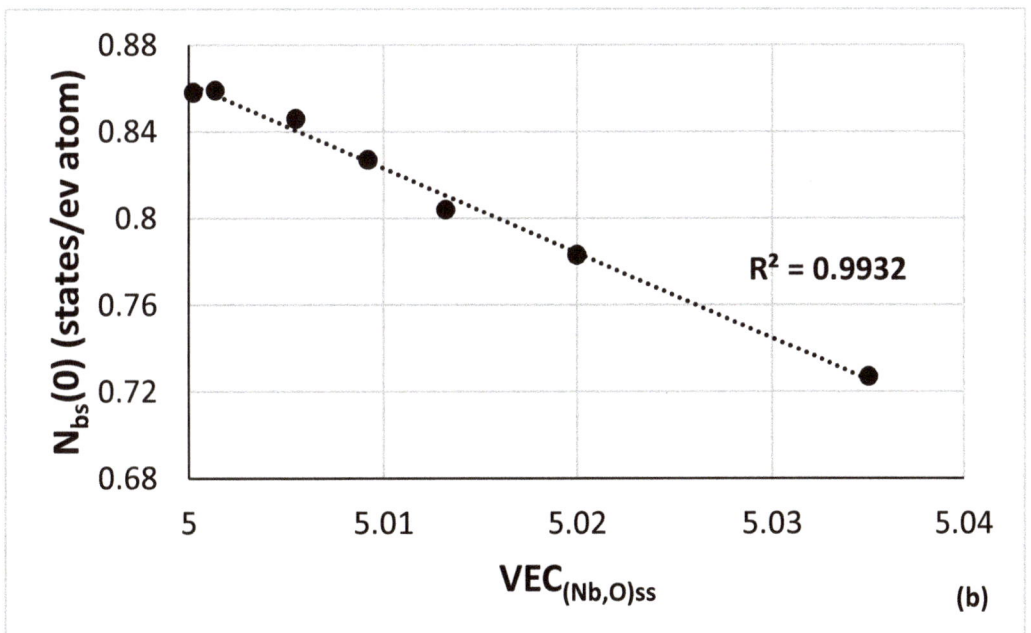

Figure 13. Data for Nb contaminated with oxygen. (a) density of electronic states at the Fermi level N(0) and (b) "band structure" density of states $N_{bs}(0)$ versus the parameter VEC of the $(Nb,O)_{ss}$.

For the Nb_{ss} of the alloy NV1-HT, the hardness is $HV_{ss\ NV1-HT} = A_{ss\ NV1-HT}[O] + B_{ss\ NV1-HT}$. If we were to assume that $B_{ss\ NV1-HT}$ is the average of the measured microhardness values of the Nb_{ss} in the bulk of NV1-HT given in [16] and in [82] and [78] (i.e., $B_{ss\ NV1-HT}$ = 507 HV), and take into account the measured oxygen content of Nb_{ss} in NV1-HT below the surface (see previous section) and the nanohardness data for different areas below the surface (Figure 12a), we can calculate the oxygen concentration of the Nb_{ss} with distance below the surface of NV1-HT. This is shown in Figure 12c. Similarly with the other two parts of Figure 12, all the data for oxygen concentration as a function of distance below the surface fit to the 4th order polynomial $[O]_{Nbss}$ (at.%) = $-4 \times 10^{-12}d^4$ + $3 \times 10^{-08}d^3 - 8 \times 10^{-05}d^2 + 0.049d + 2.2136$ with R^2 = 0.901. In Figure 12c, the blue data point corresponds to Nb_{ss} in the bulk. The hardness of the solid solution increased with oxygen contamination. The R^2 values were 0.9909, 0.9907 and 0.9792, respectively, for the linear fit of the $nanoH_{ss}$ data versus $[O]$, $[O]^{2/3}$ and $[O]^{0.5}$ (at.%). The best fit of the data is shown in Figure 12d.

For the alloy NV1, if we assume that the solid solution at 2000 µm below the surface was uncontaminated we can obtain the contribution to solid solution hardening from the alloying additions as $HV_{Nbss}{}^{2000\ \mu m} - HV°_{Nb}$. The contribution of solute additions to hardening depends on the value of $HV°_{Nb}$, it would be about 437.5 HV if we take $HV°_{Nb}$ as the average of the values in the Table 1 in [60] or about 430 HV if we take $HV°_{Nb}$ = 77.6 from [81]. We can also calculate the change in hardness of the Nb_{ss} (ΔHV) due to contamination with oxygen ($\Delta HV = HV_{Nbss\ contaminated} - HV_{Nbss}{}^{bulk}$) with distance d below the surface. This is shown in Figure 14a, where the data fit to a 4th order polynomial $\Delta HV = -2 \times 10^{-10}d^4 + 1 \times 10^{-06}d^3 - 0.0022d^2 + 1.2977d + 99.418$ with R^2 = 0.9229. The contribution to hardening of the solid solution due to contamination with oxygen increased with the concentration of the latter in the Nb_{ss}. The R^2 values were 0.9916, 0.9918 and 0.9805, respectively, for the linear fit of ΔHV_{ss} data versus $[O]$, $[O]^{2/3}$ and $[O]^{0.5}$ (at.%). The best linear fit of the data is shown in Figure 14b. Note that all the data in Figure 14b fit to a 6th order polynomial $\Delta HV_{ss} = -0.1603x^6 + 3.0517x^5 - 23.073x^4 + 87.252x^3 - 163.54x^2 + 180.57x - 4.4811$ with R^2 = 0.9989, shown with blue dashed line, and that at low oxygen contents the increase in ΔHV_{ss} is parabolic (R^2 = 0.9596), followed by linear increases (R^2 = 0.9981 and R^2 = 0.9997) with increasing $\Delta HV_{ss}/[O]^{2/3}$ as the severity of contamination with oxygen increased, in agreement with [60].

Boron Containing RM(Nb)ICs

For boron containing KZ series alloys (for "definition" of these alloys see the previous section) the hardness of the solid solution also is given by a liner relationship of the form $HV_{ss} = a\delta + b$, where a < 0 and b > 0 (see the descending part of the data (green data points) in Figure 7 in [3]). Currently, there are no data for the change in the parameter δ with oxygen concentration in the Nb_{ss}. Let us assume (i) that a linear relationship of the form $\delta = c[O] + d$ applies for the dependence of δ on contamination and (ii) that for an alloy 1 the hardness of its Nb_{ss} increases with oxygen contamination, i.e., the equation $HV_1 = A_1[O] + B_1$ applies, or $HV_{ss1} = a_1c_1[O] + a_1d_1 + b_1$. Given that $a_1 < 0$, the first term would be negative if c_1 were to be positive. Thus, based on the aforementioned assumptions, we conclude that c_1 must be negative, in other words the parameter δ of the solid solution would decrease as the [O] concentration increases. This must be tested experimentally. The last two terms (i.e., $B_1 = a_1d_1 + b_1$) give the hardness of the "uncontaminated" solid solution, i.e., Nb_{ss} with [O] = 0 at.%. Similarly with the boron free alloys, the value of B_1 is made of two parts, one (the constant b_1) is the hardness of a Nb_{ss} with the same solute elements and concentrations that give δ = 0 and this part is positive, and the other part, which in this case is negative, depends (i) on how changes in atomic size affect hardness (the constant $a_1 < 0$) owing to the alloying additions and their concentrations (excluding oxygen contamination) and (ii) on how oxygen contamination affects atomic size (the constant $d_1 > 0$).

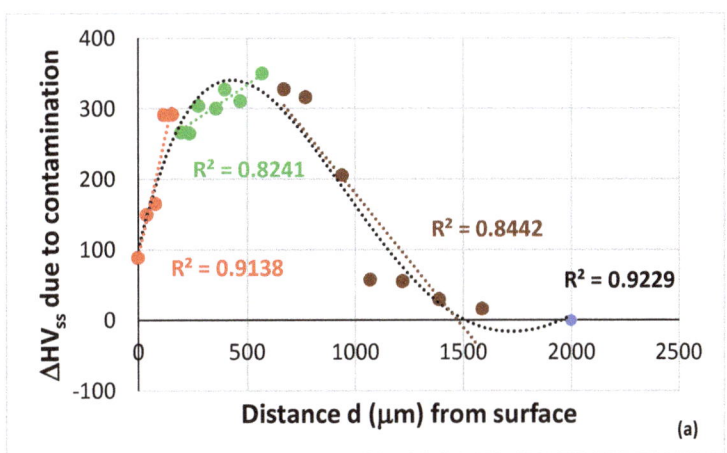
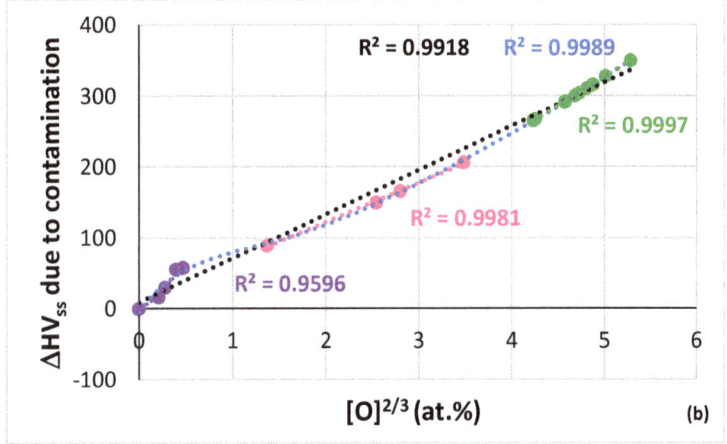
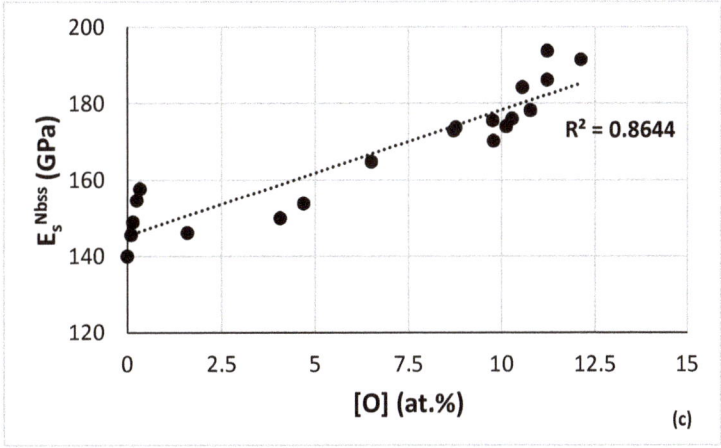

Figure 14. *Cont.*

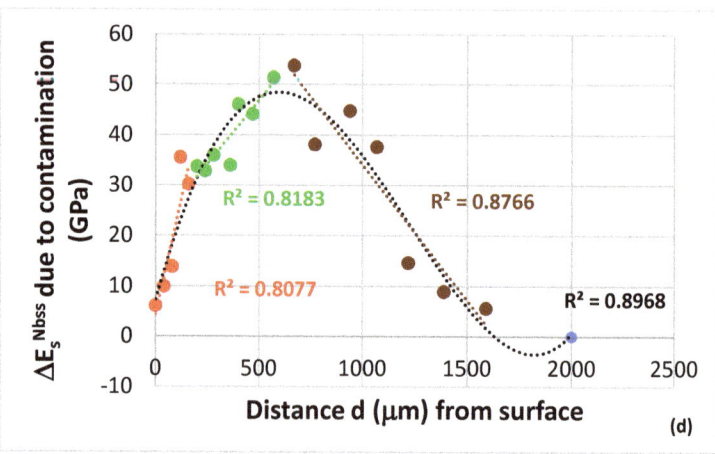

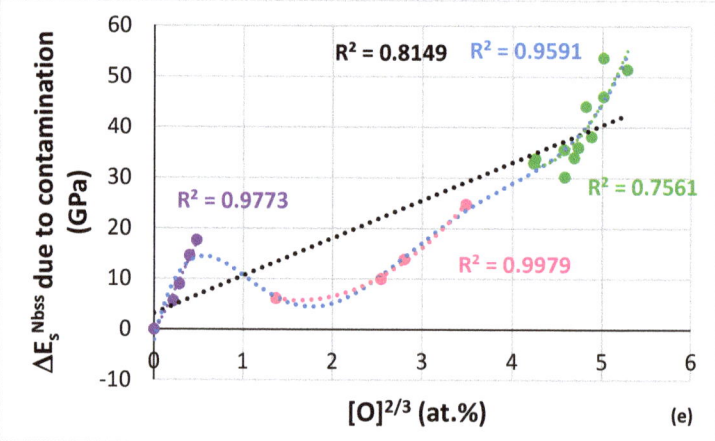

Figure 14. Data for the bcc solid solution Nb_{ss} in the alloy NV1-HT. (**a,b**) change in solid solution hardness, respectively, with distance below the surface and with oxygen contamination. (**c,e**) dependence on oxygen contamination, respectively, of the nanoindentation Young's modulus and its change with oxygen concentration. (**d**) Change in nanoindentation Young's modulus because of contamination with oxygen with distance below the surface. In (**b,e**) the black and blue dashed lines, respectively, are for linear and polynomial fit of all data.

However, in the case of boron containing KZ series alloys, the parameter δ of the Nb_{ss} depends on its boron concentration, as shown in Figure 7a, and the hardness of the solid solution decreased with increasing B concentration (Figure 15). In other words, for these alloys the experimental evidence gives HV = C[B] + D, where C < 0, and D > 0 (Figure 15) and δ = e[B] + f, where e > 0 and f > 0 (Figure 7a). Thus, for a boron containing alloy 1, $HV_{ss1} = C_{ss1}[B] + D_{ss1}$, $HV_{ss1} = a_1δ + b_1$ (where $a_1 < 0$, see previous paragraph) or $HV_{ss1} = a_1(e_1[B] + f_1) + b_1$ or $HV_{ss1} = a_1e_1[B] + a_1f_1 + b_1$. Therefore, $C_{ss1} = a_1e_1$, i.e., the constant C_{ss1} is negative, in agreement with the experimental data (Figure 15). The value of $D_{ss1} = a_1f_1 + b_1$ is made of two parts, one (the constant b_1) is the hardness of a Nb_{ss} with the same solute elements and concentrations that give δ = 0 and this part is positive, as was the case for B_1 (see above), and the other part, which in this case is negative, depends (i) on how changes in atomic size affect hardness (the constant $a_1 < 0$) owing to the alloying additions and their concentrations (excluding oxygen contamination), as was the case for

B_1, and (ii) on how alloying with boron affects atomic size (the constant $f_1 > 0$), differently with B_1.

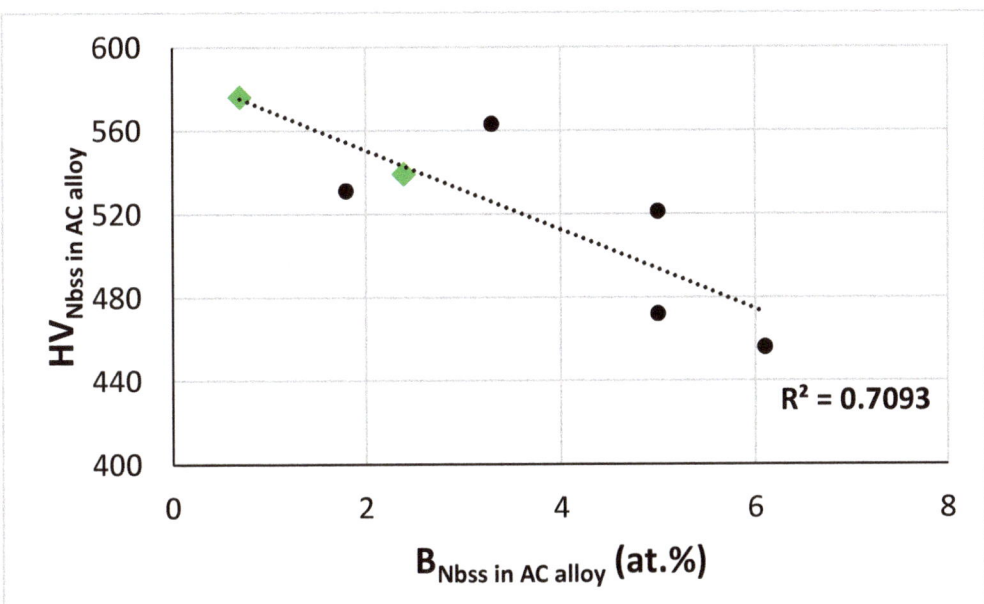

Figure 15. Hardness of the Nb_{ss} in AC boron containing KZ series alloys versus the B concentration of the solid solution. Data for the alloys TT1, TT2, TT3, TT4, TT5, TT7 and TT8. Green diamonds for Nb_{ss} in RM(Nb)ICs/RCCAs. All data $R^2 = 0.7093$.

3.2.2. Young's Modulus

The Young's modulus E_s (GPa) of the Nb_{ss} was calculated using the data from the nano-indentation experiments (see Section 3.2.1) and Equation (1)

$$E_s = \frac{E_r E_i (1 - v_s^2)}{E_i - E_r(1 - v_i^2)} \qquad (1)$$

where E_s and v_s are the Young's modulus and Poisson's ratio of the phase, and E_i, v_i are the parameters for the Berkovich indenter [83]. For the calculation of E_s the values of E_i and v_i that were given in the TriboScope manual [84] as 1140 GPa and 0.07, respectively, were used and the v_s was 0.38 [46,85]. Data for E_s are shown in Figure 12b.

The data showed that E_r^{Nbss} and E_s^{Nbss} increased to maximum values in the area 570 μm below the surface, and then decreased to the "bulk" of the HT specimen (blue data point). All the data fit to 4th order polynomials, as follows: $E_r^{Nbss} = 6 \times 10^{-12}d^4 + 2 \times 10^{-08}d^3 - 0.0001d^2 + 0.1284d + 149.81$ with $R^2 = 0.8972$ (figure not shown) and $E_s^{Nbss} = 6 \times 10^{-12}d^4 + 3 \times 10^{-08}d^3 - 0.0002d^2 + 0.1501d + 147.33$ with $R^2 = 0.8968$ (Figure 12b). First, there is a rapid increase in E_s from 146 GPa at the surface to 175 GPa about 120 μm below the surface (red data points, the R^2 value is for linear fit of data), then the change in E_s with distance decreases (green data, the R^2 value is for linear fit of data) and reaches its maximum value of 194 GPa in the area 670 μm below the surface, then the E_s decreases with distance to about 155 GPa at 1220 μm (brown data points, the R^2 value is for linear fit of data) followed with minor changes for distances greater than 1590 μm below the surface to about 140 GPa in the bulk.

The values of the Young's modulus of the alloyed and contaminated with oxygen Nb_{ss} in NV1-HT at 400 to 570 μm below the surface approached that of unalloyed γNb_5Si_3 [46]. Significant increase in the Young's modulus of Nb owing to interstitial contamination has been reported for Nb contaminated with 16 at.% C (i.e., for $(Nb,C)_{ss}$), where the increase was up to three times the Young's modulus of "uncontaminated" Nb [79].

The Young's moduli of the Nb_{ss} at the surface and the bulk of NV1-HT were higher than those of the (uncontaminated?) single phase solid solution RHEAs TiZrNbMo (142 GPa), TiZrNbMoV (141 GPa), MoNbTaTiV (139 GPa), TiZrVNb (121 GPa), NbTaTiV (117 GPa), TaNbHfZrTi (104 GPa), TiZrHfNbCr (104 GPa), TiZrHfNbV (95 GPa) anc TiZrHfNb (89 GPa), and the maximum Young's modulus (194 GPa) was lower than that of the (uncontaminated?) single phase solid solution RHEAs NbMoTaW (229 GPa), VnbMoTaW (205 GPa) and AlMoNbV (197 GPa) [86].

The Young's modulus of the Nb_{ss} in NV1-HT increased with oxygen contamination. The R^2 values were 0.8644, 0.8149 and 0.7815, respectively, for the linear fit of E_s data versus [O], $[O]^{2/3}$ and $[O]^{0.5}$ (at.%). The best linear fit of the data is shown in Figure 14c. If we take the Young's modulus of uncontaminated and pure Nb as $E°_{Nb}$ = 101.9 GPa [46] and assume that the solid solution at 2000 μm below the surface was uncontaminated, we can obtain the contribution to the Young's modulus from the alloying additions (about 38.1 GPa) and then we can calculate the change in the Young's modulus of the Nb_{ss} (ΔE_s) due to contamination with oxygen with distance below the surface, as shown in Figure 14d, where $\Delta E_s = 6 \times 10^{-12}d^4 + 3 \times 10^{-08}d^3 - 0.0002d^2 + 0.1501d + 7.3296$, with $R^2 = 0.8968$. The R^2 values were 0.8644, 0.8149 and 0.7815, respectively, for the linear fit of ΔE_s data versus [O], $[O]^{2/3}$ and $[O]^{0.5}$ (at.%), but the best fit to 6th order polynomial was for ΔE_s versus $[O]^{2/3}$, which is shown with the dashed blue line in Figure 14e for which $\Delta E_s = -0.0951x^6 + 2.0575x^5 - 16.628x^4 + 62.854x^3 - 109.2x^2 + 73.741x - 2.07$ with $R^2 = 0.9591$. Note that at low contamination level the increase in ΔE_s was linear ($R^2 = 0.9773$) and was followed with parabolic increases ($R^2 = 0.9979$ and $R^2 = 0.7561$) as the severity of contamination increased. As shown in the Figure 16 there is a linear relationship between the change in Vickers hardness (ΔHV) and the change in Young's modulus (ΔE) of the solid solution due to contamination with oxygen.

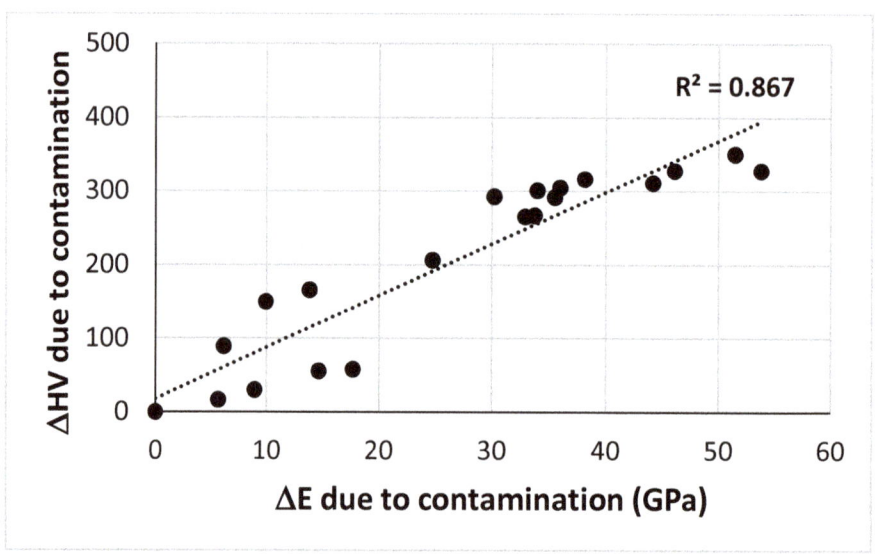

Figure 16. Change in Vickers hardness versus the change in Young's modulus of the solid solution due to contamination with oxygen. The R^2 value is for the linear fit of all data.

4. Summary

In this paper the stability of CC/HE solid solutions and the contamination with oxygen of solid solutions in (RM(lNb)ICs), RM(Nb)ICs/RCCAs and RM(Nb)ICs/RHEAs was studied. "Conventional" solid solutions were compared with CC/HE ones. "Conventional" Nb_{ss} can be Ti rich only in AC alloys, and Ti rich Nb_{ss} is not observed in HT alloys. In B containing alloys the Ti rich solid solution is usually CC/HE Nb_{ss}. The CC/HE Nb_{ss} is stable after heat treatment in alloys with simultaneous addition of Mo, W with Hf, Ge and Sn. There is a strong correlation between the δ parameters of "conventional" and CC/HE Nb_{ss} with the B or the Ge+Sn concentration in the solid solution. Similarities and differences between relationships of other solutes in alloys with B or Ge+Sn addition were noted and their implications for alloy design were discussed.

The CC/HE Nb_{ss} has low $\Delta\chi$, VEC and Ω and high ΔS_{mix}, $|\Delta H_{mix}|$ and δ parameters and is formed in alloys that also have high entropy of mixing. These parameters were compared with those of single-phase solid solution HEAs and differences in the values of ΔH_{mix}, δ, $\Delta\chi$ and Ω, and similarities in the values of ΔS_{mix} and VEC were discussed. Relationships between the alloy and "conventional" solid solution parameters in NICE also apply for CC/HE Nb_{ss}. The parameters δ_{ss} and Ω_{ss}, and VEC_{ss} and VEC_{alloy} can differentiate between types of alloying additions and their concentrations and are key regarding the formation or not of CC/HE Nb_{ss}.

After isothermal oxidation at a pest temperature (800 °C/100 h) the contaminated with oxygen Nb_{ss} in the diffusion zone is CC/HE Nb_{ss}, whereas the solid solution in the bulk of the oxidised alloys can be "conventional" Nb_{ss} or CC/HE Nb_{ss}. The parameters of "uncontaminated" and contaminated with oxygen solid solutions are linked with linear relationships. There are strong correlations between the oxygen concentration in contaminated solid solutions in the diffusion zone and the bulk of isothermally oxidised alloys at 800 °C with the parameters $\Delta\chi_{Nbss}$, δ_{Nbss} and VEC_{Nbss}, the values of which increase with increasing oxygen concentration in the solid solution.

Correlations between oxygen content and the parameters δ, $\Delta\chi$ and VEC showed that the effects of interstitial contamination on properties can be understood and/or described with all three parameters. The boron content on the other hand correlates only with δ.

The effects of contamination with oxygen of the near surface areas of a heat-treated RM(Nb)IC with high vol.% Nb_{ss} on the hardness and Young's modulus of the solid solution, and contributions to the hardness of the Nb_{ss} in B free or B containing KZ series alloys were discussed. The hardness and Young's modulus of the bcc solid solution increased linearly with its oxygen concentration and the change in hardness and Young's modulus due to contamination increased linearly with $[O]^{2/3}$.

5. Suggestions for Future Research

In RM(Nb)ICs/RCCAs, the CC bcc solid solutions with Ge+Sn and Al, Cr, Hf, Mo, Ti and W additions are (i) stable (Figure 1) and (ii) Si free [33,34]. It is suggested (a) that single phase bcc solid solution RCCAs or RHEAs with above elements would be stable at high temperatures. Furthermore, given that Ge+Sn with Al, Cr, Hf, Mo and Ti improve oxidation at pest and high temperatures [32–34,87] it is suggested (b) that these RCCAs or RHEAs would also be oxidation resistant. For B containing RM(Nb)ICs, RM(Nb)ICs/RCCAs, RM(NB)ICs/RHEAs, RCCAs and RHEAs it is suggested to study the contamination with oxygen of their solid solutions and to find out if there is a relationship between the parameter δ and the oxygen concentration, see Boron Containing RM(Nb)ICs in Section 3.2.1.

Funding: The support of this work by the University of Sheffield, Rolls-Royce Plc and EPSRC (EP/H500405/1, EP/L026678/1) is gratefully acknowledged.

Institutional Review Board Statement: Not applicable.

Informed Consent Statement: Not applicable.

Data Availability Statement: All the data for this work is given in the paper, other data cannot be made available to the public.

Acknowledgments: The support of this work by the University of Sheffield, Rolls-Royce Plc and EPSRC (EP/H500405/1, EP/L026678/1) is gratefully acknowledged. Discussions with N Vellios, P Keating, J Uttley and O Chapman are gratefully acknowledged. For the purpose of open access, the author has applied a 'Creative Commons Attribution (CC BY) licence to any Author Accepted Manuscript version arising.

Conflicts of Interest: The author declares no conflict of interest.

Appendix A

Table A1. Nominal compositions (at.%) of reference alloys used in this work.

Alloy	Nb	Ti	Si	Al	B	Hf	Cr	Mo	Ta	V	W	Fe	Ge	Sn	Ref.
EZ5	43	24	18	5	-	5	-	-	-	-	-	-	-	5	[15]
EZ6	43	24	18	-	-	5	5	-	-	-	-	-	-	5	[15]
EZ8	38	24	18	5	-	5	5	-	-	-	-	-	-	5	[15]
JG3	46	24	18	5	-	-	5	2	-	-	-	-	-	-	[87]
JG6	36	24	18	5	-	5	5	2	-	-	-	-	-	5	[87]
JN1	43	24	18	5	-	5	5	-	-	-	-	-	-	-	[44]
JN2	43	15	18	-	-	2	10	5	-	-	5	-	-	2	[51]
JN3	51	15	18	-	-	2	2	5	-	-	5	-	-	2	[51]
JN4	45	20	20	-	-	2	2	6	-	-	-	-	-	5	[51]
JZ3	41.8	12.4	17.7	4.7	-	1	5.2	-	6	-	2.7	-	4.8	3.7	[33]
JZ3+	38.7	12.4	19.7	4.6	-	0.8	5.2	-	5.7	-	2.3	-	4.9	5.7	[33]
JZ4 *	38.9	12.5	17.8	5	-	1.1	5.2	6.2	-	-	2.3	-	5.2	5.8	[34]
JZ5 *	32	20.4	19.2	4.5	-	0.9	4.7	6.3	-	-	1.1	-	5.2	5.7	[34]
KZ4	53	24	18	-	-	-	5	-	-	-	-	-	-	-	[88]
KZ5	48	24	18	5	-	-	5	-	-	-	-	-	-	-	[88]
KZ6	42	24	18	5	-	-	5	-	6	-	-	-	-	-	[89]
KZ7	53	24	18	5	-	-	-	-	-	-	-	-	-	-	[88]
OHS1	38	24	18	5	-	-	5	-	-	-	-	-	5	5	[32]
NV1	53	23	5	5	-	5	2	-	-	5	-	-	-	2	[16]
NV2	43	30	10	2	-	2	5	-	-	-	-	3	-	5	[13]
NV5	43	24	18	-	-	-	5	-	-	-	-	5	-	5	[11]
TT1	50	24	18	-	8	-	-	-	-	-	-	-	-	-	[90]
TT2	48	24	16	-	7	-	5	-	-	-	-	-	-	-	[36]
TT3	48	24	16	5	7	-	-	-	-	-	-	-	-	-	[36]
TT4 *	42.4	24.6	15.7	5	6.9	-	5.4	-	-	-	-	-	-	-	[36]
TT5	37	24	18	5	5	-	5	-	6	-	-	-	-	-	[35]
TT6	39	24	18	4	6	-	5	-	-	-	-	-	-	4	[35]
TT7	38	24	17	5	6	5	5	-	-	-	-	-	-	-	[35]
TT8	42.5	24	17	3.5	6	5	2	-	-	-	-	-	-	-	[36]
YG8	67	-	20	-	-	5	-	5	-	-	3	-	-	-	[91]
YG10	59	10	18	-	-	5	-	5	-	-	3	-	-	-	[92]

Table A1. *Cont.*

Alloy	Nb	Ti	Si	Al	B	Hf	Cr	Mo	Ta	V	W	Fe	Ge	Sn	Ref.
YG11	54	10	18	5	-	5	-	5	-	-	3	-	-	-	[92]
ZF6	43	24	18	5	-	-	5	-	-	-	-	-	5	-	[93]
ZF9	38	24	18	5	-	5	5	-	-	-	-	-	5	-	[93]
ZX5	51	24	18	5	-	-	-	-	-	-	-	-	-	2	[17]
ZX7	46	24	18	5	-	-	5	-	-	-	-	-	-	2	[17]

* actual composition.

References

1. Shah, D.M.; Anton, D.L.; Musson, C.W. Feasibility Study of Intermetallic Composites. *MRS Proc.* **1990**, *194*, 333–340. [CrossRef]
2. Bewlay, B.P.; Jackson, M.R.; Gigliotti MF, X. Niobium silicide high temperature in situ composites. Chapter 6 in Intermetallic Compounds: Principles and Practice. *Intermet. Compd. Princ. Pract.* **2002**, *3*, 541–560.
3. Tsakiropoulos, P. Alloys for application at ultra-high temperatures: Nb-silicide in situ composites. *Prog. Mater. Sci.* **2020**, *123*, 100714. [CrossRef]
4. Eshed, E.; Larianovsky, N.; Kovalevsky, A.; Popov, V., Jr.; Gorbachev, I.; Katz-Demyanetz, A. Microstructural Evolution and Phase Formation in 2nd-Generation Refractory-Based High Entropy Alloys. *Materials* **2018**, *11*, 175. [CrossRef]
5. Zhang, H.; Zhao, Y.; Huang, S.; Zhu, S.; Wang, F.; Li, D. Manufacturing and Analysis of High-Performance Refractory High-Entropy Alloy via Selective Laser Melting (SLM). *Materials* **2019**, *12*, 720. [CrossRef] [PubMed]
6. Moravcikova-Gouvea, L.; Moravcik, I.; Pouchly, V.; Kovacova, Z.; Kitzmantel, M.; Neubauer, E.; Dlouhy, I. Tailoring a Refractory High Entropy Alloy by Powder Metallurgy Process Optimization. *Materials* **2021**, *14*, 5796. [CrossRef]
7. Srikanth, M.; Annamalai, A.R.; Muthuchamy, A.; Jen, C.-P. A Review of the Latest Developments in the Field of Refractory High-Entropy Alloys. *Crystals* **2021**, *11*, 612. [CrossRef]
8. Heilmaier, M.; Krüger, M.; Saage, H.; Rösler, J.; Mukherji, D.; Glatzel, U.; Völkl, R.; Hüttner, R.; Eggeler, G.; Somsen, C.; et al. Metallic materials for structural applications beyond nickel-based superalloys (Review). *JOM* **2009**, *61*, 61–676. [CrossRef]
9. Tsakiropoulos, P. Refractory Metal (Nb) Intermetallic Composites, High Entropy Alloys, Complex Concentrated Alloys and the Alloy Design Methodology NICE—Mise-en-scène [†] Patterns of Thought and Progress. *Materials* **2021**, *14*, 989. [CrossRef]
10. Tsakiropoulos, P. On Nb Silicide Based Alloys: Alloy Design and Selection. *Materials* **2018**, *11*, 844. [CrossRef]
11. Vellios, N.; Tsakiropoulos, P. Study of the role of Fe and Sn additions in the microstructure of Nb-24Ti-18Si-5Cr silicide based alloys. *Intermetallics* **2010**, *18*, 1729–1736. [CrossRef]
12. Vellios, N.; Tsakiropoulos, P. The role of Fe and Ti additions in the microstructure of Nb-18Si-5Sn silicide-based alloys. *Intermetallics* **2007**, *15*, 1529–1537. [CrossRef]
13. Vellios, N.; Tsakiropoulos, P. The Effect of Fe Addition in the RM(Nb)IC Alloy Nb–30Ti–10Si–2Al–5Cr–3Fe–5Sn–2Hf (at.%) on Its Microstructure, Complex Concentrated and High Entropy Phases, Pest Oxidation, Strength and Contamination with Oxygen, and a Comparison with Other RM(Nb)ICs, Refractory Complex Concentrated Alloys (RCCAs) and Refractory High Entropy Alloys (RHEAs). *Materials* **2022**, *15*, 5815. [CrossRef]
14. Tsakiropoulos, P. Refractory Metal Intermetallic Composites, High-Entropy Alloys, and Complex Concentrated Alloys: A Route to Selecting Substrate Alloys and Bond Coat Alloys for Environmental Coatings. *Materials* **2022**, *15*, 2832. [CrossRef] [PubMed]
15. Zacharis, E.; Utton, C.; Tsakiropoulos, P. A Study of the Effects of Hf and Sn on the Microstructure, Hardness and Oxidation of Nb-18Si Silicide-Based Alloys-RM(Nb)ICs with Ti Addition and Comparison with Refractory Complex Concentrated Alloys (RCCAs). *Materials* **2022**, *15*, 4596. [CrossRef]
16. Vellios, N.; Keating, P.; Tsakiropoulos, P. On the Microstructure and Properties of the Nb-23Ti-5Si-5Al-5Hf-5V-2Cr-2Sn (at.%) Silicide-Based Alloy—RM(Nb)IC. *Metals* **2021**, *11*, 1868. [CrossRef]
17. Xu, Z.; Utton, C.; Tsakiropoulos, P. A Study of the Effect of 2 at.% Sn on the Microstructure and Isothermal Oxidation at 800 and 1200 °C of Nb-24Ti-18Si-Based Alloys with Al and/or Cr Additions. *Materials* **2018**, *11*, 1826. [CrossRef] [PubMed]
18. Xu, Z.; Utton, C.; Tsakiropoulos, P. A Study of the Effect of 5 at.% Sn on the Micro-Structure and Isothermal Oxidation at 800 and 1200 °C of Nb-24Ti-18Si Based Alloys with Al and/or Cr Additions. *Materials* **2020**, *13*, 245. [CrossRef] [PubMed]
19. Senkov, O.N.; Miracle, D.B.; Chaput, K.J.; Couzinie, J.-P. Development and exploration of refractory high entropy alloys—A review. *J. Mater. Res.* **2018**, *33*, 3092–3128. [CrossRef]
20. Shah, D.M. MoSi2 and other silicides as high temperature structural materials. In *Superalloys 1992*; Antolovich, S.D., Stusrud, R.W., MacKay, R.A., Anton, D.L., Khan, T., Kissinger, R.D., Klastrom, D.L., Eds.; TMS (The Minerals, Metals & Materials Society): Pittsburgh, PA, USA, 1992; pp. 409–422.
21. Shah, D.; Anton, D. Evaluation of refractory intermetallics with A15 structure for high temperature structural applications. *Mater. Sci. Eng. A* **1992**, *153*, 402–409. [CrossRef]
22. Reed, R.C. *The Superalloys: Fundamentals and Applications*; Cambridge University Press: Cambridge, UK, 2006.
23. Inoue, A. Stabilization of metallic supercooled liquid and bulk amorphous alloys. *Acta Mater.* **2000**, *48*, 279–306. [CrossRef]

24. Fang, S.; Xiao, X.; Xia, L.; Li, W.; Dong, Y. Relationship between the widths of supercooled liquid regions and bond parameters of Mg-based bulk metallic glasses. *J. Non-Cryst. Solids* **2003**, *321*, 120–125. [CrossRef]
25. Suryanarayana, C. Phase formation under non-equilibrium processing conditions: Rapid solidification processing and mechanical alloying. *J. Mater. Sci.* **2018**, *53*, 13364–13379. [CrossRef]
26. Tsakiropoulos, P. On Nb silicide based alloys: Part II. *J. Alloys Compd.* **2018**, *748*, 569–576. [CrossRef]
27. Zhang, Y. *High Entropy Materials*; Springer: Berlin/Heidelberg, Germany, 2019. [CrossRef]
28. Tsakiropoulos, P. On the Nb silicide based alloys: Part I—The bcc Nb solid solution. *J. Alloys Compd.* **2017**, *708*, 961–971. [CrossRef]
29. Tsakiropoulos, P. On the Alloying and Properties of Tetragonal Nb5Si3 in Nb-Silicide Based Alloys. *Materials* **2018**, *11*, 69. [CrossRef] [PubMed]
30. Tsakiropoulos, P. Alloying and Properties of C14–NbCr2 and A15–Nb3X (X = Al, Ge, Si, Sn) in Nb–Silicide-Based Alloys. *Materials* **2018**, *11*, 395. [CrossRef] [PubMed]
31. Tsakiropoulos, P. Alloying and Hardness of Eutectics with Nbss and Nb5Si3 in Nb-silicide Based Alloys. *Materials* **2018**, *11*, 592. [CrossRef] [PubMed]
32. Hernández-Negrete, O.; Tsakiropoulos, P. On the Microstructure and Isothermal Oxidation at 800 and 1200 °C of the Nb-24Ti-18Si-5Al-5Cr-5Ge-5Sn (at.%) Silicide-Based Alloy. *Materials* **2020**, *13*, 722. [CrossRef] [PubMed]
33. Zhao, J.; Utton, C.; Tsakiropoulos, P. On the Microstructure and Properties of Nb-12Ti-18Si-6Ta-5Al-5Cr-2.5W-1Hf (at.%) Silicide-Based Alloys with Ge and Sn Additions. *Materials* **2020**, *13*, 3719. [CrossRef]
34. Zhao, J.; Utton, C.; Tsakiropoulos, P. On the Microstructure and Properties of Nb-18Si-6Mo-5Al-5Cr-2.5W-1Hf Nb-Silicide Based Alloys with Ge, Sn and Ti Additions (at.%). *Materials* **2020**, *13*, 4548. [CrossRef] [PubMed]
35. Thandorn, T.; Tsakiropoulos, P. On the Microstructure and Properties of Nb-Ti-Cr-Al-B-Si-X (X = Hf, Sn, Ta) Refractory Complex Concentrated Alloys. *Materials* **2021**, *14*, 7615. [CrossRef] [PubMed]
36. Thandorn, T.; Tsakiropoulos, P. The Effect of Boron on the Microstructure and Properties of Refractory Metal Intermetallic Composites (RM(Nb)ICs) Based on Nb-24Ti-xSi (x = 16, 17 or 18 at.%) with Additions of Al, Cr or Mo. *Materials* **2021**, *14*, 6101. [CrossRef] [PubMed]
37. Ghadyani, M.; Utton, C.; Tsakiropoulos, P. Microstructures and Isothermal Oxidation of the Alumina Scale Forming Nb1.7Si2.4Ti2.4Al3Hf0.5 and Nb1.3Si2.4Ti2.4Al3.5Hf0.4 Alloys. *Materials* **2019**, *12*, 222. [CrossRef] [PubMed]
38. Ghadyani, M.; Utton, C.; Tsakiropoulos, P. Microstructures and Isothermal Oxidation of the Alumina Scale Forming Nb1.45Si2.7Ti2.25Al3.25Hf0.35 and Nb1.35Si2.3Ti2.3Al3.7Hf0.35 Alloys. *Materials* **2019**, *12*, 759. [CrossRef] [PubMed]
39. Zhang, Y.; Zhou, Y.J.; Lin, J.P.; Chen, G.L.; Liaw, P.K. Solid-Solution Phase Formation Rules for Multi-component Alloys. *Adv. Eng. Mater.* **2008**, *10*, 534–538. [CrossRef]
40. Tsai, M.-H.; Yeh, J.-W. High-entropy alloys: A critical review. *Mater. Res. Lett.* **2014**, *2*, 107–123. [CrossRef]
41. Zhu, J.H.; Liaw, P.K.; Liu, C.T. Effect of electron concentration on the phase stability of NbCr2-based Laves phase alloys. *Mater. Sci. Eng. A* **1997**, *239-240*, 260–264. [CrossRef]
42. Li, Z.; Tsakiropoulos, P. The Effect of Ge Addition on the Oxidation of Nb-24Ti-18Si Silicide Based Alloys. *Materials* **2019**, *12*, 3120. [CrossRef]
43. Zacharis, E.; Utton, C.; Tsakiropoulos, P. A Study of the Effects of Hf and Sn on the Microstructure, Hardness and Oxidation of Nb-18Si Silicide Based Alloys without Ti Addition. *Materials* **2018**, *11*, 2447. [CrossRef] [PubMed]
44. Nelson, J.; Ghadyani, M.; Utton, C.; Tsakiropoulos, P. A Study of the Effects of Al, Cr, Hf, and Ti Additions on the Microstructure and Oxidation of Nb-24Ti-18Si Silicide Based Alloys. *Materials* **2018**, *11*, 1579. [CrossRef] [PubMed]
45. Zhao, J.; Utton, C.; Tsakiropoulos, P. On the Microstructure and Properties of Nb-12Ti-18Si-6Ta-2.5W-1Hf (at.%) Silicide-Based Alloys with Ge and Sn Additions. *Materials* **2020**, *13*, 1778. [CrossRef] [PubMed]
46. Papadimitriou, I.; Utton, C.; Tsakiropoulos, P. The impact of Ti and temperature on the stability of Nb5Si3 phases: A first-principles study. *Sci. Technol. Adv. Mater.* **2017**, *18*, 467–479. [CrossRef] [PubMed]
47. McCaughey, C.; Tsakiropoulos, P. Type of Primary Nb5Si3 and Precipitation of Nbss in αNb5Si3 in a Nb-8.3Ti-21.1Si-5.4Mo-4W-0.7Hf (at.%) Near Eutectic Nb-Silicide-Based Alloy. *Materials* **2018**, *11*, 967. [CrossRef] [PubMed]
48. Schlesinger, M.E.; Okamoto, H.; Gokhale, A.B.; Abbaschian, R. The Nb-Si (Niobium-Silicon) system. *J. Phase Equilibria Diffus.* **1993**, *14*, 502–509. [CrossRef]
49. Wang, Z.; Wu, H.; Wu, Y.; Huang, H.; Zhu, X.; Zhang, Y.; Zhu, H.; Yuan, X.; Chen, Q.; Wang, S.; et al. Solving oxygen embrittlement of re-fractory high-entropy alloy via grain boundary. *Mater. Today* **2022**, *54*, 83–89. [CrossRef]
50. Chan, K.S. Modelling creep behaviour of niobium silicide in-situ composites. *Mater. Sci. Eng. A* **2002**, *337*, 59–66. [CrossRef]
51. Nelson, J. Study of the Effects of Cr, Hf and Sn with Refractory Metal Additions on the Microstructure and Properties of Nb-Silicide Based Alloys. Ph.D. Thesis, University of Sheffield, Sheffield, UK, 2015.
52. Knittel, S.; Mathieu, S.; Portebois, L.; Vilasi, M. Effect of tin addition on Nb-Si based in situ composites. Part II: Oxidation behaviour. *Intermetallics* **2014**, *47*, 43–52.
53. Jackson, M.R.; Bewlay, B.P.; Zhao, J.-C. Niobium Silicide Based Composites Resistant to Low Temperature Pesting. U.S. Patent 6,419,765, 16 July 2002.
54. Tsakiropoulos, P. Alloys. U.S. Patent 10,227,680, 12 May 2019.
55. Jackson, M.R.; Jones, K.D. Mechanical behaviour of Nb-Ti base alloys. In *Refractory Metals, Extraction, Processing and Applications*; Liddell, K.C., Sadoway, D.R., Bautista, R.G., Eds.; Tms: Warrendale, PA, USA, 1990; pp. 311–320.

56. Pionke, L.J.; Davis, J.W. *Technical Assessment of Nb Alloys Data Base for Fusion Reactor Applications*, Report C00-4247-2; McDonnell Douglas Corporation: St. Louis, MI, USA, 1979.
57. Inouye, H. Interactions of refractory metals with active gases in vacua and inert gas environments. In *Refractory Metals and Alloys—Metallurgy and 'Technology*; Plenum Press: Salem, OR, USA, 1968; pp. 165–195.
58. Harris, B.; Quarrell, A.G. (Eds.) *Niobium, Tantalum, Molybdenum and Tungsten*; Elsevier: Amsterdam, The Netherlands, 1961; p. 367.
59. Seigle, L.L.; Promisel, N.E. (Eds.) *The Science and Technology of Tungsten, Tantalum, Molybdenum, Niobium and their Alloys*; Pergamon: London, UK, 1964; p. 63.
60. Szkopiak, Z.C. Hardness of Nb-N and Nb-O alloys. *J. Less Common Met.* **1969**, *16*, 93–103. [CrossRef]
61. Sankar, M.; Baligidad, R.; Gokhale, A. Effect of oxygen on microstructure and mechanical properties of niobium. *Mater. Sci. Eng. A* **2013**, *569*, 132–136. [CrossRef]
62. Tottle, C.R. The physical and mechanical properties of niobium. *Inst. Met.* **1957**, *85*, 375–378.
63. Donoso, J.R.; Reed-Hill, R.E. Slow strain-rate embrittlement of niobium by Oxygen. *Met. Mater. Trans. A* **1976**, *7*, 961–965. [CrossRef]
64. Yang, P.-J.; Li, Q.-J.; Tsuru, T.; Ogata, S.; Zhang, J.-W.; Sheng, H.-W.; Shan, Z.-W.; Sha, G.; Han, W.-Z.; Li, J.; et al. Mechanism of hardening and damage initiation in oxygen embrittlement of body-centred-cubic niobium. *Acta Mater.* **2019**, *168*, 331–342. [CrossRef]
65. Enrietto, J.F.; Sinclair, G.M.; Wert, C.A. Mechanical behaviour of Columbium containing oxygen. In *Columbium Metallurgy, Metallurgical Society Conference*; Interscience Publishers: Geneva, Switzerland, 1960; Volume 10, pp. 503–519.
66. Begley, R.T.; France, L.L. Effect of oxygen and nitrogen on the workability and mechanical properties of Columbium. In *Newer Metals*; ASTM STP-272; ASTM International: West Conshohocken, PL, USA, 1959; pp. 56–67.
67. Begley, R.T.; Bechtold, J.H. Effect of alloying on the mechanical properties of Niobium. *J. Less Common Met.* **1961**, *3*, 1–12. [CrossRef]
68. Hahn, G.T.; Gilbert, A.; Jaffee, R.I. The Effects of Solutes on the Ductile-to-Brittle Transition in Refractory Metals. In Proceedings of the Refractory Metals and Alloys II, Metallurgical Society Conferences, Chicago, IL, USA, 12–13 April 1963; Interscience Publishers: Geneva, Switzerland, 1963; Volume 17, pp. 23–63.
69. *ASM Alloy Phase Diagram Database*; Villars, P.; Okamoto, H.; Cenzual, K. (Eds.) ASM International: Materials Park, OH, USA, 2020; ISBN 0-87170-682-2.
70. Klopp, W.D.; Sims, C.T.; Jaffee, R.I. *High Temperature Oxidation and Contamination of Nb*; Report Number BMI-1170; Battelle Memorial Institute: Columbus, OH, USA, 19 February 1957.
71. Bryant, R. The solubility of oxygen in transition metal alloys. *J. Less Common Met.* **1962**, *4*, 62–68. [CrossRef]
72. Fisher, E.S.; Westlake, D.G.; Ockers, S.T. Effects of hydrogen and oxygen on the elastic moduli of vanadium, niobium, and tantalum single crystals. *Phys. Status Solidi* **1975**, *28*, 591–602. [CrossRef]
73. Wriedt, H.A.; Oriani, R.A. The effect of hydrogen on the Young's modulus of tantalum, niobium and vanadium. *Scripta Metall.* **1970**, *8*, 203–208. [CrossRef]
74. Buck, O.; Thompson, D.O.; Wert, C.A. The effect of hydrogen on the low temperature internal friction of a niobium single crystal. *J. Phys. Chem. Solids* **1971**, *32*, 2331–2344. [CrossRef]
75. Jones, K.A.; Moss, C.; Rose, R.M. The effect of small oxygen additions on the elastic constants and low temperature ultrasonic attenuation of Nb single crystals. *Acta Metall.* **1969**, *17*, 365–372. [CrossRef]
76. Calhoun, C.D. *Brittle to Ductile Transition of Nb-V Alloys as Affected by Notches, Strain Rate, Nitrogen and Oxygen*; KAPLL 3119 report; GE: Boston, MA, USA, April 1965.
77. Gorr, B.; Schellert, S.; Müller, F.; Christ, H.J.; Kauffmann, A.; Heilmaier, M. Current Status of Research on the Oxidation Behaviour of Refractory High Entropy Alloys. *Adv. Eng. Mater.* **2021**, *23*, 2001047. [CrossRef]
78. Uttley, J. *Study of the Hardness of the Nb Solid Solution and Nb5Si3 Silicide in Nb Silicide Based Alloys Exposed to High Temperatures*; Final Year Project; University of Sheffield: Sheffield, UK, 2010.
79. Zinkle, S.J.; Huang, J.S. Mechanical Properties of Carbon-Implanted Niobium. *MRS Proc.* **1990**, *188*, 121–126. [CrossRef]
80. Yao, H.W.; Qiao, J.W.; Hawk, J.A.; Zhou, H.F.; Chen, M.W.; Gao, M.C. Mechanical properties of refractory high-entropy alloys: Experiments and modeling. *J. Alloy. Compd.* **2017**, *696*, 1139–1150. [CrossRef]
81. Koch, C.C.; Scarbrough, J.O.; Kroeger, D.M. Effects of interstitial oxygen on the superconductivity of niobium. *Phys. Rev. B* **1974**, *9*, 888–897. [CrossRef]
82. Chapman, O. *Study of the Hardness of the Nbss and Nb5Si3 Phases in a Nb Silicide Based Alloy*; Final Year Project; University of Sheffield: Sheffield, UK, 2010.
83. Oliver, W.C.; Pharr, G.M. An improved technique for determining hardness and elastic modulus using load and displacement sensing indentation experiments. *J. Mater. Res.* **1992**, *7*, 1564–1583. [CrossRef]
84. *Triboscope User Manual*; Hysitron Ltd: Eden Prairie, MN, USA, 2005.
85. Kumar, K.S. *Intermetallic Compounds: Principles and Practice*; Westbrook, H., Fleischer, R.L., Eds.; John Wiley & Sons: Chichester, UK, 1995; Volume 2, p. 213.
86. Shang, Y.; Brechtl, J.; Pistidda, C.; Liaw, P.K. Mechanical Behavior of High-Entropy Alloys: A Review. In *High-Entropy Materials: Theory, Experiments, and Applications*; Springer: Berlin/Heidelberg, Germany, 2021; pp. 435–522. [CrossRef]

87. Geng, J.; Tsakiropoulos, P. A study of the microstructures and oxidation of Nb-Si-CrAl-Mo in situ composites alloyed with Ti, Hf and Sn. *Intermetallics* **2007**, *15*, 382–395. [CrossRef]
88. Zelenitsas, K.; Tsakiropoulos, P. Study of the role of Cr and Al additions in the microstructure of Nb-Ti-Si in situ composites. *Intermetallics* **2005**, *13*, 1079–1095. [CrossRef]
89. Zelenitsas, K.; Tsakiropoulos, P. Study of the role of Ta and Cr additions in the microstructure of Nb-Ti-Si-Al in situ composites. *Intermetallics* **2006**, *14*, 639–659. [CrossRef]
90. Thandorn, T.; Tsakiropoulos, P. Study of the role of B addition on the microstructure of the Nb-24Ti-18Si-8B alloy. *Intermetallics* **2010**, *18*, 1033–1038. [CrossRef]
91. Grammenos, I.; Tsakiropoulos, P. Study of the role of Hf, Mo and W additions in the microstructure of Nb-20Si silicide based alloys. *Intermetallics* **2011**, *19*, 1612–1621. [CrossRef]
92. Grammenos, I. Characterisation of Creep Resistant Ultra-High Temperature Niobium Silicide-Based Alloys. PhD Thesis, University of Surrey, Guildford, UK, 2008.
93. Li, Z.; Tsakiropoulos, P. On The Microstructures and Hardness of The Nb-24Ti-18Si-5Al-5Cr-5Ge and Nb-24Ti-18Si-5Al-5Cr-5Ge-5Hf (at.%) Silicide Based Alloys. *Materials* **2019**, *12*, 2655. [CrossRef] [PubMed]

Article

Optimal Design of Annular Phased Array Transducers for Material Nonlinearity Determination in Pulse–Echo Ultrasonic Testing

Sungjong Cho [1], Hyunjo Jeong [2,*] and Ik Keun Park [3]

[1] NDT Research Center, Seoul National University of Science and Technology, Seoul 01811, Korea; cho-sungjong@seoultech.ac.kr
[2] Department of Mechanical Engineering, Wonkwang University, Iksan 54538, Korea
[3] Department of Mechanical and Automotive Engineering, Seoul National University of Science and Technology, Seoul 01811, Korea; ikpark@seoultech.ac.kr
* Correspondence: hjjeong@wku.ac.kr; Tel.: +82-063-850-6690

Received: 3 November 2020; Accepted: 3 December 2020; Published: 6 December 2020

Abstract: Nonlinear ultrasound has been proven to be a useful nondestructive testing tool for micro-damage inspection of materials and structures operating in harsh environment. When measuring the nonlinear second harmonic wave in a solid specimen in the pulse–echo (PE) testing mode, the stress-free boundary characteristics brings the received second harmonic component close to zero. Therefore, the PE method has never been employed to measure the so-called "nonlinear parameter (β)", which is used to quantify the degree of micro-damage. When there are stress-free boundaries, a focused beam is known to improve the PE reception of the second harmonic wave, so phased-array (PA) transducers can be used to generate the focused beam. For the practical application of PE nonlinear ultrasonic testing, however, it is necessary to develop a new type of PA transducer that is completely different from conventional ones. In this paper, we propose a new annular PA transducer capable of measuring β with improved second harmonic reception in the PE mode. Basically, the annular PA transducer (APAT) consists of four external ring transmitters and an internal disk receiver at the center. The focused beam properties of the transducers are analyzed using a nonlinear sound beam model which incorporates the effects of beam diffraction, material attenuation, and boundary reflection. The optimal design of the APAT is performed in terms of the maximum second harmonic reception and the total correction close to one, and the results are presented in detail.

Keywords: phased array transducer; transducer optimization; beam focusing; pulse–echo mode; harmonic generation; stress-free boundary; total correction

---

1. Introduction

Power generation facilities in nuclear power and thermal power plants that are operated at high temperature and high pressure can lead to various types of micro-damage (e.g., deterioration, residual stress, fatigue, creep, and micro-cracks) as the number of years of use increases. The management of such damage is an essential part of ensuring the soundness and safe operation of power plants. In particular, a more reliable diagnosis technology is required for major components and parts made of nickel alloy or carbon steel welding because they are susceptible to micro-damage.

Currently, in nondestructive testing of power generation facilities, conventional radiography testing (RT) is being replaced by ultrasonic testing. Among the various ultrasonic methods, phased array ultrasonic testing is widely applied to the inspection of power generation facilities and pressure vessels [1–4]. However, it is not easy to detect various types of micro-damage described above with conventional linear ultrasonic testing techniques. Nonlinear ultrasonic technology uses nonlinear acoustic effects that occur when a strong ultrasonic wave is incident inside a material. The nonlinear ultrasound, such as the second harmonic wave, is known to be sensitive to micro-damage [5–10], and mainly measures the nonlinear parameter, β, which is defined by the displacement amplitudes of the fundamental and second harmonic waves to quantify the degree of damage. Studies on nonlinear ultrasonic applications are actively being conducted [11–20], and the use of longitudinal waves dominates in most cases, although surface and Lamb waves are also used. Damage types include fatigue, deterioration, creep, and irradiation, and most studies measure the uncorrected nonlinear parameter, β', for ease and convenience of measurement.

Although nonlinear parameters are mainly measured in the through-transmission (TT) mode, pulse–echo (PE) measurements are frequently required for field applications. According to Bender et al. [21], the amplitude of the second harmonic received after reflection from the stress-free boundary of a sample in the PE mode is theoretically zero. This was the main reason the PE method has not been applied until recently. However, the zero reception of the second harmonic in the PE mode was the result of the pure plane wave. In the case of a real transducer of finite size, the second harmonic wave can be received owing to the diffraction effect, but it is extremely small and can only be measured when the specimen is thick enough [22,23]. Therefore, increasing the amplitude of the received second harmonic in applications of nonlinear PE method for thin samples is of utmost importance for obtaining the second harmonic signal with high signal-to-noise ratio and for accurate and reliable measurement of β.

It has been found that a focusing beam increases the amplitude of the received second harmonic in the PE testing of a sample with the stress-free boundary. Actually, the received second harmonic amplitude was found to significantly increase when a spherically focusing transducer was used in the water–air boundary [24,25]. The spherical focusing with a linear phased array ultrasonic transducer (PAUT) is not possible on the flat surface of a solid specimen. In addition, because the current PAUT for linear ultrasonic testing achieves beam focusing using dozens of channels and short pulses, it is still difficult to apply the nonlinear ultrasonic technique that employs a high-power toneburst type signal. It is also necessary to minimize the source nonlinearity, which is an important variable in nonlinear ultrasonic measurement. This is because when source nonlinearity occurs, it is mixed with the nonlinearity caused by damage, making it difficult to observe the damage-induced nonlinearity alone. Furthermore, the receive transducer should have a broad bandwidth capable of covering both fundamental and second harmonic wave frequencies. Therefore, for the development of PAUTs applicable to PE nonlinear testing, it is necessary to minimize the number of channels by designing a new type of PAUT that is completely different from the conventional PAUT. Before fabricating and applying such PAUT, a prototype design is required, and the focused beam properties should be fully understood. Further optimization of the PAUT is possible to achieve the maximum second harmonic reception and the uncorrected nonlinear parameter (β') close to the absolute nonlinear parameter (β).

In this paper, we propose an annular phased array transducer (APAT) and model the nonlinear acoustic fields generated by the APAT in the PE setup to determine the optimal dimensions of the transducer. Basically, the APAT for PE nonlinear testing purposes consists of the four external ring transmitters and an internal disk receiver at the center. The fundamental and second harmonic wave fields, which are focused at various positions of the 1 cm thick specimen and then reflected from the stress-free boundary, are calculated and their characteristics are examined with the received average fields. For a given specimen thickness and frequency, the optimization of the APAT is then performed in terms of the maximum possible reception of the second harmonic and the total correction as close to one as possible. The optimization results are presented in detail. The shape of the time domain waveform formed at the focal position and at the receiver position are examined through finite element (FE) simulation.

Section 2 describes the requirements of the phased array transducer to be considered when performing the nonlinear parameter measurement in the pulse–echo mode using a focused beam and introduces the conceptual design of an APAT along with its focal characteristics in linear ultrasound.

In Section 3, we outline the nonlinear acoustic model developed in our previous work [26] and define the nonlinear parameter with necessary corrections. This model combines the effects of nonlinearity, diffraction, and boundary reflection in order to calculate the fundamental and second harmonic fields in the focused beam generated by the four annular transmit elements. Section 4 compares the focused beam properties of the two types of APAT—equal width (EW) and equal area (EA). We then present details on the optimization process of the EW type APAT and provide results on the received second harmonic amplitude and the relative nonlinear parameter. The FE simulation results are also presented in Section 4 to see how well the focused and/or received waveform matches the initially incident waveform. Conclusions are drawn in Section 5.

2. Phased Array Transducer Design

Ultrasonic phased array techniques are widely used in nondestructive testing areas and in many medical applications. Some of the attractive features of PAs include electronic focusing and steering capabilities. To generate a focused beam at any specified angle and distance, time delays are calculated and applied electronically to each element, as shown in Figure 1. In general, PA types are classified as linear or annular depending on the shape and arrangement of the elements, as shown in Figure 1.

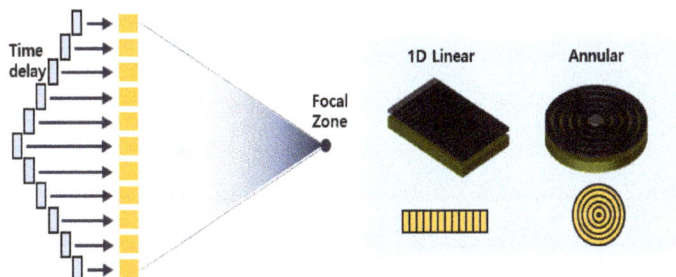

Figure 1. Schematic illustration of phased array beam focusing through time delay and two types of phased arrays.

To utilize the PA focusing technology to the measurement of nonlinear material properties, feasibility studies (e.g., senor materials and fabrication methods) and equipment availability must be preceded at the design stage of PAUT. Material nonlinearity measurements using the finite amplitude method require high power amplifiers, and the number of such amplifiers increases as the number of PA elements increases. Therefore, it is desirable to achieve beam focusing with minimum number of elements.

The generation of second harmonic waves in solids typically requires very high input voltages at the transmitter. Therefore, it is important to minimize the source nonlinearity caused by the transmit element. Most transmit transducers use single crystal $LiNbO_3$ instead of commercial transducers made of piezoelectric materials such as PZT for the purpose of harmonic generation with minimal source nonlinearity from the transmitter. The $LiNbO_3$ piezoelectric element shows a narrowband spectrum around its fundamental resonant frequency when no backing material is used. Thus, a transducer made of $LiNbO_3$ cannot receive both fundamental and second harmonic components at the same time. To solve this problem, the practical approach is to use a separate receive transducer of a broad bandwidth.

The next thing to consider is the calibration of the receiver. For a quantitative evaluation of the material damage, an absolute nonlinear parameter (β), not the uncorrected nonlinear parameter (β'), needs to be measured. To measure β, the receive transducer must be calibrated [27–29]. To summarize the above, the PAUT for pulse–echo nonlinearity measurement requires a minimum number of transmit elements to generate a focused beam, separation of transmit and receive elements, and calibration of the receive element.

Taking these requirements into account, the arrangement of the transmit and receive elements for the conceptual design of linear and annular PAs is shown in Figure 2. Compared with conventional PAs, the central element is used only for reception, and the other elements are used for transmission, where the transmit and receive elements have the same central axis. This design allows the application of conventional receiver calibration method. In the conceptual design and wave field simulation with a beam focusing, the number of transmit elements is limited to four. Because the central element is used for reception, the beam focusing behavior is analyzed along the central axis. The beam focusing simulation in this section was conducted using the CIVA program [30–32], a nondestructive simulation platform. In the CIVA simulation, aluminum was selected as the propagation medium, and the following acoustic properties including the nonlinear parameter were used [28]: longitudinal wave velocity, $c = 6422$ m/s; density, $\rho = 2700$ kg/m^3; fundamental wave frequency, $f = 5$ MHz; and nonlinear parameter, $\beta = 5.5$. The specifications of the PAUTs used in the simulation are summarized in Table 1. The specifications of the annular phased array were imported from the optimal design of the equal width (EW) type annular phased array described in Section 4. The diameter of the receive element is 3.2 mm, and the diameter of the innermost transmit element is 10 mm. Specifications of the EW phase array can be found in Table 2. The dimensions of the linear PA along the width direction are the same as the cross-sectional dimensions of the annular PA, and the length of the linear PA was taken arbitrarily as 10 mm. In Section 4, the effect of the sizes of the transmit elements (i.e., equal width type and equal area type) on the received amplitude of the fundamental wave and the second harmonic was compared.

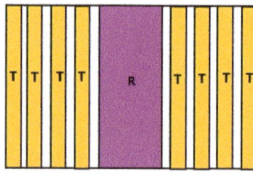 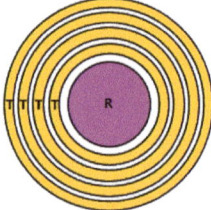

Figure 2. Configuration of linear and annular phase arrays.

Table 1. Specifications of linear and annular phased arrays used in the simulation.

Array Pattern	Linear	Annular
Number of elements	8	4
Gap between elements	0.5 mm	0.5 mm
Element width	1 mm	1 mm
Element length	10 mm	-
Total array width	21 mm	21 mm

Beam focusing simulation results of the linear and annular phased arrays are shown in Figure 3. The simulations were performed in the linear ultrasound range. In the case of the linear PA, the focusing is hardly found, whereas, in the case of the annular PA, a distinct focusing can be seen at three focal lengths. This is due to the geometry of the annular PA, which is much more efficient in forming the focused beam. If we look at the results of the annular PA in more detail, the simulated focal spot sizes at −6 dB along the cross-axis of the beam are about 1 mm for all three focal lengths. If we treat the annular PA as a single transmit element of diameter D, the measured spot size is slightly larger than the estimated focal spot size using the equation $d = \lambda L/D$ where λ is the wavelength, L is the focal length, and D is the total width of the array. On the other hand, the focal spot size along the on-axis of the beam increases with increasing focal length, and this is clearly seen in the simulation results of Figure 3. Increased focal spot size also means decrease of peak amplitude. The difficulty in creating well-shaped focal zones in the annular PA can be attributed to the hollow structure and a small number of elements [33–35]. The current annular PA structure with four transmit elements is hollow in the center, but the overall beam focusing behavior is similar to that observed in the conventional linear PAs with dozens of elements. In particular, since it has a good focusing performance at the focal length of 10 mm, it can be used for pulse–echo nonlinear measurement of relatively thin specimens with a thickness of about 10 mm.

Table 2. Dimensions of two types of annular phased arrays (unit: mm).

	EW			EA	
Radius		Radius		Radius	Radius
r_{1in}	5	r_{1out}	6	r_{1in} 5	r_{1out} 6
r_{2in}	6.5	r_{2out}	7.5	r_{2in} 6.5	r_{2out} 7.3
r_{3in}	8	r_{3out}	9	r_{3in} 7.8	r_{3out} 8.5
r_{4in}	9.5	r_{4out}	10.5	r_{4in} 9	r_{4out} 9.6

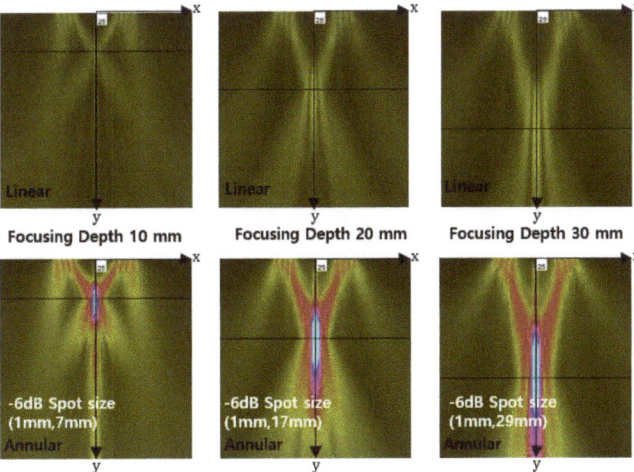

Figure 3. Beam focusing simulation results of the linear and annular phased arrays.

3. Annular PA Wave Fields and Definition of β

3.1. Theory

Second harmonic generation in the nonlinear PE testing with the stress-free boundary condition is schematically illustrated in Figure 4. The APAT consisting of four ring transmitters and a central disk receiver are also included. In Figure 4, $p_{1,i}$ is the acoustic pressure of the fundamental wave emitted from the ring transmitters, and $p_{2,i}$ is the generated second harmonic wave owing to the forcing of $p_{1,i}$. $p_{1,r}$ denotes the reflected fundamental wave when the wave $p_{1,i}$ hits the boundary, $p_{2,r1}$ is the reflected wave when the wave $p_{2,i}$ hits the boundary, and $p_{2,r2}$ is the second harmonic wave generated by the reflected $p_{1,r}$. Therefore, the total reflected second harmonic, $p_{2,r}$, can be obtained by adding $p_{2,r1}$ and $p_{2,r2}$. Both the reflected fundamental and second-harmonic waves are received by the center receiver.

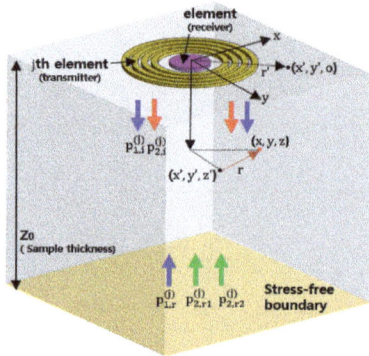

Figure 4. Schematic illustration of second harmonic generation process in a nonlinear PE testing with the stress-free boundary.

3.2. Sound Beam Solution

The second harmonic wave is produced due to material nonlinearity when the finite amplitude fundamental wave radiates from the transmitter and propagates in the solid. The related acoustic fields for a single element circular transducer have been derived previously [26,36,37]. In this section, we briefly present the mathematical equations to calculate the received acoustic fields when a phased array transducer composed of four ring elements radiates a finite amplitude longitudinal wave. The incident fundamental wave $p_{1,i}$ and the generated second harmonic $p_{2,i}$ are given by Equations (1) and (2). Pressure p is used here as a field variable.

$$p_{1,i}(x_1, y_1, z_1) = -2ik \int_{-\infty}^{+\infty}\int_{-\infty}^{+\infty} p_1(x', y', 0)\, G_1(x, y, z|x', y', 0)\, dx'\, dy' \tag{1}$$

$$p_{2,i}(x_1, y_1, z_1) = \frac{2\beta k^2}{\rho c^2} \int_0^z \int_{-\infty}^{+\infty}\int_{-\infty}^{+\infty} p_1^2(x', y', z')\, G_2(x, y, z|x', y', z')\, dx'\, dy'\, dz' \tag{2}$$

where the Green's function is given by the following equation:

$$G_1(x, y, z|x', y', 0) = \frac{1}{4\pi r} \exp(ikr) \tag{3}$$

$$G_2(x, y, z|x', y', z') = \frac{1}{4\pi R} \exp(i2kR) \tag{4}$$

Here, $r = \sqrt{(x-x')^2 + (y-y')^2 + z^2}$ and $R = \sqrt{(x-x')^2 + (y-y')^2 + (z-z')^2}$. For p_1, the source function is $p_1(x', y', z' = 0)$, and the integration is applied over the transducer surface element $ds' = dx'dy'$ at the source plane $z' = 0$,

$$p_1(x', y', z' = 0) = p_0, \quad a^2 \le x'^2 + y'^2 \le b^2 \tag{5}$$

where p_0 is the uniform acoustic pressure and a and b are the inner and outer radii of the ring transmitter. $p_{1,i}$ and $p_{1,r}$ of the mth transmission element can be obtained by calculating $p_{1,i}^{(m)}$ and $p_{1,r}^{(m)}$ under different boundary conditions of Equation (5), and the total fields are found by adding them together. The reflected fundamental pressure $p_{1,r}$ is given by

$$p_{1,r}(x, y, z) = R_1 p_{1,i}(x, y, z) \tag{6}$$

The total pressure of the reflected second harmonic $p_{2,r}$ is obtained as the sum of $p_{2,r1}$ and $p_{2,r2}$ given by

$$p_{2,r1}(x, y, z) = -2ik \int_{-\infty}^{\infty}\int_{-\infty}^{\infty} R_2 p_{2,i}(x', y', z_0) G_2(x, y, z|x'y', z_0)\, dx'dy' \tag{7}$$

$$p_{2,r2}(x, y, z) = \frac{2\beta k^2}{\rho c^2} \int_{z_0}^{z}\int_{-\infty}^{\infty}\int_{-\infty}^{\infty} \{p_{1,r}(x', y', z')\}^2 G_2(x, y, z|x', y', z')\, dx'dy'dz' \tag{8}$$

In Equations (6) and (7), R_1 and R_2 are the reflection coefficients for the fundamental and second harmonic waves at the solid–air interface and are given by $R_1 = R_2 = -1$. The reflected second harmonic fields for mth element can be obtained by calculating $p_{2,r1}^{(m)}$ and $p_{2,r2}^{(m)}$, and the total fields are found by adding contributions from all elements.

Next, to calculate the received pressure at a distance z by the receiver of area S_R, the concept of the average pressure can be defined and calculated as follows:

$$\tilde{p}_n(z) = \frac{1}{S_R} \int_{S_R} p_n(x, y, z) dS_R \quad n = 1, 2 \tag{9}$$

3.3. Time Delay

Consider an array of N elements radiating into a solid to produce a sound beam with a focal length F, as shown in Figure 5. The focusing time delays can be calculated as follows [38]:

$$\Delta t_N = \frac{\sqrt{\left(\frac{r_{2N}+r_{2N-1}}{2}\right)^2 + F^2} - F}{c} = \frac{\sqrt{r_N^2 + F^2} - F}{c} \tag{10}$$

where Δt_N is the required time delay for element $N = 0, 1, \ldots, N$. Note that in Equation (10) each calculated time has a positive value, which is a time delay. The delay of the time-domain signal is equivalent to multiplying the frequency domain signal by a phase term that is linear in frequency and proportional to its delay. If $F(\omega)$ is the Fourier transform of the time domain signal $f(t)$, then the Fourier transform of the time-shifted signal $f(t - \Delta t_N)$ can be obtained as $\exp(i\omega \Delta t_N) F(\omega)$, where Δt_N is the delay time.

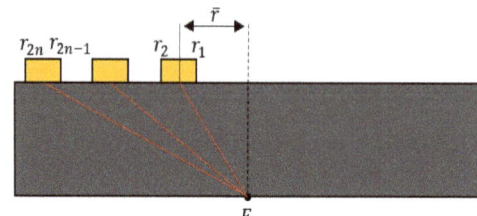

Figure 5. Geometrical parameters for calculating the focusing time delay of phased array.

3.4. Definition of β with Total Correction

Equation (9) can be expressed more conveniently in terms of the plane wave solutions modified by the correction terms owing to the effects of attenuation, diffraction, and boundary reflection, i.e.,

$$\widetilde{p}_{1,r} = \left[p_1^{plane}(z) \right][C_{T1}] \tag{11}$$

$$\widetilde{p}_{2,r} = \left[p_2^{plane}(z) \right][C_{T2}] \tag{12}$$

Here, $p_1^{plane} = p_0 exp(ikz)$ and $p_2^{plane} = \frac{\beta k p_0^2 z}{2\rho c^2} exp(2ikz)$ where k is the wave number, ρ is the density, and c is the wave velocity. In addition, the average pressure is calculated at the initial source position, i.e., at the total propagation distance $z = 2z_0$. In Equations (11) and (12), C_{Tn} is the correction due to attenuation, diffraction, and boundary reflection in the fundamental ($n = 1$) and second harmonic ($n = 2$) waves and is defined as follows:

$$C_{T1} = R_1 M_1 \widetilde{D}_1 \tag{13}$$

$$C_{T2} = \left[R_2 M_{21} \widetilde{D}_{21} + R_1^2 M_{22} \widetilde{D}_{22} \right] \tag{14}$$

where M_1, M_{21}, and M_{22} and $\widetilde{D}_1$, $\widetilde{D}_{21}$, and $\widetilde{D}_{22}$ are the attenuation corrections and diffraction corrections in $\widetilde{p}_{1,r}$, $\widetilde{p}_{2,r1}$, and $\widetilde{p}_{2,r2}$. The detailed expressions for these corrections can be found elsewhere [26]. If we put $\frac{C_{T1}^2}{C_{T2}} = C_T$, where C_T is called the "total correction", combining Equations (11) and (12) yields the nonlinear parameter β_f in fluids

$$\beta_f = \frac{2\rho c^2}{kz} \frac{\widetilde{p}_{2,r}}{\widetilde{p}_{1,r}^2} C_T = \beta'_f C_T \tag{15}$$

The nonlinear parameter β_s in solids can be obtained by replacing β_f with $\frac{1}{2}\beta_s$ in Equation (15). Then, using the relationship between pressure and displacement, β_s can be determined in terms of the received average displacement by

$$\beta_s = \beta = \frac{8}{k^2 z} \frac{\widetilde{u}_{2,r}}{\widetilde{u}_{1,r}^2} C_T = \beta' C_T \tag{16}$$

Since the amplitude of the actually measured wave deviates from the plane wave, C_T appearing in Equation (16) is to correct the attenuation, diffraction, and boundary reflection effects in the received amplitudes of the fundamental and second harmonic waves. Hence, β' is called the "uncorrected" nonlinear parameter.

4. Optimization of Phase Array Dimensions

The received second harmonic wave, $\widetilde{u}_{2,r}$, in the pulse–echo testing is in general much smaller than the through-transmission method, consequently the uncorrected nonlinear parameter, β', becomes also very small. To recover the correct nonlinear parameter, β, a large value of the total correction,

C_T, should be multiplied. For accurate and reliable determination of β, especially for thin specimens, it is necessary to maximize $\tilde{u}_{2,r}$ and reduce the dependence on C_T by optimizing the design of APAT. These two parameters depend on many variables including sample thickness, frequency, and shape and size of the transmit and receive elements. Here, the thickness of the specimen is fixed, so it is not a design variable for optimization. It is also assumed that the frequency is fixed. Then, the optimization of APAT can be considered a process of determining the size, arrangement, and shape of the transmit and receive elements. The optimization of APATs can be approached in terms of two objective functions: the second harmonic reception and the total correction. The optimized transducer should provide the largest possible second harmonic reception and the total correction value as close to one as possible.

In the simulation-based optimization here, the received amplitudes of the fundamental and second harmonics and the total correction were calculated through wave field analysis for various combinations of shape and size of the transmit and receive elements. Then, the optimized APAT is finally obtained by comparing the received second harmonic amplitude and the total correction value from various simulation cases.

As a final step, the waveform of the received signal was obtained through finite element analysis (FEA). The purpose of FEA is to check the distortion of waveforms received through focusing, and to validate the analytical model used for wave field calculation and optimization of annular phased arrays.

4.1. Focused Beam Field Analysis Results

APATs can be divided into two types: equal width (EW) and equal area (EA). The wave fields for these two types are analyzed and the focusing properties are compared. The source displacement used in the analysis is $u_0 = 10^{-9}$ m, and the fundamental frequency is $f_1 = 5$ MHz. The attenuation effect is not considered. For the EW type, the element width is 1 mm and the kerf is 0.5 mm. For the EA type, the area of the innermost element is the same as the first element of the EW type, and the size of the remaining elements is determined from the area of the first element. The kerf is 0.5 mm. The number of elements in both types is four. The focal length (F) is set to 10 mm. The central receiver has a fixed diameter of 3.2 mm. The target thickness of the specimen is assumed to be 10 mm. The dimensions used in the wave field analysis are given in Table 2.

Simulation results and comparisons between the EW and EA types of APAT for fundamental and second harmonic waves are shown in Figure 6. More specifically, the 2D beam profile and the on-axis variation of displacement are presented. When considering only the beam profile, there seems to be no difference in beam focusing between the two types. However, comparison of on-axis profiles shows a noticeable difference between the two types. The maximum amplitude of the EW type is slightly larger than that of the EA type at $F = 10$ mm in both wave types. The EA type is found to focus at a distance slightly shorter than $F = 10$ mm in the fundamental wave. Based on this, it can be said that the EW type has better focusing performance than the EA type. Therefore, optimization is performed using the EW type.

The focusing behavior of the fundamental wave along the lateral and axial directions is shown in Figure 3. Similar behavior is also observed here, as shown in Figure 6a,c.

Compared to the fundamental wave, the second harmonic wave shows a narrower beamwidth at the focal length due to the twice as large frequency, as shown in Figure 6b,d. As a result, the second harmonic wave forms a sharp focus at the specified focal length. In fact, the amplitude of the received wave is determined by a receiver of finite size. Thus, in the case of a focused beam, it is important to determine the size of the receiver according to the focal spot size in the lateral direction of the beam in order to receive maximum amplitude. In Figure 6b,d, the focal spot sizes at −6 dB along the lateral direction are estimated to be less than 1 mm, which can be treated as a point focus. Therefore, for maximum reception of the second harmonic amplitude, a point reception device such as a laser interferometer may be the best choice, but, considering the actual situation, a broadband ultrasonic receiver made of the smallest possible size piezoelectric element may be more suitable.

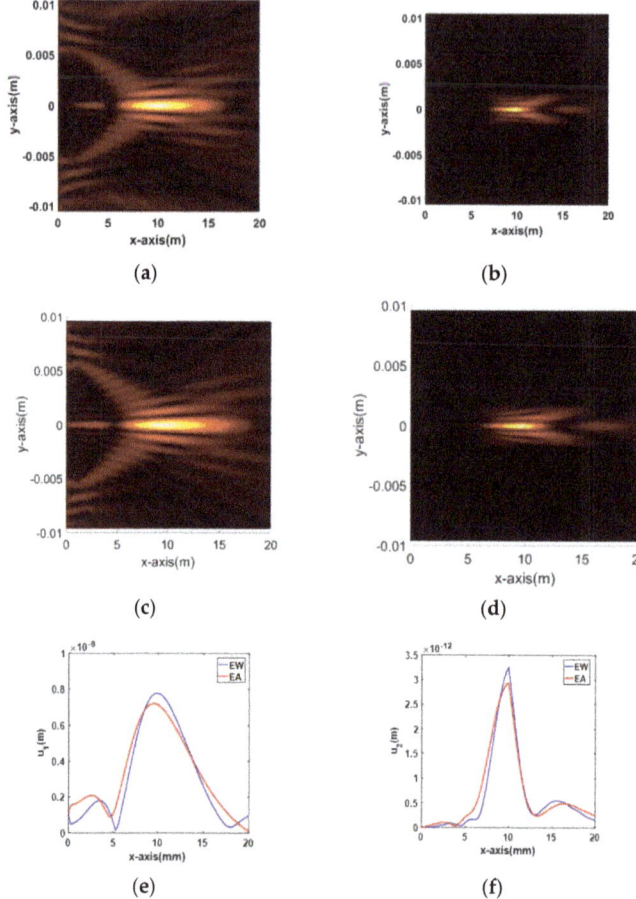

Figure 6. Simulation results and comparisons between the EW and EA types of APAT for fundamental and second harmonic waves: (**a**) fundamental, EW; (**b**) second harmonic, EW; (**c**) fundamental, EA; (**d**) second harmonic, EA; (**e**) on-axis, fundamental wave; and (**f**) on-axis, second harmonic.

4.2. Optimization for Second Harmonic Generation and Total Correction

In the EW type APAT, the focal length and element width affect the received amplitude of the second harmonic and the total correction. Here, the effect of these two parameters is examined. The focal position is set at three distances, namely 10, 15, and 20 mm, corresponding to the reflection boundary of the 10 mm thick specimen, the center of the specimen equal to 1.5 times the specimen thickness, and twice the specimen thickness equal to the receiver position. The width of the element is set to 1, 1.5, 2, and 2.5 mm. Twelve simulation cases are listed in Table 3. According to the simulation results in Section 4.1, among commercially available broadband transducers, a receiver with a minimum diameter of 3.2 mm is used.

Table 3. Simulation cases for various combinations of focal length and element width.

Group	Case Number	Focal Length (mm)	Element Width
A	1		1
	2	10	1.5
	3		2
	4		2.5
B	5		1
	6	15	1.5
	7		2
	8		2.5
C	9		1
	10	20	1.5
	11		2
	12		2.5

Using various combinations of focal length and element width, the received fundamental and second harmonic amplitudes were calculated. The simulation results for 12 cases are shown in Figure 7. It also includes the results of the through-transmission (TT) mode calculations when a transmitter and a receiver both 12.7 mm in diameter are used. The propagation distance for the single element TT mode is 10 mm, which is the thickness of the specimen. The received amplitude is largest in Case C-1 for both the fundamental and the second harmonic waves. This result shows that the optimal design of APAT with the beam focusing can produce a received second harmonic amplitude that is about 50% larger than the single element TT case.

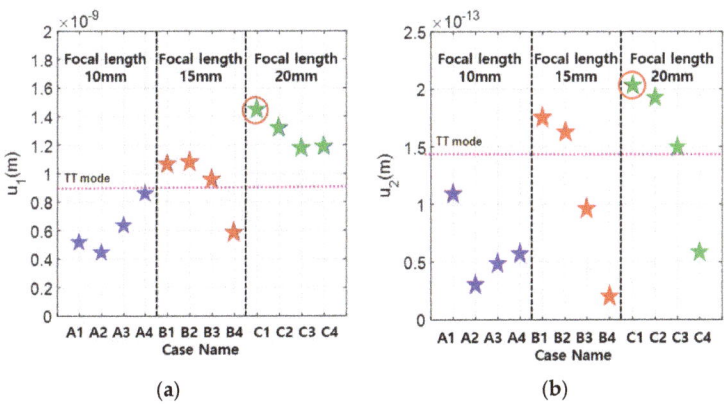

Figure 7. The received displacements calculated using the simulation parameters in Table 3: (**a**) fundamental wave; and (**b**) second harmonic wave.

The uncorrected β' was calculated using the received amplitude data in Figure 7, and the results are shown in Figure 8. Since the nonlinear parameter is given by $\beta = [\beta'][C_T]$, the total correction C_T should be as close to one as possible in the optimization process. This means that the uncorrected β' should be as close to β as possible. It can be seen that $\beta' = 1.56$ in Case C-1, where the received second harmonic amplitude is the largest, and $\beta' = 2.53$ in Case B-1. Therefore, the optimal case for β' or C_T is Case A-1 giving $\beta' = 6.55$, which is about 19% larger than $\beta = 5.5$. These results show that the APAT specifications optimized for one objective function may not satisfy the other objective function.

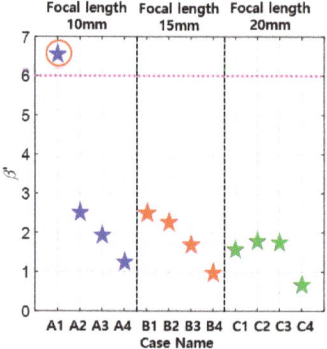

Figure 8. The calculated β' using the data in Figure 7.

Next, the influence of kerf was analyzed by changing the kerf size to 0.1, 0.3, and 0.5 mm in the EW type APAT. When designing an APAT, the interelement spacing (kerf) should be less than half the wavelength to suppress the occurrence of grating lobes. All three kerf sizes selected here meet this condition. Three different focal lengths were used as before. Various combinations of focal length and kerf size are shown in Table 4.

Table 4. Simulation cases for various combinations of focal length and kerf size.

Case Numbers	Focal Length (mm)	Kerf (mm)
A11		0.1
A12	10	0.3
A13		0.5
B11		0.1
B12	15	0.3
B13		0.5
C11		0.1
C12	20	0.3
C13		0.5

The simulation results for various combinations of data in Table 4 are shown in Figure 9, showing the received fundamental and second harmonic amplitudes and the uncorrected β'. From the viewpoint of the maximum second harmonic reception, the best case is B11 or C11, and, from the viewpoint of β' or C_T, Case A13 may be better. Considering both of these goals, all three cases in Group A are good. These results show that the APAT specifications optimized for one objective may not satisfy the other objective. Therefore, in the optimization of the APAT specification, the objective function—second harmonic reception, total correction, or both—needs to be clearly defined.

4.3. Summary of Optimization Results

In relation to the measurement of nonlinear parameters of materials in the pulse–echo mode, the optimal design of the annular phased array transmitter consisting of four equal width (EW) elements was considered. With the specimen thickness, frequency, and receiver size fixed, the optimization of the APAT was performed from two viewpoints: received second harmonic amplitude and total correction. The received amplitudes of the fundamental and second harmonics and the total correction were calculated through wave field analysis for various combinations of the element width, kerf, and focal length of the transmitter. Then, the optimized specifications of the APAT were obtained by comparing the received second harmonic amplitude and the total correction from various simulation cases. In the optimal design process of APAT, the results of the through-transmission (TT) method by a single transmitter and a single receiver were used as a reference.

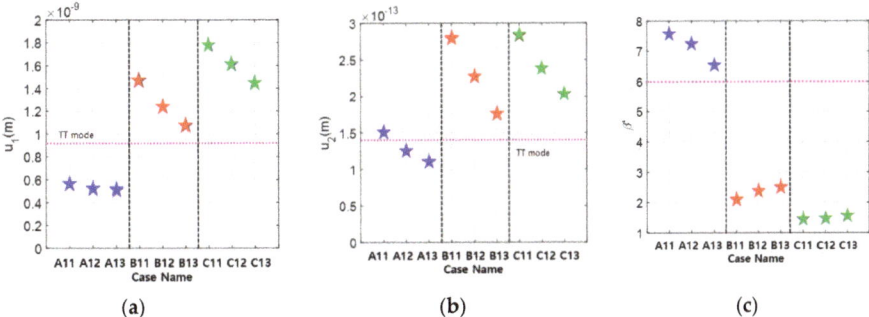

Figure 9. Received displacements and uncorrected nonlinear parameter calculated using the simulation parameters in Table 4: (**a**) fundamental wave; (**b**) second harmonic wave; and (**c**) uncorrected nonlinear parameter.

The optimization results are summarized in Table 5, where the three optimized APAT designs—A13, B13, and C13 in Table 4—are given together with the TT results. The kerf sizes of these types are all 0.5 mm, and the focal length is 10, 15, and 20 mm, respectively. From the viewpoint of the maximum second harmonic reception only, the best case is B13 or C13, and, from the viewpoint of uncorrected nonlinear parameter β' or the total correction C_T only, A13 is better. Considering both of these conditions, A13 is just fine. These results show that the APAT specifications optimized for one objective may not satisfy the other objective. Therefore, in the optimization of the APAT specification, the objective function—second harmonic reception, total correction, or both—should be clearly specified.

Table 5. Summary of optimized APAT specifications and simulation results.

Specification	Single (TT)	PA (PE)	PA (PE)	PA (PE)
Transmitter (mm)	Diameter = 12.7	Element width = 1 kerf = 0.5		
Focal length (mm)	-	10	15	20
Receiver dia. (mm)	12.7	3.2	3.2	3.2
u_1(m)	8.82×10^{-10}	5.20×10^{-10}	1.07×10^{-10}	1.45×10^{-9}
u_2(m)	1.46×10^{-13}	1.10×10^{-13}	1.76×10^{-13}	2.03×10^{-13}
β'	6.04	6.54	2.48	1.56
C_T	0.91	0.84	2.22	3.53

Note: PA = Phased array; TT = Through–transmission; PE = Pulse-echo.

We already developed the measurement procedure to determine material nonlinearity in the pulse–echo method using a single element transducer and a dual element transducer [29,36,37]. The current work is the extension of our previous work on the dual element transducer approach. The single annular transmit element was simply replaced by the four annular transmit elements to create a focused beam at a specific location in the specimen. If the annular phased array transducer with four element transmitter and a single element receiver can be made and used, similar measurement procedures can be applied, including receiver calibration.

4.4. FE Simulation Results

The analytical acoustic model introduced in Section 3 is a method of calculating the wave field in the frequency domain and provides the received displacement value at a specific frequency. To obtain the received waveform in the time domain, displacement must be calculated at hundreds of frequency values and then inversely Fourier-transformed. Therefore, the analytic method is not suitable for time domain waveform calculation. In ultrasonic modeling, one of the most efficient ways to directly calculate the waveform is the finite element method.

In the case of performing a nonlinear experiment using the optimized APAT of Sections 4.1 and 4.2, a tone burst waveform of tens of cycles is used as an input signal and is received by the receive transducer after being focused on a specific position in the specimen. Therefore, it may be necessary to ensure that the time-delayed signal emitted by each element of the APAT is arrived in-phase at the focal position and then received in the same waveform as the initially incident signal without distortion.

In this section, the waveform of the received signal was obtained through FEA. The purpose of FEA is to check the distortion of the received waveform after being focused on a position in the specimen, and to validate the analytical model used for wave field calculation and optimization of annular phased arrays. COMSOL Multiphysics FE program was used to simulate the nonlinear wave fields calculation. The specimen used is an aluminum with quadratic material nonlinearity [39]. The quadratic nonlinear material is defined by the third-order elastic constants l, m, and n, which is also called "Murnaghan material" in the built-in option of COMSOL program. Simulation was carried out using the second-order axisymmetric model. The source displacement used in the FE simulation was $u_0 = 10^{-7}$ m, which is two orders of magnitude larger than that used in the analytical simulation. This is for easy visualization of the relatively small second harmonic component in the received signal of the FE simulation.

The three types of EW APAT in Table 4—A13, B13, and C13—were used in the FE simulation. The kerf sizes of these types are all 0.5 mm, and the focal length is 10, 15, and 20 mm, respectively. The FE simulation results are presented in Figure 10, showing the received signal waveforms for the three cases. The results show that the peak amplitude of the received waveform, measured in the order of C13 > B13 > A13, is in good qualitative agreement with the analytical simulation results shown in Figure 9a. In the case of C13, where the focal position and the receiver position are the same (Figure 10c), the signal radiated from each element appears to arrive in-phase at the specified focal position. In addition, the overall shape of the waveform seems to match the input signal well without any significant difference. In Cases A13 and B13, where the focal position and the receiver position are different, there is a slight difference in the leading and trailing edges of the received waveform compared to the input waveform. This difference may occur because the focal position and the reception position are not the same, and it occurs slightly larger in Case A13, where the difference between these two positions is larger. Since the difference in waveform is very small, it is believed that it will have little effect on the measurement of nonlinear parameters. If the reception time delay is applied to the received waveform, the waveform difference can be reduced.

However, it was determined that there is little effect on the measurement of nonlinear parameters because the distortion of the waveform was not large in the simulation results.

The frequency spectrum of the received signal in Figure 10 is shown in Figure 11. To easily visualize the second harmonic component, the spectral values were multiplied by 10. The second harmonic component can be clearly seen in all three cases. The peak magnitude of each spectrum is in the order of C13 > B13 > A13, which also agrees well with the analytical calculation results. It should be noted that a spectral component of large size is present in the very low frequency region. This is known as a zero-frequency component or a quasi-static component which is produced by nonlinear acoustic wave propagation in an elastic solid of quadratic nonlinearity. Although its existence has been proven through theory and FE simulation [39,40], it is not easy to experimentally observe this component because a wideband receiver that covers down to zero frequency cannot be easily found.

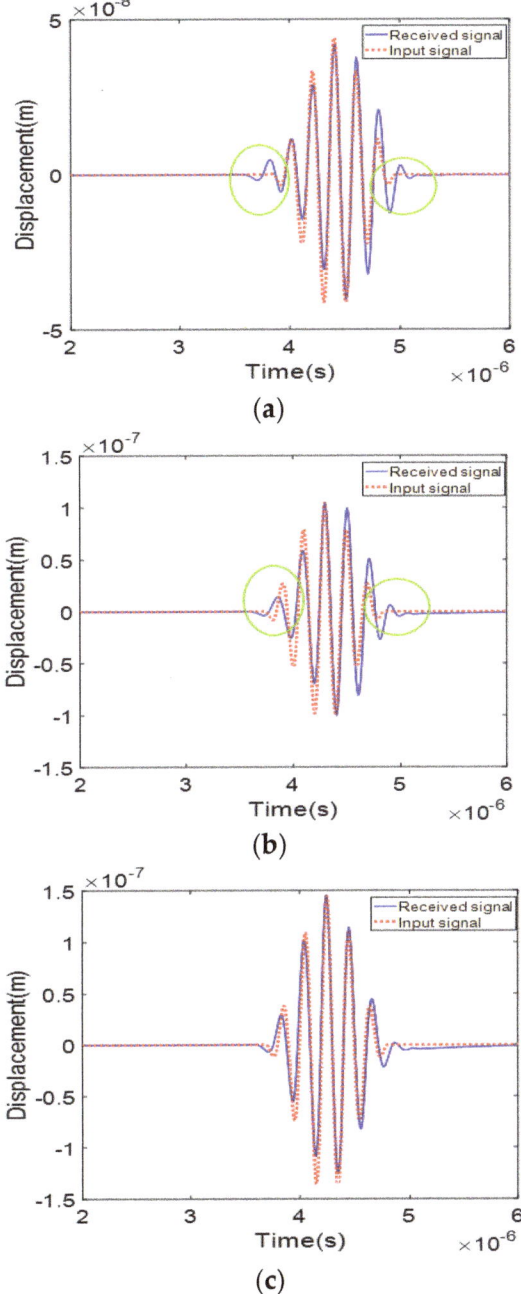

Figure 10. The received signal waveforms for three different APAT cases in Table 4: (**a**) Case A13; (**b**) Case B13; and (**c**) Case C13.

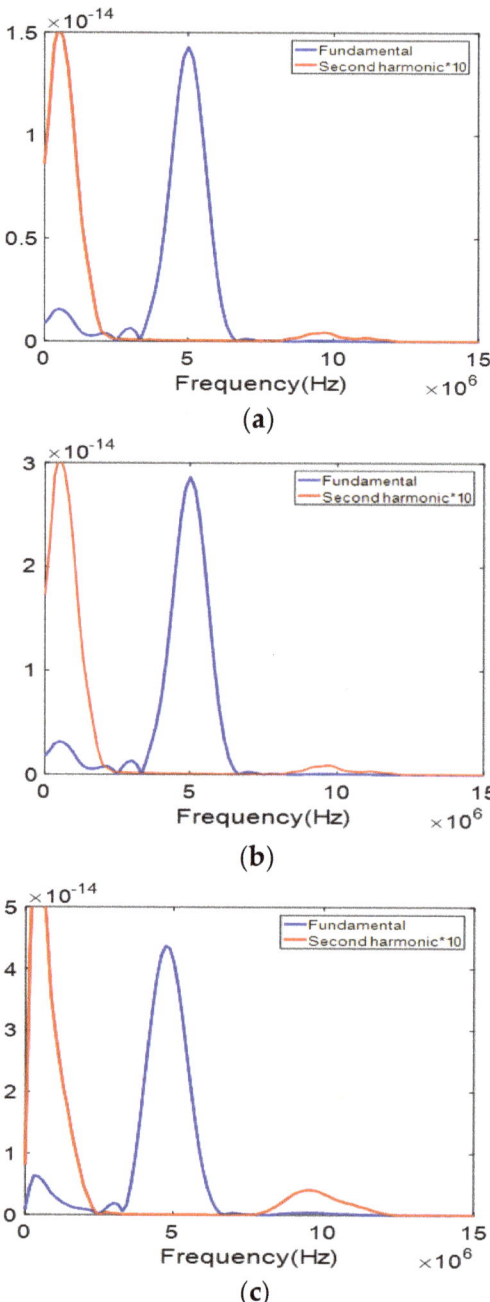

Figure 11. The received signal spectrum for three different APAT cases in Figure 10: (**a**) Case A13; (**b**) Case B13; and (**c**) Case C13.

5. Conclusions

In this paper, we present the analytical model, optimization method, and optimized design results of annular phased array transmitter for efficient second harmonic generation and nonlinear parameter determination in the pulse–echo nonlinear ultrasonic testing. The annular phased array transducer consisting of four-element transmitter and a single-element receiver was optimized in terms of second harmonic reception and total correction. The performance of various combinations of transmitter design variables and focal lengths were tested through wave field analysis, and the optimized specifications of the transmitter were determined and presented. In the future, the fabrication and experimental verification of optimized annular phased array transducers through acoustic performance testing and nonlinear parameter measurement is required. In addition, when using the focused beam of the annular phased array transducer, applying a reception time delay will further enhance the received second harmonic amplitude in the pulse echo mode. These additional studies are expected to develop the pulse–echo nonlinear ultrasonic tests as more practical nondestructive evaluation and diagnosis techniques.

Author Contributions: Writing—original draft preparation, S.C.; Writing—review and supervision, H.J.; and review and editing, I.K.P. All authors have read and agreed to the published version of the manuscript.

Funding: This work was supported by the Korea Institute of Energy Technology Evaluation and Planning (KETEP), the Ministry of Trade, Industry and Energy (MOTIE) of the Republic of Korea, and National Research Foundation of Korea(NRF) (Grant Nos. 20181510102130 and 2019R1F1A1045480).

Conflicts of Interest: The authors declare no conflict of interest.

References

1. Ciorau, P.; Pullia, L.; Hazelton, T.; Daks, W. Phased array ultrasonic technology (PAUT) contribution to detection and sizing of microbially influenced corrosion (MIC) of service water systems and shut down coolers heat exchangers in OPG CANDU stations. In Proceedings of the 8th International Maintenance Conference for CANDU, Toronto, ON, Canada, 16–18 November 2008.
2. Carboni, M.; Cantini, S.; Gilardoni, C. Validation of the rotating UT probe for in-service inspections of freight solid axles by means of the MAPOD approach. In Proceedings of the 5th European-American Workshop on Reliability of NDE, Berlin, Germany, 7–10 October 2013.
3. Hagglund, F.; Robson, M.; Troughton, M.J.; Spicer, W.; Pinson, I.R. A novel phased array ultrasonic testing (PAUT) system for on-site inspection of welded joints in plastic pipes. In Proceedings of the 11th European Conference on Non-Destructive Testing (ECNDT), Prague, Czech Republic, 6–10 October 2014.
4. Hwang, Y.I.; Park, J.; Kim, H.J.; Song, S.J.; Cho, Y.S.; Kang, S.S. Performance comparison of ultrasonic focusing techniques for phased array ultrasonic inspection of dissimilar metal welds. *Int. J. Precis. Eng. Manuf.* **2019**, *20*, 525–534. [CrossRef]
5. Cantrell, J.H.; Yost, W.T. Nonlinear ultrasonic characterization of fatigue microstructures. *Int. J. Fatigue* **2001**, *23*, 487–490. [CrossRef]
6. Cantrell, J.H.; Yost, W.T. Acoustic nonlinearity and cumulative plastic shear strain in cyclically loaded metals. *J. Appl. Phys.* **2013**, *113*, 153506. [CrossRef]
7. Matlack, K.H.; Kim, J.Y.; Wall, J.J.; Qu, J.; Jacobs, L.J.; Sokolov, M.A. Sensitivity of ultrasonic nonlinearity to irradiated, annealed, and re-irradiated microstructure changes in RPV steels. *J. Nucl. Mater.* **2014**, *448*, 26–32. [CrossRef]
8. Matlack, K.H.; Kim, J.Y.; Jacobs, L.J.; Qu, J. Review of second harmonic generation measurement techniques for material state determination in metals. *J. Nondestr. Eval.* **2015**, *34*, 273. [CrossRef]
9. Wang, X.; Wang, X.; Hu, X.L.; Chi, Y.B.; Xiao, D.M. Damage assessment in structural steel subjected to tensile load using nonlinear and linear ultrasonic techniques. *Appl. Acoust.* **2019**, *144*, 40–50. [CrossRef]
10. Kim, J.; Kim, J.G.; Kong, B.; Kim, K.M.; Jang, C.; Kang, S.S.; Jhang, K.Y. Applicability of nonlinear ultrasonic technique to evaluation of thermally aged CF8M cast stainless steel. *Nucl. Eng. Technol.* **2020**, *52*, 621–625. [CrossRef]

11. Pruell, C.; Kim, J.Y.; Qu, J.; Jacobs, L.J. Evaluation of fatigue damage using nonlinear guided waves. *Smart Mater. Struct.* **2009**, *18*, 035003. [CrossRef]
12. Walker, S.V.; Kim, J.Y.; Qu, J.; Jacobs, L.J. Fatigue damage evaluation in A36 steel using nonlinear Rayleigh surface waves. *NDT E Int.* **2012**, *48*, 10–15. [CrossRef]
13. Xiang, Y.; Deng, M.; Xuan, F.Z.; Liu, C.J. Experimental study of thermal degradation in ferritic Cr-Ni alloy steel plates using nonlinear Lamb waves. *NDT E Int.* **2011**, *44*, 768–774. [CrossRef]
14. Ruiz, A.; Ortiz, N.; Medina, A.; Kim, J.Y.; Jacobs, L.J. Application of ultrasonic methods for early detection of thermal damage in 2205 duplex stainless steel. *NDT E Int.* **2013**, *54*, 19–26. [CrossRef]
15. Matlack, K.H.; Wall, J.J.; Kim, J.Y.; Qu, J.; Jacobs, L.J.; Viehrig, H.W. Evaluation of radiation damage using nonlinear ultrasound. *J. Appl. Phys.* **2012**, *111*, 054911. [CrossRef]
16. Apple, T.M.; Cantrell, J.H.; Amaro, C.M.; Mayer, C.R.; Yost, W.T.; Agnew, S.R.; Howe, J.M. Acoustic harmonic generation from fatigue-generated dislocation substructures in copper single crystals. *Philos. Mag.* **2013**, *93*, 2802–2825. [CrossRef]
17. Balasubramaniam, K.; Valluri, J.S.; Prakash, R.V. Creep damage characterization using a low amplitude nonlinear ultrasonic technique. *Mater. Charact.* **2011**, *62*, 275–286. [CrossRef]
18. Viswanath, A.; Rao, B.P.C.; Mahadevan, S.; Parameswaran, P.; Jayakumar, T.; Raj, B. Nondestructive assessment of tensile properties of cold worked AISI type 304 stainless steel using nonlinear ultrasonic technique. *J. Mater. Process. Technol.* **2011**, *211*, 538–544. [CrossRef]
19. Nucera, C.; Lanza di Scalea, F. Nonlinear wave propagation in constrained solids subjected to thermal loads. *J. Sound Vib.* **2014**, *333*, 541–554. [CrossRef]
20. Shui, G.; Wang, Y.S.; Gong, F. Evaluation of plastic damage for metallic materials under tensile load using nonlinear longitudinal waves. *NDT E Int.* **2013**, *55*, 1–8. [CrossRef]
21. Bender, F.A.; Kim, J.Y.; Jacobs, L.J.; Qu, J. The generation of second harmonic waves in an isotropic solid with quadratic nonlinearity under the presence of a stress-free boundary. *Wave Motion* **2013**, *50*, 146–161. [CrossRef]
22. Vander Meulen, F.; Haumesser, L. Evaluation of B/A nonlinear parameter using an acoustic self-calibrated pulse-echo method. *Appl. Phys. Lett.* **2008**, *92*, 214106. [CrossRef]
23. Best, S.R.; Croxford, A.J.; Neild, S.A. Pulse-echo harmonic generation measurements for non-destructive evaluation. *J. Nondestr. Eval.* **2014**, *33*, 205–215. [CrossRef]
24. Saito, S. Nonlinearly generated second harmonic sound in a focused beam reflected from free surface. *Acoust. Sci. Technol.* **2005**, *26*, 55–61. [CrossRef]
25. Zhang, S.; Li, X.; Jeong, H.; Cho, S.; Hu, H. Theoretical and experimental investigation of the pulse-echo nonlinearity acoustic sound fields of focused transducers. *Appl. Acoust.* **2017**, *117*, 145–149. [CrossRef]
26. Jeong, H.; Zhang, S.; Barnard, D.; Li, X. A novel and practical approach for determination of the acoustic nonlinearity parameter using a pulse-echo method. *AIP Conf. Proc.* **2016**, *1706*, 060006.
27. Dace, G.E.; Thompson, R.B.; Buck, O. Measurement of the acoustic harmonic generation for materials characterization using contact transducers. In *Review of Progress in Quantitative Nondestructive Evaluation*; Springer: Berlin/Heidelberg, Germany, 1992; Volume 11, pp. 2069–2076.
28. Jeong, H.; Barnard, D.; Cho, S.; Zhang, S.; Li, X. Receiver calibration and the nonlinearity parameter measurement of thick solid samples with diffraction and attenuation corrections. *Ultrasonics* **2017**, *81*, 147–157. [CrossRef] [PubMed]
29. Jeong, H.; Cho, S.; Shin, H.; Zhang, S.; Li, X. Optimization and validation of dual element ultrasound transducers for improved pulse-echo measurements of material nonlinearity. *IEEE Sen. J.* **2020**, *20*, 13596–13606. [CrossRef]
30. Sy, K.; Brédif, P.; Iakovleva, E.; Roy, O.; Lesselier, D. Development of methods for the analysis of multi-mode TFM images. *J. Phys. Conf. Ser.* **2018**, *1017*. [CrossRef]
31. Aizpurua, I.; Ayesta, I.; Castro, I. Characterization of anisotropic weld structure for nuclear industry. In Proceedings of the 11th European Conference on NDT, Prague, Czech Republic, 6–10 October 2014.
32. Jeong, H. Time Reversal-based beam focusing of an ultrasonic phased array transducer on a target in anisotropic and inhomogeneous welds. *Mater Eval.* **2014**, *72*, 589–596.
33. Shattuck, D.P.; Von Ramm, O.T. Compound scanning with a phased array. *Ultrason. Imaging* **1982**, *4*, 93–107. [CrossRef]
34. Von Ramm, O.T.; Smith, S.W. Beam steering with linear arrays. *IEEE Trans. Biomed. Eng.* **1983**, *30*, 438–452. [CrossRef]

35. McNab, A.; Stumpf, I. Monolithic phased array for the transmission of ultrasound in NDT ultrasonics. *Ultrasonics* **1986**, *24*, 148–155. [CrossRef]
36. Jeong, H.; Cho, S.; Zhang, S.; Li, X. Acoustic nonlinearity parameter measurements in a pulse-echo setup with the stress-free reflection boundary. *J. Acoust. Soc. Am.* **2018**, *143*, EL237–EL242. [CrossRef] [PubMed]
37. Cho, S.; Jeong, H.; Zhang, S.; Li, X. Dual element transducer approach for second harmonic generation and material nonlinearity measurement of solids in the pulse-echo method. *J. Nondestr. Eval.* **2020**, *39*, 62. [CrossRef]
38. Liaptsis, G.; Liaptsis, D.; Wright, B.; Charlton, P. Focal law calculations for annular phased array transducers. *e-J. Nondestr. Test.* **2015**, *20*, 1–7.
39. Nagy, P.B.; Qu, J.; Jacobs, L.J. Finite-size effects on the quasistatic displacement pulse in a solid sample with quadratic nonlinearity. *J. Acoust. Soc. Am.* **2013**, *134*, 1760–1774. [CrossRef] [PubMed]
40. Qu, J.; Jacobs, L.J.; Nagy, P.B. On the acoustic-radiation-induced strain and stress in elastic solids with quadratic nonlinearity (L). *J. Acoust. Soc. Am.* **2011**, *129*, 3449–3452. [CrossRef] [PubMed]

Publisher's Note: MDPI stays neutral with regard to jurisdictional claims in published maps and institutional affiliations.

© 2020 by the authors. Licensee MDPI, Basel, Switzerland. This article is an open access article distributed under the terms and conditions of the Creative Commons Attribution (CC BY) license (http://creativecommons.org/licenses/by/4.0/).

Article

Absolute Measurement of Material Nonlinear Parameters Using Noncontact Air-Coupled Reception

Hyunjo Jeong [1],*, Sungjong Cho [2], Shuzeng Zhang [3] and Xiongbing Li [3]

[1] Department of Mechanical Engineering, Wonkwang University, Iksan, Jeonbuk 54538, Korea
[2] Nondestructive Testing (NDT) Research Center, Seoul National University of Science and Technology, Seoul 01811, Korea; cho-sungjong@seoultech.ac.kr
[3] School of Traffic and Transportation Engineering, Central South University, Changsha 410075, China; sz_zhang@csu.edu.cn (S.Z.); lixb213@csu.edu.cn (X.L.)
* Correspondence: hjjeong@wku.ac.kr; Tel.: +82-(0)63-850-6690

Abstract: Nonlinear ultrasound is often employed to assess microdamage or nonlinear elastic properties of a material, and the nonlinear parameter is commonly used to quantify damage sate and material properties. Among the various factors that influence the measurement of nonlinear parameters, maintaining a constant contact pressure between the receiver and specimen is important for repeatability of the measurement. The use of an air-coupled transducer may be considered to replace the contact receiver. In this paper, a method of measuring the relative and absolute nonlinear parameters of materials is described using an air-coupled transducer as a receiver. The diffraction and attenuation corrections are newly derived from an acoustic model for a two-layer medium and the nonlinear parameter formula with all corrections is defined. Then, we show that the ratio of the relative nonlinear parameter of the target sample to the reference sample is equal to that of the absolute nonlinear parameter, and this equivalence is confirmed by measurements on three systems of aluminum samples. The proposed method allows the absolute measurement of the nonlinear parameter ratio or the nonlinear parameter without calibration of the air-coupled receiver and removes restrictions on the selection of reference samples.

Keywords: nonlinear parameter; noncontact reception; air-coupled receiver; aluminum samples; corrections

Citation: Jeong, H.; Cho, S.; Zhang, S.; Li, X. Absolute Measurement of Material Nonlinear Parameters Using Noncontact Air-Coupled Reception. *Materials* **2021**, *14*, 244. https://doi.org/10.3390/ma14020244

Received: 10 December 2020
Accepted: 31 December 2020
Published: 6 January 2021

Publisher's Note: MDPI stays neutral with regard to jurisdictional claims in published maps and institutional affiliations.

Copyright: © 2021 by the authors. Licensee MDPI, Basel, Switzerland. This article is an open access article distributed under the terms and conditions of the Creative Commons Attribution (CC BY) license (https://creativecommons.org/licenses/by/4.0/).

1. Introduction

The acoustic nonlinear parameter becomes a powerful tool in the nondestructive evaluation field as a measure of material nonlinearity and damage state in structural components. This parameter can quantitatively be obtained by harmonic generation measurements. The most widely used technique is the finite amplitude method, in which a high power wave of a monochromatic frequency propagates through a nonlinear medium introduces distortions, resulting in the generation of higher harmonics. Harmonic generation measurements for evaluating nonlinear parameters can be conducted using several wave types, different generation and detection methods, and a variety of experimental set-ups. The general experimental procedure is similar in all cases, where an ultrasonic tone burst at frequency ω is launched from the emitting transducer, it propagates some distance through the material, and the response is measured by the receiving transducer—specifically, the amplitudes of the fundamental and second harmonic waves are extracted from the frequency response of the received signal.

Contact piezoelectric transducers are most commonly used as emitting transducers of longitudinal waves in the through-transmission setup. Various types of detectors can be used as receiving transducers of both fundamental and second harmonic wave components. Detection of second harmonic generation measurements using longitudinal waves has been conducted with contact piezoelectric transducers [1–6], capacitive transducers [7–14], and laser interferometers [15–18]. An absolute measurement of material nonlinearity is possible using either capacitive transducers [9,12] or contact piezoelectric transducers using a calibration procedure [19,20] in which the absolute displacement amplitude of the fundamental and second harmonic waves can be measured.

The various methods used to detect nonlinear signals suffer from significant limitations. Piezoelectric contact transducers, while being easy to use in many ways, are heavily influenced by contact conditions between the transducer and sample surface, so that application of a consistent force is crucial to measurement repeatability.

Noncontact detection methods such as capacitive receivers and laser interferometers are more desirable from the practical point of view, but they also have some drawbacks. Laser interferometry requires a mirror-finished sample surface and relies on complicated optics to maximize sensitivity. Careful preparation of sample surfaces is also very important in the capacitive receiver technique, requiring an optically flat and parallel sample surface over the entire receiver area and a small gap spacing of only a few microns [21,22].

Compared to existing noncontact detection methods such as capacitive receivers or laser interferometers, air-coupled transducers are easy to handle, significantly less expensive, and robust relative to surface conditions. As far as we know, only relative measurements were possible with air-coupled transducers, mainly because they are difficult to calibrate for use in nonlinear measurements. Existing calibration techniques such as self-reciprocity methods are not directly applicable to the air-coupled receivers because of high ultrasonic attenuation loss in air. Consequently, most second harmonic generation measurements were limited to relative measurements. Recently, air-coupled transducers have been applied to second harmonic generation measurements as an efficient detection tool for Rayleigh and Lamb waves [23]. The air-coupled transducer detects a longitudinal wave in air that is leaked from the propagating Rayleigh wave or Lamb wave in the sample. Torello et al. [24] reported a hybrid acoustic modeling and experimental approach to air-coupled transducer calibration and the use of this calibration in a model-based optimization to determine the absolute nonlinear parameter of representative materials. More recently, Li et al. [25] proposed a comparative approach where four separate experimental setups are used to obtain the sensitivity or the transfer function of an air-coupled ultrasonic receiver and to measure material nonlinear parameter.

The purpose of this paper is to develop a new technique for absolute measurement of material nonlinearity using air-coupled receivers without separate receiver calibration. First, an acoustic model for a two-layer medium composed of solid specimen and air is considered, and the diffraction and attenuation corrections are derived from the wave field analysis. These corrections convert the measured fundamental and second harmonic amplitudes to the plane wave amplitudes in nonlinear parameter calculations. Next, it is shown that the ratio of the relative nonlinear parameter of the target specimen to the reference specimen (β'/β'_{ref}) is the same as the absolute nonlinear parameter ratio of the two specimens (β/β_{ref}). This equivalence allows an absolute comparison of the material nonlinearity between different materials from the measurement of relative nonlinear parameter. Furthermore, the absolute nonlinear parameter of a target specimen (β) can be obtained by measuring the relative nonlinear parameter ratio, β'/β'_{ref}, if the absolute nonlinear parameter of the reference specimen (β_{ref}) is available. A nonlinear ultrasonic testing system including an air-coupled receiver is constructed, and the relative and absolute nonlinear parameters are measured for aluminum specimens of three different types: Al2024, Al6061, and Al7075. The equivalence between the relative and absolute parameter ratios is verified through experimental results.

Section 2 describes the nonlinear acoustic model for a two-layer medium composed of solid and air and defines the nonlinear parameter formula with necessary corrections. The equivalence of relative nonlinear parameter ratio and absolute nonlinear parameter ratio is also demonstrated. Simulation results on the received displacement and diffraction and attenuation corrections are provided in Section 3. Section 4 introduces the experimental setup and specimens, and Section 5 presents the experimental results. Conclusions are drawn in Section 6.

2. Sound Beam Fields and Nonlinear Parameter

In order to calculate the received wave fileds and to define the nonlinear parameter, we need an appropriate model equation for finite amplitude radiation that takes into account the combined effects of nonlinearity, diffraction, and attenuation. For this purpose, a Westervelt-type equation similar to the Westervelt equation for sound beams of fnite amplitude in fuids can be used for longitudinal wave motion in isotropic solids [26,27]. Such equation can be obtained from the Westervelt equation in a manner similar to the derivation of the KZK-type equation from the KZK equation [28].

Applying the quasilinear theory to the Westervelt-type equation yields the governing equations for the fundamental and second harmonic waves for axisymmetric sound source. The Green's function approach is a convenient method for constructing the integral solutions to these equations. Then, the solutions can be obtained by integrating over the product of the Green's function and the appropriate source function to sum up the contributions from all source points. To calculate the received ultrasonic fields by a finite radius receiver, the concept of field averaging can be used.

In general, the accurate determination of the nonlinear parameter in a wide range of experimental conditions requires attenuation and diffraction corrections to the plane wave solution. The attenuation correction for the fundamental wave is well known [29]. The attenuation correction for the second harmonic wave generated by the focring of the propagating fundamental wave is also well known [30]. For wave propagation in a multi-layered medium, attenuation correction can be defined for each layer, and the total attenuation correction for the entire medium can be obtained by simply multiplying the correction in each layer.

A closed form of the diffraction correction for the fundamental wave was found by Rogers and Van Buren [31] when the transmitter and receiver are of the same size. However, the integral solutions can also be used to numerically calculate the diffraction correction of the fundamental wave for a more general transmitter-receiver combination [32]. The diffraction correction is defined as the magnitude of the fundamental or second harmonic wave divided by the corresponding plane wave solution. Similarly, the diffraction correction for the second harmonic wave can be found numerically from the magnitude of the second harmonic wave and the plane wave solution of the second harmonic wave [32]. The concept of diffraction correction for a single medium can be extended to a two-layer medium that is covered in this study.

In this work, the diffraction corrections for the fundamental and second harmonic waves are developed using the integral solutions of the Westervelt-type equation [26,27]. The diffraction correction for a single medium is extended to a two-layer medium consisting of a solid layer (specimen) and an air layer. The analytical diffraction correction can be efficiently used for a wide range of two-layer media and transmitter-receiver geometries.

First, using the approach mentioned above, we construct the displacement field expressions for the fundamental and second harmonic waves propagating in the solid specimen depicted in Figure 1a. The propagation in a single medium is then extended to the analysis of propagation in a two-layer medium composed of the solid specimen and air, and the displacement fields received by an air-coupled transducer are obtained, as shown in Figure 1b. Finally, we will define the diffraction corrections for the fundamental and second harmonic waves in the two-layer medium, and derive a nonlinear parameter expression modified by the attenuation and diffraction corrections.

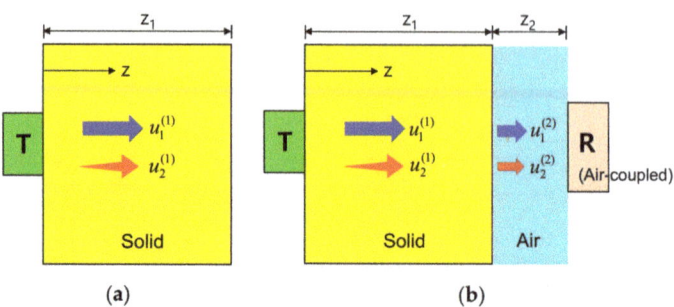

Figure 1. (a) Fundamental wave propagation and second harmonic generation in a solid specimen, and (b) Transmission at the solid-air interface, propagation in the air, and reception by an air-coupled ultrasonic receiver.

2.1. Sound Beam Solutions for a Single Medium

Referring to Figure 1a, the fundamental displacement field in medium 1, $u_1^{(1)}$, in the forward propagation ($0 \leq z \leq z_1$) region and the generated second harmonic displacement field, $u_2^{(1)}$, due to the finite amplitude radiation at the transmitter can be expressed as: [26]

$$u_1^{(1)}(x,y,z) = -2ik^{(1)} \int_{-\infty}^{+\infty} \int_{-\infty}^{+\infty} u_1(x',y',0)\, G_1^{(1)}(x,y,z|x',y',0)\, dx'dy' \quad (1)$$

$$u_2^{(1)}(x,y,z) = \frac{\beta k^{(1)}}{2(c^{(1)})^2} \int_0^z \int_{-\infty}^{+\infty} \int_{-\infty}^{+\infty} \left[u_1^{(1)}(x',y',z')\right]^2 G_2^{(1)}(x,y,z|x',y',z')\, dx'dy'dz' \quad (2)$$

Here, β is the nonlinear parameter of the solid specimen, and defined as $\beta = -\left(3 + \frac{C_{111}}{C_{11}}\right)$ where C_{11} and C_{111} are the second and third order elastic constants, respectively. In Equations (1) and (2), $k^{(1)}$ and $c^{(1)}$ are the wave number and wave velocity of medium 1, respetively. The Green's function can be obtained as

$$G_n^{(1)}(x,y,z|x',y',z') = \frac{1}{4\pi r_n} \exp(ink^{(1)} r_n), \; n = 1, 2 \quad (3)$$

where $r_1 = \sqrt{(x-x')^2 + (y-y')^2 + z^2}$ and $r_2 = \sqrt{(x-x')^2 + (y-y')^2 + (z-z')^2}$ is the distance from the sound source point $(x',y',0)$ to the target point (x,y,z). The attenuation effect is not included in the Green function and will be treated separately. A constant displacement U_0 is prescribed over the surface S' of a circular piston transducer of radius a:

$$\begin{cases} u_1(x',y',z'=0) = U_0, \\ u_2(x',y',z'=0) = 0 \end{cases} \quad x'^2 + y'^2 \leq a^2 \quad (4)$$

Note that the range of wave propagation distance in Equations (1) and (2) is $0 \leq z \leq z_1$. Equation (1) provides an exact solution to calculate the fundamental wave field, and represents a superposition of spherical waves radiating from the point sources distributed on the source plane $z' = 0$. When the second harmonic wave field is calculated using Equation (2), the Green's function used in this integral includes contribution of the element $dV' = dx'dy'dz'$ of the virtual source formed by the fundamental wave field. If a displacement source such as Equation (4) is defined, these equations can be used to obtain the fundamental and second harmonic displacement fields.

Equations (1) and (2) provide the exact displacement solutions for the fundamental and second harmonic waves for a planar, circular P-wave transducer radiating at normal incidence in medium 1, and can be expressed in the following form:

$$u_1^{(1)}(x,y,z) = \left[U_1^{plane}\right]\left[D_1^{(1)}(a,x,y,z)\right] \tag{5}$$

$$u_1^{(2)}(x,y,z) = \left[U_2^{plane}\right]\left[D_2^{(1)}(a,x,y,z)\right] \tag{6}$$

where the first square bracket on the right-hand side represents the pure plane wave solutions, and the second bracket the diffraction corrections. The plane wave solutions are given by:

$$U_1^{plane} = U_0 \exp(ik^{(1)}z) \tag{7}$$

$$U_2^{plane} = \frac{(k^{(1)})^2 \beta z U_0^2}{8} \exp(2ik^{(1)}z) \tag{8}$$

and the diffraction corrections are defined as $D_1^{(1)} = u_1^{(1)}(x,y,z)/U_1^{plane}$ and $D_2^{(1)} = u_2^{(1)}(x,y,z)/U_2^{plane}$.

2.2. Sound Beam Solutions for a Two-Layer Medium

Prior to further acoustic modeling and simulation, it is necessary to understand the transmission, generation, propagation, and reception of the fundamental and second harmonic waves in a two-layer medium consisting of solid and air. The fundamental wave and the second harmonic wave generated by the propagating fundamental wave both propagate in the solid medium (Figure 1a) and these two waves are transmitted at the solid-air interface and then propagate into the air (Figure 1b). Due to the small intensity of the transmitted fundamental wave resulting from the very low transmission coefficient and the very high attenuation loss in the air, it is difficult to meet the conditions for generating a new second harmonic wave in the air. Therefore, the generation of the second harmonic that may be newly generated by the transmitted fundamental wave in the air is ignored. Similarly, it is assumed that the second harmonic wave in the solid propagates only as the second harmonic in the air without generating other waves. These are schematically shown in Figure 1b.

Approximate methods such as the multi-Gaussian beam (MGB) model may be used to formulate the wave fields in the two-layer medium. The MGB model is based on the paraxial approximation and is known to be very computationally efficient [29]. This model accurately calculates the sound beam fields of an ultrasonic transducer at distances of approximately one transducer diameter or greater from the transducer face. In the two-layer medium covered here, however, the distance from the solid-air interface to the air-coupled receiver is only a few millimeters, which is very short. Therefore, in this case, the paraxial MGB solutions may fail in providing accurate beam field results, especially the diffraction corrections. Therefore, exact solutions are required for modeling a planar, circular P-wave transducer radiating at normal incidence in the solid-air interface. The integral solutions of Equations (1) and (2) are extended to the wave field analysis in the second medium.

Based on this observation, the propagating fundamental and second harmonic waves in the second medium (air) can be found by using the results of Equations (1) and (2) at $z = z_1$ as new sound sources for radiation into the second medium. Then, the integral solutions become:

$$u_1^{(2)}(x,y,z) = T_{12} \times \left[-2ik^{(2)} \int_{-\infty}^{+\infty}\int_{-\infty}^{+\infty} u_1^{(1)}(x',y',z_1)\, G_1^{(2)}(x,y,z|x',y',z_1)\, dx'dy'\right] \tag{9}$$

$$u_2^{(2)}(x,y,z) = T_{12} \times \left[-2ik^{(2)} \int_{-\infty}^{\infty}\int_{-\infty}^{\infty} u_2^{(1)}(x',y',z_1)\, G_2^{(2)}(x,y,z|x',y',z_1)\, dx'dy'\right]. \tag{10}$$

In the above equation, T_{12} is the transmission coefficient given by $T_{12} = \frac{Z_2}{Z_1+Z_2}$ where Z_1 and Z_2 are the acoustic impedeances of the medium 1 and 2, respectively. The linear transmission is assumed for the nonlinear second harmonic wave in Equation (10), but this is a good approximation for the solid-air interface. Note that the range of wave propagation distance in these equations is $z_1 \leq z \leq z_2$. These equations can also be written in the quasi-plane wave form similar to Equations (5) and (6). The diffraction effects of the fundamental and second harmonic waves propagating in the second medium are defined as $D_1^{(2)} = u_1^{(2)}(x,y,z)/u_1^{plane}$, $D_2^{(2)} = u_2^{(2)}(x,y,z)/u_2^{plane}$, where the plane wave solutions are given by $u_1^{plane} = U_0 T_{12} \exp(ik^{(1)}z_1 + ik^{(2)}z)$, $u_2^{plane} = \frac{(k^{(1)})^2 \beta z_1 U_0^2}{8} T_{12} \exp(i2k^{(1)}z_1 + i2k^{(2)}z)$. The detailed expressions for the plane wave solutions and diffraction corrections are given later together with the attenuation corrections.

Finally, to calculate the received displacement at distance z_2 by a circular air-coupled transducer of radius b, the concept of average field can be used and calculated as follows:

$$\tilde{u}_n^{(m)}(z_2) = \frac{1}{\pi b^2} \int_S u_n^{(m)}(x,y,z_2)\, dS, n = 1,2 \qquad (11)$$

where $u_n^{(m)}(x,y,z_2)$ is computed from Equations (9) and (10). Substituting Equations (9) and (10) into Equation (11) and performing some manipulation, the received displacement fields in the second medium can be written in the following form:

$$\tilde{u}_1(z_2) = \left[U_0 T_{12} \exp(ik^{(1)}z_1 + ik^{(2)}z_2) \right] [D_1(a,b,z_1,z_2)] \qquad (12)$$

$$\tilde{u}_2(z_2) = \left[\frac{(k^{(1)})^2 \beta z_1 U_0^2}{8} T_{12} \exp(i2k^{(1)}z_1 + i2k^{(2)}z_2) \right] [D_2(a,b,z_1,z_2)] \qquad (13)$$

where the first term in the bracket in each equation represents the plane wave solution and D_n denotes the diffraction correction for the fundamental ($n = 1$) and second harmonic ($n = 2$) waves. The detailed expressions of u_n^{plane} and D_n in Equations (12) and (13) are given by

$$u_1^{plane} = U_0 T_{12} \exp(ik^{(1)}z_1 + ik^{(2)}z_2) \qquad (14)$$

$$u_2^{plane} = \frac{(k^{(1)})^2 \beta z_1 U_0^2}{8} T_{12} \exp(i2k^{(1)}z_1 + i2k^{(2)}z_2) \qquad (15)$$

$$D_1(a,b,z_1,z_2) = \frac{\tilde{u}_1(z_2)}{u_1^{plane}} \qquad (16)$$

$$D_2(a,b,z_1,z_2) = \frac{\tilde{u}_2(z_2)}{u_2^{plane}}. \qquad (17)$$

The diffraction corrections given by Equations (16) and (17) are defined as the received average displacement of the fundamental or second harmonic wave propagated through the entire medium divided by the amplitude of the plane wave involved in the same propagation process. In the current solid-air medium, the fundamental wave is generated from the start of radiation and continues to propagate only as the fundamental wave in medium 1 (solid) and 2 (air). However, the second harmonic wave is generated and propagated by the propagating fundamental wave in the first medium, and then propagates as a fundamental wave of frequency $2f$ in the second medium.

The attenuation correction for the fundamental and second harmonic waves is well known in the second harmonic generation process in a single medium and can be considered separately since it only affects the amplitude of the propagating wave. Thus, the attenuation correction that occurs over the entire propagation process can be obtained by successively multiplying the attenuation correction of each layer. With the inclusion of the attenuation corrections, the final expressions of the received displacement fields in the second medium are obtained as:

$$\tilde{u}_1(z_2) = \left[U_0 T_{12} \exp(ik^{(1)}z_1 + ik^{(2)}z_2)\right][D_1(a,b,z_1,z_2)]\left\{\left[M_1^{(1)}(\alpha_1^{(1)},z_1)\right]\left[M_1^{(2)}(\alpha_1^{(2)},z_2)\right]\right\} \qquad (18)$$

$$\tilde{u}_2(z_2) = \left[\frac{(k^{(1)})^2 \beta z_1 U_0^2}{8} T_{12} \exp(i2k^{(1)}z_1 + i2k^{(2)}z_2)\right][D_2(a,b,z_1,z_2)]\left\{\left[M_2^{(1)}(\alpha_1^{(1)},\alpha_2^{(1)},z_1)\right]\left[M_2^{(2)}(\alpha_2^{(2)},z_2)\right]\right\} \qquad (19)$$

where $\alpha_1^{(m)}$ and $\alpha_2^{(m)}$ are the attenuation coefficients at the fundamental and second harmonic frequencies in the mth medium, respectively. The detailed expressions of M_n, $n = 1, 2$ appearing in the above equations are given by:

$$\left[M_1^{(1)}\right]\left[M_1^{(2)}\right] = M_1 = \left[\exp(-\alpha_1^{(1)}z_1)\right]\left[\exp(-\alpha_1^{(2)}z_2)\right] \qquad (20)$$

$$\left[M_2^{(1)}\right]\left[M_2^{(2)}\right] = M_2 = \left[\frac{\exp(-2 \times \alpha_1^{(1)}z_1) - \exp(-\alpha_2^{(1)}z_1)}{(\alpha_2^{(1)} - 2\alpha_1^{(1)})z_1}\right]\exp\left[-\alpha_2^{(2)}z_2\right]. \qquad (21)$$

The attenuation correction of the fundamental wave in each layer is found to be the same as the simple exponential attenuation law in linear acoustics. The attenuation correction of the second harmonic wave in the first solid layer shows a somewhat complicated behavior, since it will grow with propagation distance due to the nonlinear interaction effects, and will also decrease due to attenuation effects.

2.3. Definition of Nonlinear Parameter

The nonlinear parameter, β, can be found from Equations (18) and (19) by cancelling U_0 in both terms:

$$\beta = \left[\frac{8U_2(z_2)}{(k^{(1)})^2 z_1 U_1^2(z_2)}\right]\left[\frac{M_1^2\left(\alpha_1^{(1)}, \alpha_1^{(2)} z_1, z_2\right)}{M_2\left(\alpha_1^{(1)}, \alpha_2^{(1)}, \alpha_2^{(2)}, z_1, z_2\right)}\right]\left[\frac{|D_1(a,b,z_1,z_2)|^2}{|D_2(a,b,z_1,z_2)|}\right] \qquad (22)$$

where U_1 and U_2 are the displacement amplitudes of the received fundamental and second harmonic waves, respectively. The first square bracket in Equation (22) represents the uncorrected nonlinear paramter, and the second and third brackets represent the attenuation correction and diffraction correction, respectively. The nonlinear parameter β is defined using the amplitude of the fundamental and second harmonic waves of the pure plane wave. However, the amplitude of the measured wave in the real environment deviates from the pure plane wave due to material attenuation and finite size transducers. Therefore, attenuation and diffraction corrections are required to convert the actually measured wave amplitude into the plane wave amplitude.

The absolute measurement of the received displacement generally requires the use of a calibrated receiving transducer. In case of an air-coupled receiver, however, the reciprocity-based calibration [19,20] is very difficult to perform because of high attenuation in the air. Due to the difficulty of obtaining calibration measurements, it is possible to use an electrical output signal instead of the received displacement in the measurement of nonlinear parameters. If the amplitudes of received fundamental and second harmonic waves are denoted by the electrical current output signals denoted by $A_1(\omega)$ and $A_2(\omega)$, respectively, then the relative nonlinear parameter, β', with the inclusion of attenuation and diffraction corrections can be expressed as:

$$\beta' = \left[\frac{8A_2}{(k^{(1)})^2 z_1 A_1^2}\right]\left[\frac{M_1^2}{M_2}\right]\left[\frac{|D_1|^2}{|D_2|}\right]. \qquad (23)$$

Now we define the relative nonlinear parameter ratio of a target sample to a reference sample as follows:

$$\frac{\beta'}{\beta'_{ref}} = \frac{\left[\frac{8A_2}{(k^{(1)})^2 z_1 A_1^2}\right] \left[\frac{M_1^2}{M_2}\right] \left[\frac{|D_1|^2}{|D_2|}\right]}{\left(\left[\frac{8A_2}{(k^{(1)})^2 z_1 A_1^2}\right] \left[\frac{M_1^2}{M_2}\right] \left[\frac{|D_1|^2}{|D_2|}\right]\right)_{ref}}. \tag{24}$$

By including attenuation and diffraction corrections in the denominator of Equation (24), the restrictions on the type and thickness of the reference specimen can be removed and any specimen can be used. In Equation (24), the amplitude of the electrical current signal can be expressed in terms of the absolute displacement amplitude by using the relationship between these two quantities, $A(\omega) = U(\omega)/H(\omega)$, where $H(\omega)$ is the transfer function of the receive transducer [33]. Then, it can be easily shown from Equation (25) below that the relative nonlinear parameter ratio is equal to the absolute nonlinear parameter ratio, since the ratio of the transfer function H_1^2/H_2 is a characteristic of the receive transducer and does not depend on the type of specimen.

$$\frac{\beta'}{\beta'_{ref}} = \frac{\left[\frac{8U_2}{(k^{(1)})^2 z_1 U_1^2}\right] \left[\frac{H_1^2}{H_2}\right] \left[\frac{M_1^2}{M_2}\right] \left[\frac{|D_1|^2}{|D_2|}\right]}{\left(\left[\frac{8U_2}{(k^{(1)})^2 z_1 U_1^2}\right] \left[\frac{H_1^2}{H_2}\right] \left[\frac{M_1^2}{M_2}\right] \left[\frac{|D_1|^2}{|D_2|}\right]\right)_{ref}} = \frac{\beta}{\beta_{ref}} \tag{25}$$

Therefore, the ratio of the absolute nonlinear parameters between two different materials can be obtained by measuring the ratio of the relative nonlinear parameters, which then allows a quantitative comparison of the material nonlinearity between different materials. Equation (25) also demonstrates that the absolute nonlinear parameter of a target specimen (β) can be obtained by measuring the relative nonlinear parameter ratio of a target sample to a reference sample (β'/β'_{ref}) if the absolute nonlinear parameter of the reference specimen (β_{ref}) is available. This observation can be written as

$$\beta = \left(\frac{\beta'}{\beta'_{ref}}\right) \beta_{ref}. \tag{26}$$

3. Simulation Results

In the measurement of the relative or absolute nonlinear parameter of a solid specimen using an air-coupled transducer, the diffraction and attenuation corrections are important factors affecting the measurement results. In particular, the diffraction correction equations for a two-layer medium of solid-air is complicated and computationally heavy due to multiple integrations involved. Therefore, the calculation of accurate diffraction correction is the main purpose of the wave field simulation.

The acoustic parameters of the two-layered medium used in the calculation of wave fileds are listed in Table 1. The acoustic properties of Al6061 including the nonlinear parameter β were taken from the measurement [33]. For air, the wave speed and density are the values at temperature 20 °C and the value of β was taken from [34]. The frequency-dependent attenuation coefficient of air was taken from [35]. In addition, the source displacement used is 1.0 m, and the fundamental frequency used is 2 MHz. The diameters of transmit and receive transducers are all 12.7 mm. The propagation distance or the layer thickness of layers 1 and 2 is 12 cm and 0.5 cm, respectively.

Table 1. Acoustic parameters of the two-layered medium used in the wave field calculation.

Materials	Wave Speed (m/s)	Density (kg/m³)	Attenuation (Np/m)
Al6061	6422	2700	$\alpha_1 = 0.5$, $\alpha_2 = 3\alpha_1$
Air	346	1.29	$1.83 \times 10^{-11} f^2$

In the wave field calculation, the received displacement, diffraction correction, and attenuation correction for the fundamental wave and second harmonic waves were obtained over the entire propagation process of the two-layer medium. In the calculation of the received displacement, both the transmission coefficient and the attenuation effect were neglected.

Figure 2a,b respectively shows the received displacement amplitude of the fundamental and second harmonic waves, calculated using Equations (9) and (10). The behavior of the received displacement in the solid specimen is not unfamiliar and is similar to the results observed in previous studies [36]. The distance between the solid-air interface and the receiver is 0.5 cm, which is very short, the transmission coefficient was assumed to be 1, and the attenuation in the air layer was neglected, so the displacement received in the air layer remains almost the same and is equal to the displacement at the end of the solid specimen. Thus, the diffraction correction at the receiver position can be replaced by the value at the end of the solid specimen.

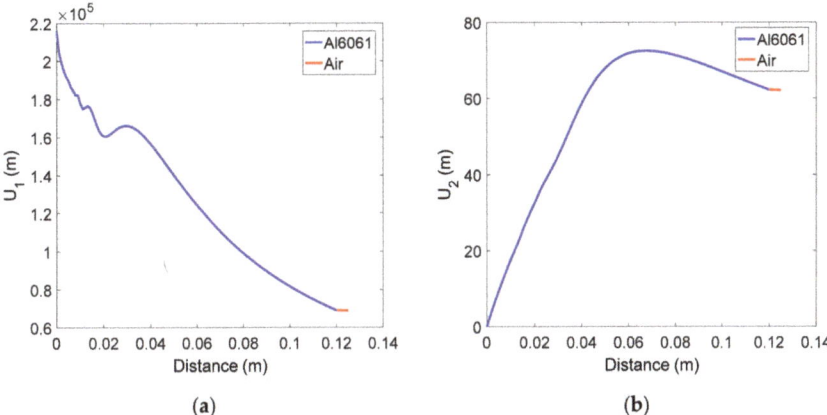

Figure 2. Variation of received average displacement in the two-layer medium of solid-air: (**a**) Fundamental wave, and (**b**) Second harmonic wave.

Diffraction corrections are defined by Equations (16) and (17) as the division of the received average displacement at each propagation distance by the displacement of the plane wave at the same distance without considering the effects of attenuation and transmission interface. Diffraction corrections were calculated for the Al6061, 12 cm thick sample and the results are shown in Figure 3 as a function of the propagation distance. Figure 3a is the plot of $\tilde{D}_1$ of the fundamental wave, and Figure 3b is the plot of $\tilde{D}_2$ of the second harmonic wave. Since the amplitude of the plane fundamental wave is a constant at all propagation distances, the overall behavior of $\tilde{D}_1$ looks the same as that of the received fundamental wave shown in Figure 2a. Since the amplitude of the plane second harmonic wave increases linearly with the propagation distance, $\tilde{D}_2$ decreases with the propagation distance in the solid layer and then looks like Figure 3b. $\tilde{D}_2$ is smaller than $\tilde{D}_1$ at the same propagation distance. As pointed out earlier, the diffraction correction in the air layer can be replaced by the value at the end of the solid layer, in which case the two-layer medium composed of solid and air can be treated as a single solid medium.

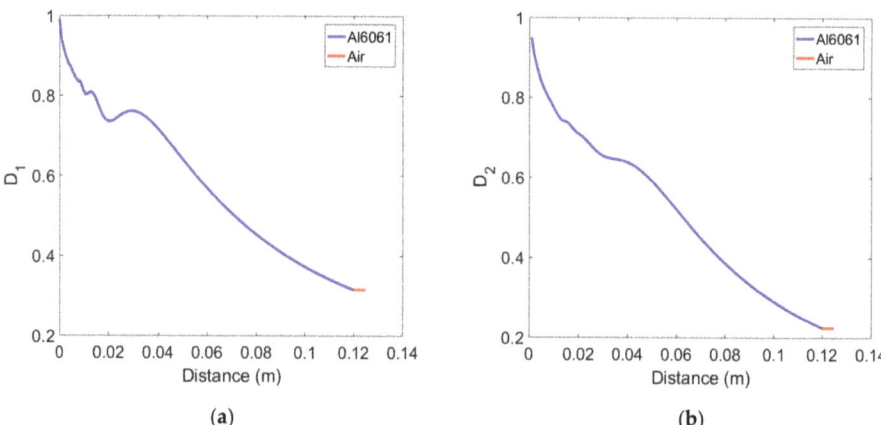

Figure 3. Variation of diffraction correction in the two-layer medium composed of solid and air: (**a**) Fundamental wave, and (**b**) Second harmonic wave.

Attenuation corrections in each layer of the solid-air medium are defined by Equations (20) and (21). Using these equations, attenuation corrections were calculated for the Al6061, 12 cm thick sample, and the results for the fundamental and second harmonics as a function of propagation distance are shown in Figure 4a,b, respectively. In the first solid layer, the fundamental wave exponentially decreases, and the second harmonic shows a slightly greater decrease in amplitude at the same propagation distance than the fundamental wave. The attenuation correction in the second air layer shows a very rapid decrease compared to the first solid layer in both the fundamental wave and the second harmonic wave, which is due to the very high attenuation coefficient of the air layer. The amplitude reduction of the second harmonic due to the attenuation is very severe compared to the fundamental wave, which is caused by the attenuation coefficient proportional to the square of the frequency.

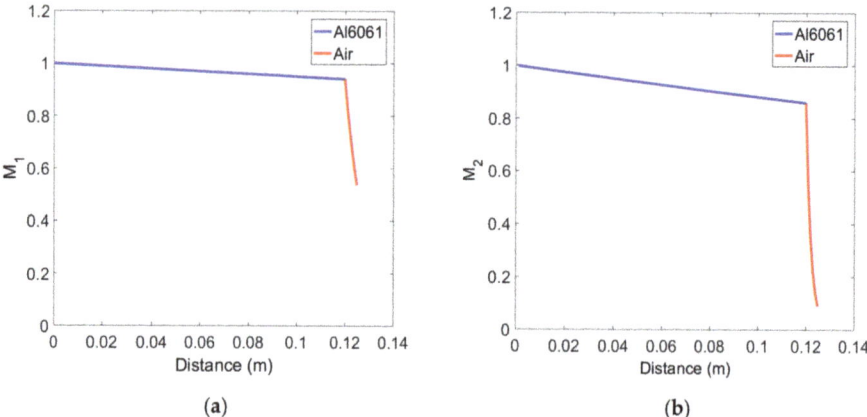

Figure 4. Variation of attenuation correction in the two-layer medium of solid-air: (**a**) Fundamental wave, and (**b**) Second harmonic wave.

4. Measurement of Relative Nonlinear Parameter β'

4.1. Experiment

Figure 5 shows the experimental setup for second harmonic generation measurement in the through-transmission mode using an air-coupled transducer as a receiver. The transmit transducer is a single crystal LiNbO$_3$ (Boston Piezo-Optics, Bellingham, MA, USA) of 12.7 mm diameter and 2 MHz center frequency. The receive transducer is an air-coupled transducer (NCT2-D13, Ultran Group, State College, PA, USA) of 12.7 mm diameter and 2 MHz center frequency. The two transducers are aligned with each other for maximum output signal capture. The entire propagation region is composed of two layers: solid specimen and air. The air gap from the solid specimen end to the receiver face is fixed at 5 mm. In the transmission side, a toneburst of 20 cycles tuned to the fundamental frequency (2 MHz) is supplied by a function generator (33250A, Agilent Technologies, Inc., Santa Clara, CA, USA), and then amplified by a linear amplifier (2100L, Electronics & Innovation, Ltd., Rochester, NY, USA) to provide a high-power monochromatic toneburst for harmonic generation in the solid specimen. In the reception side, the receive transducer is directly connected with a current probe (Tektronix CT-2, Tektronix, Inc., Wilsonville, OR, USA) and digitized using a digital oscilloscope (LT 332, LeCroy, Inc., Chestnut Ridge, NY, USA).

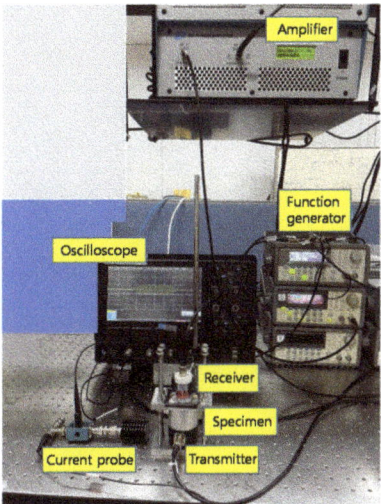

Figure 5. Experimental setup for harmonic generation measurement in the through-transmission mode using a contact transmitter and an air-coupled receiver.

The purpose of this experiment is to accurately obtain the relative nonlinear parameter, β', of aluminum specimens by applying the attenuation and diffraction corrections to the measured current output. Then, the ratio of relative nonlinear parameters of two different materials can be obtained accordingly, which is equal to the ratio of absolute nonlinear parameters.

4.2. Specimen

Three types of aluminum specimens: Al2024, Al6061 and Al7075, were selected, and five different thicknesses in each Al type were prepared: 4 cm, 6 cm, 8 cm, 10 cm, and 12 cm. The shape of the specimen is a block of circular cross section, 5 cm in diameter.

It has been known that the measured value of nonlinear parameter is affected by the parallelism of the top and bottom surfaces of the specimen. The surface roughness of the specimen also has a direct influence on the accuracy of the measurement of nonlinear parameters [37]. In order to minimize the effect of surface roughness, it is necessary to maintain the same surface roughness on each specimen as much as possible. The prepared specimens were machined so that the upper and lower surfaces were parallel. The surface roughness of each specimen was maintained at the same level as possible using a metal abrasive. Table 2 shows the surface roughness of each specimen. The surface roughness was measured 5 times per specimen and then the mean value was calculated. The mean roughness for all specimens is 0.11. Figure 6 shows the representative aluminum specimens after surafce machining.

Table 2. Surface roughness of aluminum specimens (Unit: μm).

Materials	Specimen Thickness (cm)				
	4	6	8	10	12
Al2024	0.12	0.10	0.13	0.09	0.12
Al6061	0.11	0.09	0.16	0.10	0.10
Al7075	0.11	0.10	0.14	0.12	0.10

Figure 6. Aluminum specimens after surface machining.

5. Results and Discussion

Figure 7a,b show the input waveform and its frequency spectrum. Here, the toneburst waveform consisting of twenty cycles produces a narrowband spectrum with a center frequency of 2 MHz. In nonlinear ultrasonic measurements, the second harmonic amplitude is generally two or three orders of magnitude lower than the fundamental wave amplitude. When using the air-coupled receiver, it may be more difficult to receive the second harmonic component because of low transmission efficiency at the solid-air interface and high attenuation loss in the air. Therefore, the possibility of receiving the second harmonic component was confirmed through spectrum analysis of the output signal received from each specimen. Figure 7c,d show the received signal and its frequency spectrum measured on Al6061, 12 cm thick specimen. From the frequency spectrum, we can see that not only the second harmonic component but also the third harmonic component are well received.

Next, the relative nonlinear parameter β' was measured for each specimen. During the experiment, it was observed that the output signal was very noisy due to the influence of external variables such as vibration of the experiment table and alignment of the probes. Therefore, in order to increase the signal-to-noise ratio of the output signal, 500 summed averages were used. Figure 8 shows the relative nonlinear parameter β' obtained from the three types of aluminum specimens before and after corrections for attenuation and diffraction. The β' before all corrections appears differently according to the thickness, and it tends to increase with the thickness. Since the changes in the fundamental and second harmonic amplitudes according to the thickness change have already been reflected in the calculation of β', it can be thought that increasing β' with increasing thickness is due to the diffraction and attenuation effects. After applying all corrections, β' in each alloy system shows an almost constant value regardless of thickness. These results demonstrate the importance of accurate corrections in measuring relative nonlinear parameters.

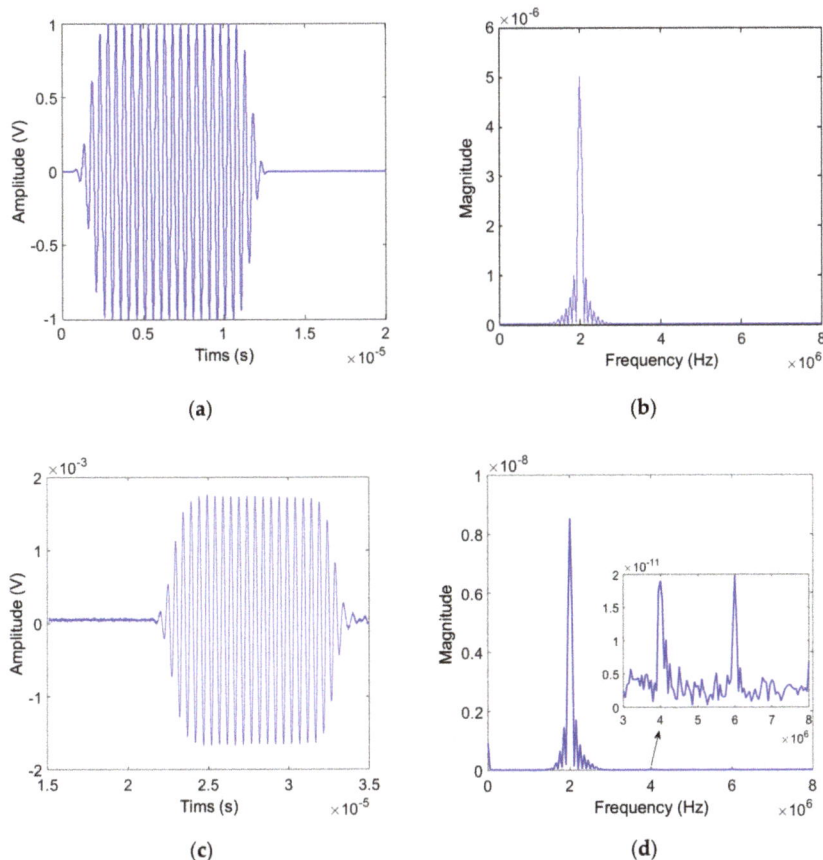

Figure 7. Input signal (**a**) Waveform and (**b**) Spectrum, and output signal (**c**) Waveform and (**d**) Spectrum.

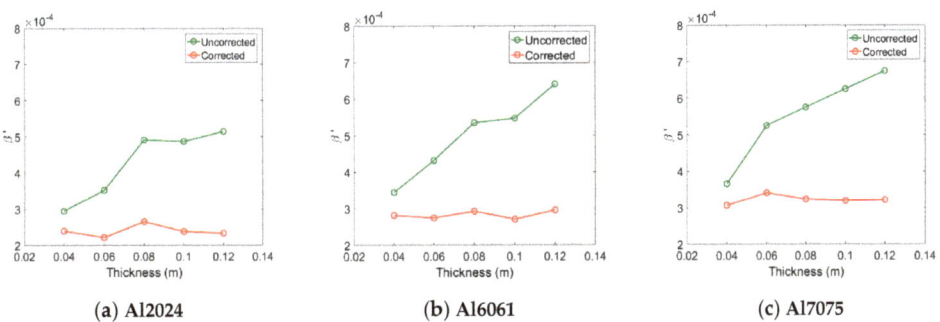

Figure 8. Measured β' before and after all corrections for three different aluminum systems.

When the three types of aluminum have the same thickness, the magnitude of the measured β' value before correction is approximately in the order of $\beta'_{7075} > \beta'_{6061} > \beta'_{2024}$. In each type of aluminum, the β' value after correction is almost constant regardless of the thickness, and the magnitude of the average β' value is also in this order, as shown in Table 3 below. The order of magnitude of the β' after correction coincides with the order of magnitude of the absolute nonlinear parameter (β) after the correction shown in Table 4.

Table 3. Measurement results of the relative nonlinear parameter β'.

Material	β'_{mean}	β'_{max}	β'_{min}
Al2024	2.39×10^{-4}	2.65×10^{-4}	2.21×10^{-4}
Al6061	2.82×10^{-4}	2.95×10^{-4}	2.70×10^{-4}
Al7075	3.22×10^{-4}	3.41×10^{-4}	3.06×10^{-4}

Table 4. Measurement results of the absolute nonlinear parameter β.

Material	β_{mean}	β_{max}	β_{min}
Al2024	4.78	5.13	4.66
Al6061	5.32	5.43	5.21
Al7075	6.46	6.68	6.31

Next, the uncorrected β' values measured for each aluminum system were normalized by the β' of the 4 cm specimen in each system and compared with the theoretical predictions. The comparisons between the two results are shown in Figure 9, and the overall agreement is found to be pretty good. The measured β' for each aluminum system was corrected for attenuation and diffraction, and then normalized by the β' of the 4 cm specimen in each system. The results are also shown in Figure 9. As can be expected, the normalized β' values in each alloy system are all close to one regardless of the specimen thickness. In fact, in each alloy system, the best fit line to the five normalized data is a horizontal line passing close to one. These results validate the proposed method of measuring material nonlinear parameters using an air-coupled transducer. These results also demonstrate the importance of accurate corrections in measuring relative nonlinear parameters.

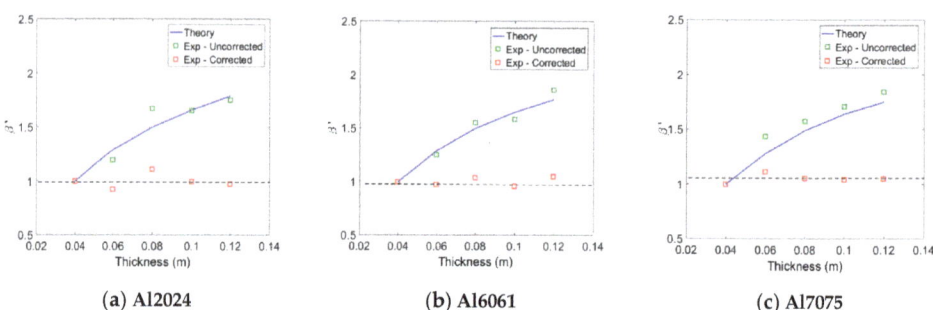

(a) Al2024 (b) Al6061 (c) Al7075

Figure 9. Comparison of normalized β' between the experiments and theory.

The mean, maximum, and minimum values were calculated for the corrected β' of each alloy system shown in Figure 8, and the results are listed in Table 3. The calculated mean β' for each system in Table 3 will be used as the β'_{ref} of the reference specimen or the β' of the target specimen in the calculation of the β'/β'_{ref} ratio in Table 5. Table 4 shows the absolute nonlinear parameter (β) measured by the through-transmission method for the same specimens in Table 3 using a contact transmitter and a contact receiver. In this case, it is necessary to use the calibrated receiver, and detailed measurement procedures are given in Ref. [33]. The mean β value for each system in Table 4 will be used as the reference value or target value in the calculation of the β/β_{ref} ratio. Below, we will compare the β'/β'_{ref} ratio and the β/β_{ref} ratio.

Table 5. Comparison of the relative nonlinear parameter ratio and absolute nonlinear parameter ratio.

Ratio	β'/β'_{ref}	β/β_{ref}	Difference (%)
Al6061/Al2024	1.18	1.11	6.3
Al7075/Al2024	1.35	1.35	0
Al2024/Al6061	0.85	0.90	5.6
Al7075/Al6061	1.14	1.21	5.8
Al2024/Al7075	0.74	0.74	0
Al6061/Al7075	0.88	0.82	7.3

Equation (25) indicates that the relative nonlinear parameter ratio and the absolute nonlinear parameter ratio are the same regardless of the type of reference specimen. Therefore, the equivalence of these two ratios was confirmed in the following way. Relative nonlinear parameter ratios were calculated for various alloy system combinations using the mean values of β' in Table 3. Similarly, absolute nonlinear parameter ratios were calculated using the mean values of β in Table 4. The results of these ratios for various combinations of specimen systems are presented in Table 5. Comparing these results, the overall agreement between the two ratios for all possible target and reference specimen combinations is within about 7%. In this study, we demonstrated the equivalence between the relative nonlinear parameter ratio and the absolute nonlinear parameter ratio using the reference samples whose acoustic impedances are not very different from the target samples. However, even if the difference in acoustic impedance between the reference sample and the target sample is large, the method can be applied in principle because only the diffraction and attenuation corrections need to be accurately calculated and applied.

Equation (26) shows that the absolute nonlinear parameter of a target specimen (β) can be obtained from the relative nonlinear parameter ratio of a target sample to a reference sample (β'/β'_{ref}) by multiplying the absolute nonlinear parameter of the reference specimen (β_{ref}). Based on this equation, the absolute β of each aluminum type was calculated using the β'/β'_{ref} in Table 5 and β_{ref} in Table 4. Table 6 contains the results of this calculation, and when compared with the directly measured β of Table 4, the overall agreement is within 7%. These results show that the absolute nonlinear parameter can also be obtained within the same level of error as the absolute nonlinear parameter ratio previously observed in Table 5.

Table 6. Absolute nonlinear parameter obtained from the relative nonlinear parameter ratio of two materials.

Ratio	β'/β'_{ref}	$\frac{\beta'}{\beta'_{ref}} \times \beta_{ref} = \beta$	Difference (%)
Al2024/Al6061	0.85	4.52	5.7
Al2024/Al7075	0.74	4.78	0
Al6061/Al2024	1.18	5.64	6.0
Al6061/Al7075	0.88	5.68	6.8
Al7075/Al2024	1.35	6.45	0.2
Al7075/Al6061	1.14	6.06	6.6

In principle, there are no restrictions on the selection of reference specimens when measuring β'/β'_{ref}, but a standardized reference specimen is more preferable. In linear ultrasonic testing, reference standards are mainly used to to establish a general level of consistency in measurements and to calibrate instruments, and a wide variety of standard calibration blocks of different designs, sizes and system of units are available [38]. However, such reference blocks are not yet available in nonlinear ultrasonic testing. The development of standardized reference blocks related to the measurement of nonlinear parameters will be another subject of future work.

6. Conclusions

This paper has covered the measurement of absolute nonlinear parameters of solid specimens using an air coupled transducer. The relative nonlinear parameter (β') was measured for three types of aluminum specimens with various thicknesses, and the following conclusions can be drawn:

- The ratio of the relative nonlinear parameters (β'/β'_{ref}) and the ratio of the absolute nonlinear parameters (β/β_{ref}) matched well within 6–7% for different target and reference specimen combinations.
- The absolute nonlinear parameter (β) obtained from the β'/β'_{ref} also agreed with the directly measured β.
- The proposed method does not require calibration of the air-coupled receiver, and there are no restrictions on the type and thickness of the reference specimen.
- The measurement of the β of a target specimen requires the β of the reference specimen.
- The received signal from the air-coupled transducer can be affected by the surface roughness of the specimen, alignment of the transmitter and receiver, and vibration of the experiment table.
- The use of low frequencies is relatively inefficient in second harmonic generation, and it is not easy to apply them to thin specimens.

Author Contributions: Manuscript preparation and writing, H.J.; Experiment and data reduction, S.C.; Theory development and simulation, S.Z. and X.L. All authors have read and agreed to the published version of the manuscript.

Funding: This work was supported by Wonkwang University in 2018.

Data Availability Statement: The data presented in this study are available on request from the corresponding author.

Conflicts of Interest: The authors declare no conflict of interest.

Abbreviations

a	Transmitter radius
b	Receiver radius
$A(\omega)$	Magnitude spectrum
$\alpha_n^{(m)}$	Attenuation coefficient of nth harmonic in mth layer
β	Absolute nonlinear parameter
β'	Relative nonlinear parameter
β_{ref}	Absolute nonlinear parameter of reference specimen
β'_{ref}	Relative nonlinear parameter of reference specimen
$c^{(m)}$	Longitudinal wave velocity in mth layer
C_{11}	Second order stiffness component
C_{111}	Third order stiffness component
$D_n^{(m)}$	Diffraction correction of nth harmonic in mth layer
f	Frequency
$G_n^{(m)}$	Green function of nth harmonic in mth layer
$H(\omega)$	Transfer function of receiving transducer
$k_n^{(m)}$	Wave number of nth harmonic in mth layer
$M_n^{(m)}$	Attenuation correction of nth harmonic in mth layer
r_n	Distance from source to target point for nth harmonic
T_{12}	Transmission coefficient at the interface of medium 1 and medium 2
$u_n^{(m)}$	Displacement of nth harmonic in mth layer
$\tilde{u}_n^{(m)}$	Received average displacement of nth harmonic in mth layer
$U_n^{(m)}$	Displacement amplitude of nth harmonic in mth layer

U_0	Source displacement
z_m	Thickness of mth layer
Z_m	Acoustic impedance of mth layer
(x, y, z)	Cartesian coordinates

References

1. Jeong, H.; Nahm, S.-H.; Jhang, K.-Y.; Nam, Y.-H. A nondestructive method for estimation of the fracture toughness of CrMoV rotor steels based on ultrasonic nonlinearity. *Ultrasonics* **2003**, *41*, 543–549. [CrossRef]
2. Kim, J.-Y.; Jacobs, L.J.; Qu, J.; Littles, J.W. Experimental characterization of fatigue damage in a nickel-base superalloy using nonlinear ultrasonic waves. *J. Acoust. Soc. Am.* **2006**, *120*, 1266–1273. [CrossRef]
3. Matlack, K.H.; Wall, J.J.; Kim, J.-Y.; Qu, J.; Jacobs, L.J.; Viehrig, H.-W. Evaluation of radiation damage using nonlinear ultrasound. *J. Appl. Phys.* **2012**, *111*, 054911. [CrossRef]
4. Matlack, K.H.; Kim, J.-Y.; Wall, J.J.; Qu, J.; Jacobs, L.J.; Sokolov, M.A. Sensitivity of ultrasonic nonlinearity to irradiated, annealed, and re-irradiated microstructure changes in RPV steels. *J. Nucl. Mater.* **2014**, *448*, 26–32. [CrossRef]
5. Frouin, J.; Sathish, S.; Matikas, T.E.; Na, J.K. Ultrasonic linear and nonlinear behavior of fatigued Ti-6Al-4V. *J. Mater. Res.* **1998**, *14*, 1295–1298. [CrossRef]
6. Viswanath, A.; Rao, B.P.C.; Mahadevan, S.; Jayakumar, T.; Baldev, R. Microstructural characterization of M250 grade maraging steel using nonlinear ultrasonic technique. *J. Mater. Sci.* **2010**, *45*, 6719–6726. [CrossRef]
7. Gauster, W.B.; Breazeale, M.A. Ultrasonic measurement of the nonlinearity parameters of copper single crystals. *Phys. Rev.* **1968**, *168*, 655–661. [CrossRef]
8. Gauster, W.B.; Breazeale, M.A. Detector for measurement of ultrasonic strain amplitudes in solids. *Rev. Sci. Instrum.* **1966**, *37*, 1544–1548. [CrossRef]
9. Thompson, R.B.; Buck, O.; Thompson, D.O. Higher harmonics of finite amplitude ultrasonic waves in solids. *J. Acoust. Soc. Am.* **1976**, *59*, 1087–1094. [CrossRef]
10. Mostavi, A.; Kamali, N.; Tehrani, N.; Chi, S.-W.; Ozevin, D.; Indacochea, J.E. Wavelet based harmonics decomposition of ultrasonic signal in assessment of plastic strain in aluminum. *Measurement* **2017**, *106*, 66–78. [CrossRef]
11. Barnard, D.J.; Dace, G.E.; Buck, O. Acoustic harmonic generation due to thermal embrittlement of Inconel 718. *J. Nondestruct. Eval.* **1997**, *16*, 67–75. [CrossRef]
12. Breazeale, M.A.; Philip, J. Determination of third-order elastic constants from ultrasonic harmonic generation measurements. In *Physical Acoustics*; Mason, W.P., Thurston, R.N., Eds.; Academic Press: New York, NY, USA, 1984; Volume XVII, pp. 1–60.
13. Cantrell, J.H.; Salama, K. Acoustoelastic characterization of materials. *Int. Mater. Rev.* **1991**, *36*, 125–145. [CrossRef]
14. Yost, W.T.; Cantrell, J.H. Anomalous nonlinearity parameters of solids at low acoustic drive amplitudes. *Appl. Phys. Lett.* **2009**, *94*, 021905. [CrossRef]
15. Hurley, D.C.; Fortunko, C.M. Determination of the nonlinear ultrasonic parameter β using a Michelson interferometer. *Meas. Sci. Technol.* **1997**, *8*, 634–642. [CrossRef]
16. Hurley, D.C.; Balzar, D.; Purtscher, P.T. Nonlinear ultrasonic assessment of precipitation hardening in ASTM A710 steel. *J. Mater. Res.* **2000**, *15*, 2036–2042. [CrossRef]
17. Moreau, A. Detection of acoustic second harmonics in solids using a heterodyne laser interferometer. *J. Acoust. Soc. Am.* **1995**, *98*, 2745–2752. [CrossRef]
18. Yost, W.T.; Cantrell, J.H.; Kushnick, P.W. Fundamental aspects of pulse phase-locked loop technology-based methods for measurement of ultrasonic velocity. *J. Acoust. Soc. Am.* **1992**, *91*, 1456–1468. [CrossRef]
19. Dace, G.E.; Thompson, R.B.; Buck, O. Measurement of the acoustic harmonic generation for materials characterization using contact transducers. In *Review of Progress in Quantitative Nondestructive Evaluation*; Thompson, D.O., Chimenti, D.E., Eds.; Plenum Press: New York, NY, USA, 1992; pp. 2069–2076.
20. Chakrapani, S.K.; Barnard, D.J. A calibration technique for ultrasonic immersion transducers and challenges in moving towards immersion based harmonic imaging. *J. Nondestruct. Eval.* **2019**, *38*, 76. [CrossRef]
21. Cantrell, J.H.; Breazeale, M.A. Ultrasonic investigation of the nonlinearity of fused silica for different hydroxyl-ion contents and homogeneities between 300 and 3K. *Phys. Rev. B* **1978**, *17*, 4864–4870. [CrossRef]
22. Matlack, K.H.; Kim, J.-Y.; Jacobs, L.J.; Qu, J. Review of second harmonic generation measurement techniques for material state determination in metals. *J. Nondestruct. Eval.* **2015**, *34*, 273. [CrossRef]
23. Thiele, S.; Kim, J.-Y.; Qu, J.; Jacobs, L.J. Air-coupled detection of nonlinear Rayleigh surface waves to assess material nonlinearity. *Ultrasonics* **2014**, *54*, 1470–1475. [CrossRef]
24. Torello, D.; Thiele, S.; Matlack, K.; Kim, J.-Y.; Qu, J.; Jacobs, L.J. Diffraction, attenuation, and source corrections for nonlinear Rayleigh wave ultrasonic measurements. *Ultrasonics* **2015**, *56*, 417–426. [CrossRef] [PubMed]
25. Zhang, S.; Dai, Z. Determining the Sensitivity of Air-Coupled Piezoelectric Transducers Using a Comparative Method—Theory and Experiments. 2019. Available online: https://ieee-dataport.org/documents/determining-sensitivity-air-coupled-piezoelectric-transducers-using-comparative-method (accessed on 9 December 2020).
26. Cho, S.; Jeong, H.; Zhang, S.; Li, X. Dual element transducer approach for second harmonic generation and material nonlinearity measurement of solids in the pulse-echo method. *J. Nondestruct. Eval.* **2020**, *39*, 62. [CrossRef]

27. Jeong, H.; Cho, S.; Shin, H.; Zhang, S.; Li, X. Optimization and validation of dual element ultrasound transducers for improved pulse-echo measurements of material nonlinearity. *IEEE Sens. J.* **2020**, *20*, 13596–13606. [CrossRef]
28. Norris, A.N. Finite-amplitude waves in solids. In *Nonlinear Acoustics*; Hamilton, M.F., Blackstock, D.T., Eds.; Acoustical Society of America: Melville, NY, USA, 2008; pp. 263–277.
29. Schmerr, L.W., Jr.; Song, S.-J. *Ultrasonic Nondestructive Evaluation System*; Springer: Berlin/Heidelberg, Germany, 2007.
30. Cantrell, J.H. Fundamental and applications of nonlinear ultrasonic nondestructive evaluation. In *Ultrasonic and Electromagnetic NDE for Structure and Material Characterization*; Kundu, T., Ed.; CRC Press: Boca Raton, FL, USA, 2013; pp. 35–456.
31. Rogers, P.H.; Van Buren, A.L. An exact expression for the Lommel diffraction correction. *J. Acoust. Soc. Am.* **1974**, *55*, 724–728. [CrossRef]
32. Jeong, H.; Zhang, S.; Cho, S.; Li, X. Development of explicit diffraction corrections for absolute measurements of acoustic nonlinearity parameters in the quasilinear regime. *Ultrasonics* **2016**, *70*, 199–203. [CrossRef] [PubMed]
33. Jeong, H.; Barnard, D.; Cho, S.; Zhang, S.; Li, X. Receiver calibration and the nonlinearity parameter measurement of thick solid samples with diffraction and attenuation corrections. *Ultrasonics* **2017**, *81*, 147–157. [CrossRef]
34. Beyer, R.T. The Parameters B/A. In *Nonlinear Acoustics*; Hamilton, M.F., Blackstock, D.T., Eds.; Acoustical Society of America: Melville, NY, USA, 2008; pp. 25–39.
35. Bond, L.J.; Chiang, C.-H.; Fortunko, C.M. Absorption of ultrasonic waves in air at high frequencies (10–20 MHz). *J. Acoust. Soc. Am.* **1992**, *92 Pt 1*, 2006–2015. [CrossRef]
36. Jeong, H.; Zhang, S.; Li, X. A novel method for extracting acoustic nonlinearity parameters with diffraction corrections. *J. Mech. Sci. Technol.* **2016**, *30*, 643–652. [CrossRef]
37. Chakrapani, S.K.; Howard, A.; Barnard, D. Influence of surface roughness on the measurement of acoustic nonlinearity parameter of solids using contact piezoelectric transducers. *Ultrasonics* **2018**, *84*, 112–118. [CrossRef]
38. Krautkramer, J.; Krautkramer, H. *Ultrasonic Testing of Materials*; Springer: Berlin/Heidelberg, Germany, 1990.

Article

Criticality Hidden in Acoustic Emissions and in Changing Electrical Resistance during Fracture of Rocks and Cement-Based Materials

Gianni Niccolini [1], Stelios M. Potirakis [2], Giuseppe Lacidogna [1,*] and Oscar Borla [1]

[1] Department of Structural, Geotechnical and Building Engineering, Politecnico di Torino, C.so Duca degli Abruzzi 24, 10129 Torino, Italy; gianni.niccolini@polito.it (G.N.); oscar.borla@polito.it (O.B.)
[2] Department of Electrical and Electronics Engineering, University of West Attica, 12244 Egaleo, GR-12244 Athens, Greece; spoti@uniwa.gr
* Correspondence: giuseppe.lacidogna@polito.it

Received: 25 October 2020; Accepted: 1 December 2020; Published: 9 December 2020

Abstract: Acoustic emissions (AE) due to microcracking in solid materials permit the monitoring of fracture processes and the study of failure dynamics. As an alternative method of integrity assessment, measurements of electrical resistance can be used as well. In the literature, however, many studies connect the notion of criticality with AE originating from the fracture, but not with the changes in the electrical properties of materials. In order to further investigate the possible critical behavior of fracture processes in rocks and cement-based materials, we apply natural time (NT) analysis to the time series of AE and resistance measurements, recorded during fracture experiments on cement mortar (CM) and Luserna stone (LS) specimens. The NT analysis indicates that criticality in terms of electrical resistance changes systematically precedes AE criticality for all investigated specimens. The observed greater unpredictability of the CM fracture behavior with respect to LS could be ascribed to the different degree of material homogeneity, since LS (heterogeneous material) expectedly offers more abundant and more easily identifiable fracture precursors than CM (homogenous material). Non-uniqueness of the critical point by varying the detection threshold of cracking events is apparently due to finite size effects which introduce deviations from the self-similarity.

Keywords: acoustic emission; electrical resistance; damage monitoring; criticality; natural time analysis

1. Introduction

A crucial question of scientists and civil engineers concerns the use of the electrical properties of geological and engineering materials as potential precursors of structural collapses and earthquakes [1–9]. The behavior of electrical properties has been used for several years in induced polarization, resistivity and electromagnetic methods in the context of likely mechanisms, for both piezoelectric and non-piezoelectric materials [10,11]. There are many experimental techniques available for rocks, ionic crystals and concrete-like materials, including the observation of changes in electrical properties, e.g., electrical resistance or resistivity [12–16], and of electrical and electromagnetic signals as functions of the applied external load. A combination of techniques involves also the observation of electrical properties as functions of applied electromagnetic (EM) fields (to test for damage-induced voltage-current non-linearity) and of EM frequency, including variations in relevant environmental parameters (temperature, water content, etc.) [17,18].

During mechanical loading of materials, fracture-induced electrical currents, acoustic emissions (AE) and electromagnetic emissions (EME) allow the real-time monitoring of damage evolution [8–12,19,20]. While the origin of AE from materials experiencing damage is well understood [21,22], different models have been proposed to explain the genesis of electrical and electromagnetic signals related to irreversible phenomena, such as the formation of electrical charges due to the breakage of bonds [23], the discharge model [24], the capacitor model [25], the surface oscillation model [26] and the moving charge dislocation model [27]. During compression tests, stress-induced polarization currents—attributed either to the well-known piezoelectric effect for polycrystalline natural rocks with piezoelectric properties (granite, quartzite), or to the moving segments of charged edge dislocations for non-piezoelectric ionic crystals (LiF) [10,11]—are detected in the sample. In complex materials (granite) containing quartz inclusions, AE originating from the microfracturing process stimulate damping vibrations of the quartz grains—polarized due to the high stress levels—which act as sources of EME. As far as non-piezoelectric materials (pure ionic crystals) are concerned, the motion of segments of charged dislocations—piling up during crack initiation and propagation—with respect their compensating point defects are held responsible for generating EME.

Recent accumulated laboratory evidence indicates that the generation of freshly formed fracture surfaces, due to opening cracks, is accompanied by simultaneous EME and AE, whereas AE signals that are not associated with EME signals are due to frictional noises during the slip between pre-formed fracture surfaces [28,29]. For this reason, EME is being increasingly considered as a precursor signal, since it is argued that it is produced only during the generation of new fresh surfaces/rupture of bonds, due to cracking in the material. In this regard, it has been observed that the larger the stress drops, the more intense the EME activity [30].

Furthermore, it should be stressed the compatibility of recently performed laboratory fracture experiments on rocks and ionic crystals with the processes occurring in the Earth's crust during the earthquake preparatory stage. The experimental evidence reveals that the final stage of the failure process coincided in time with the maximum of AE and quiescence in EME, while strong avalanche-like EME precedes this phase. Then, an EME silence is observed just before the final collapse in the laboratory, as well as at the geophysical scale before the seismic shock [8–11].

Stress-induced currents and EME are detected also from cement-based materials under compressive loading, where electrical double layer formation and motion of ions have been proposed as the possible causes of the observed EME: layers of ions in the water inside the capillary pores accelerate —as the pore solution moves upon loading—with respect to oppositely charged layers in the solid region, resulting in a time-varying dipole moment which generates EME [31–34]. Since the electrical conduction of rocks and mortar is largely electrolytic [35], the electrical resistance in relatively dry materials should reasonably increase as a result of microcracking, which breaks the existing conductive network within the material. Since growing microcracks are AE sources, a correlation between electrical resistance changes and AE bursts are eventually expected.

The present goal is to investigate the correlation between electrical resistance and AE measurements, carried out in air-dry surface cement mortar and rock specimens subjected to fracture tests. Here, the application of the well-known AE technique aims to verify the reliability of the electrical resistance measurement, which would enable damage monitoring with simple and inexpensive equipment.

In order to further investigate the possible critical behavior of fracture processes through their observables, the AE and the electrical resistance, we proceed here to the analysis of AE and electrical resistance time series using first classical definitions of damage in Kachanov's sense [36] and then natural time (NT) analysis [37–39], a recently proposed method in the framework of critical phenomena [8,40–43]. The following sections describe the experimental setup and the fracture experiments performed, as well as damage measurements based on the acquired AE and electrical signals; then, a brief description of the key concepts and basic formulas of the NT analysis method; after that, the NT analysis of the AE and electrical resistance signals, and finally a summary of the main findings.

2. Experimental Setup—AE Signals and Electrical Resistance Changes

A schematic diagram of the experimental setup and pictures of the test specimens are shown in Figures 1 and 2. The experiments were carried out on three rods of Luserna stone (with fixed height,

50 mm, and variable diameter, 52 mm for two specimens and 25 mm for the remaining one), and two blocks of cement mortar (section 40 × 40 mm², height 160 mm) enriched with ferric oxide to improve the electrical conductivity. Using different shapes (cylindrical vs. prismatic) necessitates some corrections to make the results comparable, as the post-peak behavior of prisms under compression is more ductile than that of cylinders. Such shape effects were avoided by making cement mortar specimens more slender (slenderness equal to 160/40 = 4) in order to induce a brittle collapse once the peak stress was reached.

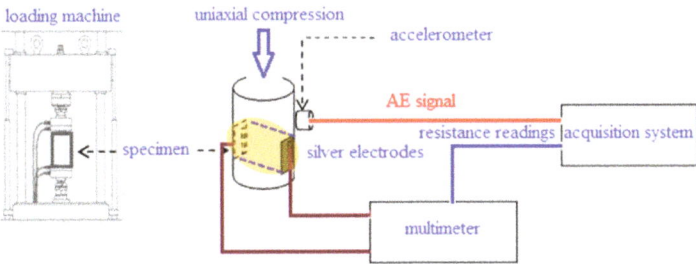

Figure 1. Experimental setup (the yellow region between the electrodes, marked by a dotted outline, defines the volume through which most of electric charges are expected to flow). Adapted from [44].

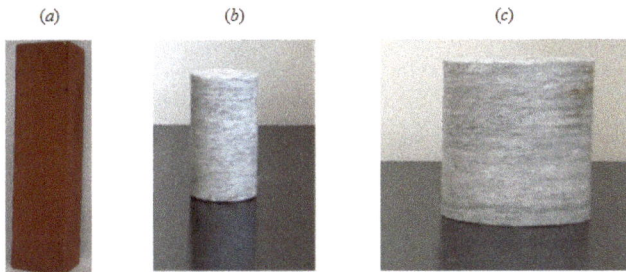

Figure 2. Blocks of cement mortar (**a**) and rods of Luserna stone (**b**,**c**).

Since mortar (as well as Luserna stone) has a high content of electrical insulator (silicon dioxide with electrical resistivity equal to 10^{14} Ω cm), an increase of 10% of ferric oxide (electrical resistivity equal to 10^{9} Ω cm) enhances its electrical conductivity. The chemical composition of both materials is reported in Table 1.

Table 1. Chemical composition of mortar enriched with ferric oxide and Luserna stone. Weight percentage of silicon dioxide and ferric oxide added (mortar) is highlighted.

Mortar		Luserna Stone	
Element	% of Weight	Element	% of Weight
SiO_2	59.7	SiO_2	72.0
CaO	21.4	Al_2O_3	14.4
Fe_2O_3	8.4	K_2O	4.1
Al_2O_3	3.3	Na_2O	3.7
SO_3	1.1	CaO	1.8
K_2O	1.0	FeO	1.7
MgO	0.7	Fe_2O_3	1.2
Na_2O	0.4	other oxides	1.1
other oxides	4.0		

Both specimens were put under uniaxial compression till macroscopic fracture, using a 500 kN servo-hydraulic loading machine equipped with electronic control in order to conduct tests at constant displacement rate (1 µm s^{-1} applied to Luserna stone and 2 µm s^{-1} to cement mortar). Such low rates were used to induce a relatively more ductile response from the specimen, as being characterized by a number of AE signals—suitable for statistical analysis—greater than that generally observed at higher loading (or strain) rates. The electrical resistance of the specimens was measured by the two-electrode technique, using an Agilent 34411 A multimeter capable of measuring resistances as high as 1 GΩ. After placing a constant voltage V in series with the multimeter and the unknown resistance, the current flowing through the specimen was measured, thus yielding the resistance from the Ohm's law. Brass screws and copper wires served as electrical contacts, placed on opposite faces of the specimen where a 30 × 30 mm^2 area was coated with a conducting silver paint in order to minimize the contact resistance (see Figure 1, and [44] for further details).

The electrical resistance R_0 of each virgin specimen (shown in Figure 2a–c) was measured at the beginning of the test. Then, the electrical resistance R of the damaged specimen was measured up to fracture, at a sampling rate of 25 Hz. The reported values were obtained by averaging over 100 samples and expressed in terms of R/R_0.

Acoustic emissions were measured by a calibrated accelerometer (charge sensitivity 9.20 pC/m s^{-2}) at a frequency range from a few hertz to 10 kHz to detect low-frequency signals generated by larger fractures, generally occurring as the failure is approaching [45,46] and held responsible for the breaking of material's conductive network. Previous experimental campaigns with transducers working in different frequency ranges (0.1–10 vs. 50–500 kHz) demonstrated a systematic reduction of the AE signal frequencies over damage accumulation [45].

AE signals were transmitted from the accelerometer to a 20 dB low-noise amplifier and then acquired at the audio sampling rate of 44.1 kHz by a sound card. Each AE signal was recorded through a hit-based method, where the HDT value was 68.1 µs (i.e., three times the delay 22.7 µs between the consecutive execution of sample taking): the end of the signal was recognized when the sum of signal readings per 68.1 µs or more descended under the threshold level. The HLT value was 68.1 µs (to avoid repeated recognitions of the same signal) and PDT was 45.4 µs.

The AE signals were characterized by the time of occurrence and the magnitude, expressed as $A_{dB} = 20 \log_{10} (A_{max}/ 1 \text{ µm s}^{-2})$, where A_{max} is the peak acceleration on the specimen surface produced by the AE wave [47]. Signal attenuation effects were not considered relevant because of the small distances involved (even if such effects, including scattering, become increasingly relevant during the damage process). The detection threshold was set to 40 dB in order to filter electrical disturbances and noisy signals, whereas a post-process FFT signal analysis was used to identify and filter the mechanical noise of the loading machine. Therefore, the AE time series, the accumulated number of AE events, the load history and the relative resistance R/R_0 are plotted in Figure 3.

Load vs. time diagrams of Figure 3a,b illustrate a brittle response by both mortar specimens. Despite the absence of relevant deviations from the linear elastic behavior, the increase in electrical resistance and AE activity revealed progressive damage accumulation within the specimen.

Except for one case depicted in Figure 3d, the load vs. time diagrams of the Luserna stone specimens were, on the contrary, characterized by a more ductile behavior, with numerous intermediate load drops (see Figure 3c,e) correlated to clusters of AE events and changes in electrical resistance, all signs revealing the preparatory damage stage of the specimen failure.

The different mechanical and electrical behaviors of the two tested materials were ascribed to their physical-mechanical properties: the Luserna stone has a heterogeneous and porous crystalline structure, whereas the cementitious specimen is an artificial material, compacted during the manufacturing process, and thus more homogeneous.

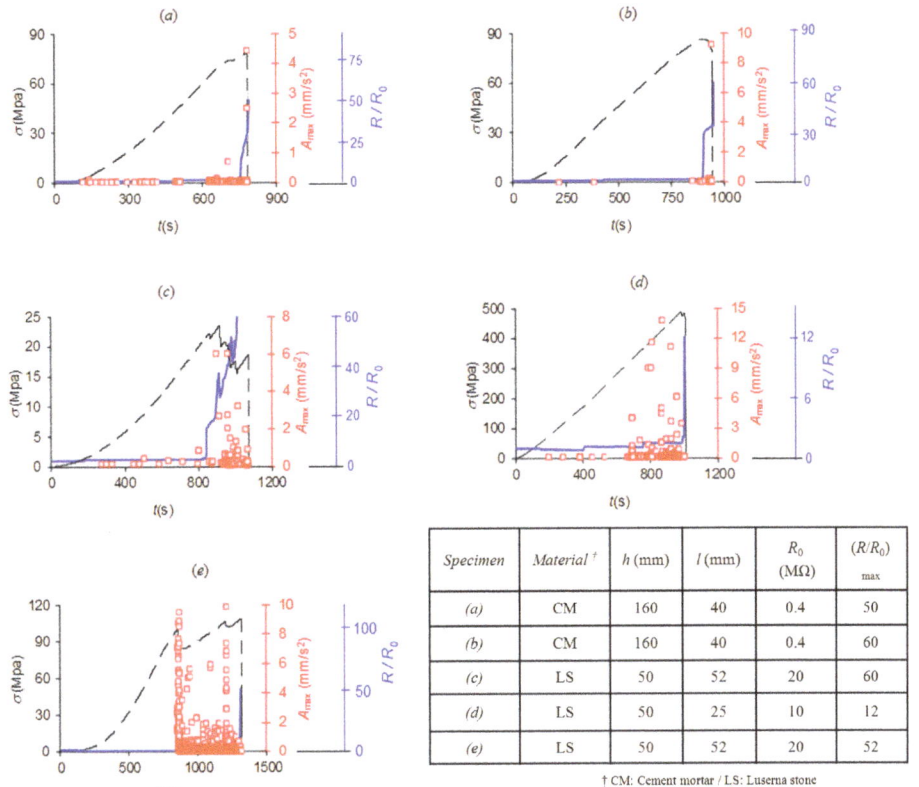

Figure 3. Variations over time of the applied stress (black dashed line), electrical resistance in R_0 units (blue solid line) and AE activity (red squares representing the peak acceleration of each AE event). The insert table provides information about the specimens corresponding to the data presented in (**a**–**e**).

3. Damage Measurements Based on AE and Electrical Resistance Time Series

The phenomenon of damage in rocks and concrete-like materials often consists of surface discontinuities in the form of inner microcracks (e.g., decohesions of interfaces between cement, sand and aggregates in concrete), or of volume discontinuities in the form of voids (responsible for measurable macroscopic volume changes). In the sense of continuum mechanics, such a discontinuous state is represented by a continuous damage variable D, representing the surface density of microcracks and cavities in any plane of a representative volume element V, i.e., the smallest volume on which a mean value of a defect characteristic may represent a field of discontinuous properties. If S is a cross-sectional area (with normal **n**) of V, with a total area $\overline{S}$ of the defect traces, $D_n = \overline{S}/S$ represents the local damage with respect to the **n**-direction. Considering only isotropic damage, i.e., in the case of defects without preferred orientation, damage is completely characterized by a scalar variable: $D_n = D$, ∀**n** for each volume element, although the accuracy of isotropic damage models generally decreases since material symmetries change during the loading process For rocks and concrete-like materials, the linear size of the representative volume element is of the order of 10.0 to 100.0 mm, which is that of the test specimens. Therefore, the evolving damage state of the current specimens is representable by a scalar function [48–50].

A full set of methods of non-direct damage measurement have been developed through its effects on strain properties. It is possible to consider the variable D as an internal variable in the sense of the thermodynamics of irreversible processes, where energy is released or dissipated in the damage processes of the creation of a new discontinuity (fracture) surface.

Considering a freshly formed crack of area S as a source of acoustic emissions, the accumulated damage can be expressed as a sum of acoustic emissions:

$$D \propto \sum_i 10^{m_i} \quad (1)$$

where $m = A_{dB}/20$ is the magnitude of an AE event, and it is proportional also to the logarithm of the source crack area S. The following relationships have been established [47]:

$$m \propto \frac{2}{3} c \log_{10} A_{max} \propto \log_{10} S \quad (2)$$

where the factor c depends on the type of transducer: $c = 1$ if the sensor acts as a strain-meter, $c = 1.5$ for a velocity transducer, and $c = 3$ in the case of an accelerometer. Here, being $c = 3$, the accumulated state of damage in terms of released energy by AE is given by the sum of the squared AE peak amplitudes, $D \propto \sum_i A_{max,i}^2$. The cumulative damage is normalized to one, where $D = 1$ is the maximum damage at the moment of failure.

In recent years, the correlation of electrical resistance with damage in solids has been investigated as well. As the failure is approached, the opening of micro- and macrocracks produces more void space in the material and, consequently, higher electrical resistance.

Considering the undamaged specimen before loading, the electrical resistance R_0 between the electrodes is given in terms of resistivity ρ by:

$$R_0 = \rho l / S \quad (3)$$

where l is the distance between the electrodes, i.e., the distance between opposite faces of the specimen, and S is the cross-sectional area of the cylindroid defined by the electrodes' surfaces (see the dashed blue contour in Figure 1) and through which, roughly, electrical charges flow.

Since an accurate calculation of the electrical resistance of the damaged specimen is extremely difficult, some simplifying assumptions appear to be necessary.

The electrical resistance of the considered volume, which experiences damage during the test, can be expressed as:

$$R = \frac{\rho' l'}{S'} = \frac{\rho l}{S(1-D)} \quad (4)$$

where changes in length and resistivity (the latter due to changes in porosity) are neglected, $l' = l$ and $\rho' = \rho$, the damage is assumed to be uniformly distributed along the axial length l and the cross-section S of the cylindroid. In this simplified model, changes in electrical resistance are entirely attributed to changes in the effective current-conducting area of the cylindroid, which is identified by its undamaged cross-sectional area, $S' = (1-D)S$. Thus, combining Equations (3) and (4) gives the simplest definition of the damage variable based on electrical resistance changes [46]:

$$D = 1 - R_0/R = \Delta R/R \quad (5)$$

where $\Delta R \equiv R - R_0$ is the increment in the electrical resistance between the undamaged state and a generic damaged state. Equation (5) correctly gives the initial value $R = R_0$ when $D = 0$, and $R \to \infty$ when $D = 1$ (namely infinite resistance at the specimen rupture).

The time series of electrical resistance measurements, $r_i(t) \equiv R_i(t)/R_0$, is transformed into a series of energy events—manageable with the NT analysis method—making the following considerations:

- Each increase in electrical resistance is due to the creation of a new discontinuity surface in the conductive network of the specimen. According to Equation (5), each electrical resistance measurement R_i is related to the effective current-conducting cross-sectional area S_i of the cylindroid by:

$$R_i = \rho l / S_i, \quad i = 0, 1, 2, \ldots \quad (6)$$

- The increase in electrical resistance $\Delta R_1 \equiv R_1 - R_0$, between the virgin state and the damaged state of the cylindroid, is related to the resulting surface of the freshly formed microcracks intersecting the cylindroid, expressed by $\Delta S_1 \equiv S_0 - S_1$. By exploiting Equation (6), it becomes:

$$\Delta S_1 = (\rho l / R_0)(1 - R_0 / R_1) = (\rho l / R_0)(1 - 1/r_1) \quad (7)$$

- The subsequent increase $\Delta R_2 \equiv R_2 - R_1$ is related to the corresponding crack surface advancement $\Delta S_2 \equiv S_1 - S_2$ by:

$$\Delta S_2 = (\rho l / R_1)(1 - R_1 / R_2) = (\rho l / R_0 r_1)(1 - r_1 / r_2) \quad (8)$$

- At the generic step, $\Delta R_i \equiv R_i - R_{i-1}$ is related to $\Delta S_i \equiv S_{i-1} - S_i$ by:

$$\Delta S_i = (\rho l / R_0 r_{i-1})(1 - r_{i-1} / r_i) \quad i = 1, 2, \ldots \quad r_0 \equiv 1 \quad (9)$$

If G_C is the toughness of the material, the amount of energy dissipated over this cracking step is calculated by means of fracture mechanics:

$$W_i = G_C \Delta S_i \quad (10)$$

- Therefore, the experimental time-varying electrical resistance values $r_i \equiv R_i / R_0$ are transformed into a time series of point-like energy events W_i, expressed as functions of r_i:

$$r_i \to \Delta S_i \propto W_i \propto \frac{1}{r_{i-1}}\left(1 - \frac{r_{i-1}}{r_i}\right) \quad (11)$$

Trends of accumulated damage expressed by Equations (1) and (5) and by

$$D \propto \sum_i \Delta S_i \propto \sum_i \frac{1}{r_{i-1}}\left(1 - \frac{r_{i-1}}{r_i}\right) \quad (12)$$

(where Equation (11) is inserted) are depicted in Figure 4.

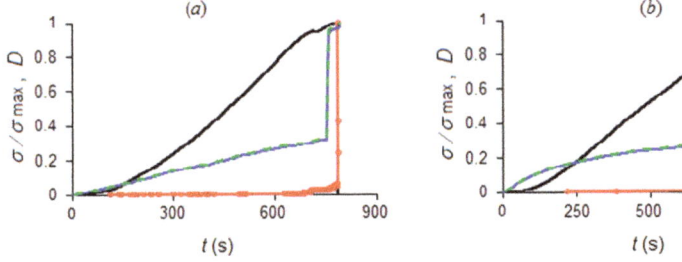

Figure 4. *Cont.*

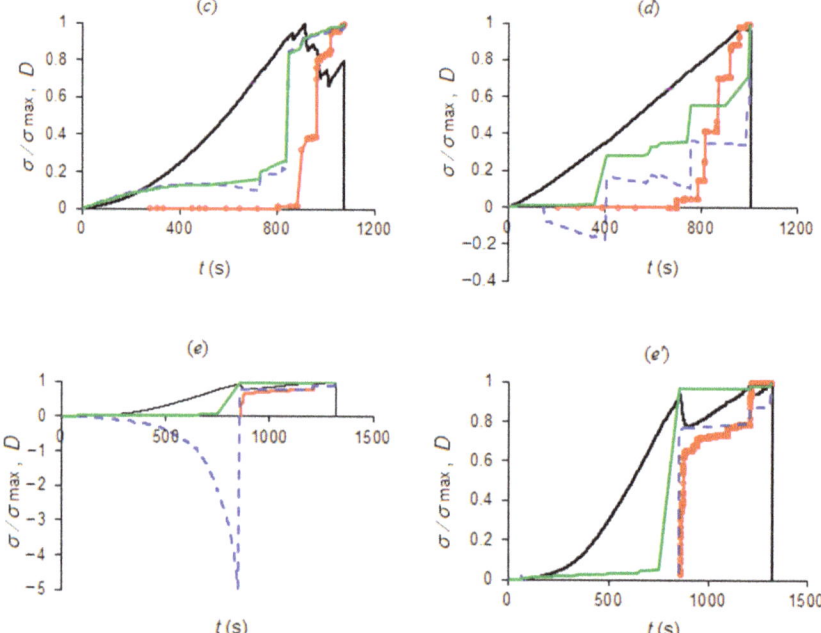

Figure 4. Normalized applied stress (black line) and damage evolution in terms of acoustic emissions (dotted red line); electrical resistance (dashed blue line); and crack surface advancements by indirect measurement (light green line). The diagrams refer to the specimens corresponding to the data presented in (**a**–**e**) (see inset table of Figure 3). The last picture (**e′**) is a detail of the positive half-plane (damage $D > 0$): negative values are due to temporary decrements in the electrical resistance of Luserna specimen (**e**).

Due to the dependence of rock resistivity on the porosity (also known as Archie's law [51], and here not explicitly considered), different trends in electrical resistivity were observed. The mortar specimens show a constant electrical resistance value up to the failure, whereas Luserna stone specimens are characterized by an initial decrease in the electrical resistance—presumably due to compaction caused by compressive loading—as emphasized by the negative part of the blue curves in Figure 4, representing damage in terms of electrical resistance according to Equation (5). They differ from the green curves, depicting damage in terms of freshly formed crack surfaces by Equation (12), where negative terms $\Delta R_i < 0$ are discarded. Such contributions, i.e., those with $1 - r_{i-1}/r_i < 0$, are due to specimens' compaction during the initial loading stages and are removed from the time series of the energy events, as being unrelated to the cracking process.

As it appears both in Figures 3 and 4, while the mortar specimens are characterized by a significant electrical resistance variation only at the failure, significant electrical resistance changes, caused by internal cracking of the rock, were observed in Luserna stone in coincidence with stress drops.

Despite the fact that bursts of AE activity and significant changes in electrical resistance are clearly precursors of specimen failure, the critical point does not seem easily identifiable. The purpose of the following sections is to find the approach to criticality hidden in the specimens' responses.

4. The Method of Natural Time Analysis

Although the NT method has been introduced for the analysis of SES (Seismic Electric Signals) and seismicity [52–54], it has been applied to a variety of signals providing optimal time-frequency enhancement [55]. The cornerstone of the NT analysis is the definition of a new time domain, the "natural time", in which the "events", i.e., the significant values of a time series, are equispaced, while their "energy" is retained, thus defining a new time series (χ_k, Q_k), where $\chi_k = k/N$ is the NT, Q_k is the energy of the k-th event, and N is the total number of successive events. In other words, the transformation of a time series from the conventional time to the NT focuses on the (normalized) order of occurrence of the events and ignores the time intervals between them [38].

According to the NT method, the approach to criticality is manifested by satisfaction of the following set of criteria: convergence of the parameter $\kappa_1 = \sum_{k=1}^{N} p_k \chi_k^2 - \left(\sum_{k=1}^{N} p_k \chi_k\right)^2$ to the value 0.070, while simultaneously the entropy in NT, $S_{nt} = \sum_{k=1}^{N} p_k \chi_k \ln \chi_k - \left(\sum_{k=1}^{N} p_k \chi_k\right) \ln\left(\sum_{k=1}^{N} p_k \chi_k\right)$, and the entropy under time-reversal, S_{nt-}, satisfy the condition $S_{nt}, S_{nt-} < S_u = \left(\frac{\ln 2}{2}\right) - \frac{1}{4} (\approx 0.0966)$, where $p_k = \frac{Q_k}{\sum_{n=1}^{N} Q_n}$ is the energy of the k-th event normalized by the total energy and S_u denotes the entropy in NT of a "uniform" noise [56,57]. This set of criteria has been successfully applied to a variety of unprocessed EM signals which are possibly earthquake (EQ) related, such as the SES [57–59] and the MHz fracto-EME [8,9,60,61], to reveal the approach of the underlying dynamical system to criticality. Both for models of dynamical systems (such as the Ising model and several models of self-organized criticality) and real systems, the value $\kappa_1 = 0.070$ is being considered to quantify the extent of systems' organization at the commencement of the critical stage [38].

Moreover, NT analysis is also applied to quantities, usually daily-valued ones, calculated from measured time series, such as recordings of ground-based magnetometers [56,62–64], and very low frequency (VLF) receiver recordings of sub-ionospheric propagation [65]. However, these quantities form time series of limited length (limited number of data), as happens with foreshock seismicity time series. In such cases, the set of criteria checked for the reveal of the approach to criticality is different and follows the paradigm of NT analysis of foreshock seismicity [52,54,57,66]. Finally, the identification of the approach to criticality in more complex systems calls for the investigation of the evolution of the entropy change ΔS ($= S_{nt} - S_{nt-}$) under time reversal [67,68], where the latter reference presents a methodology applicable (such as that in [69]) in specifically designed experiments to investigate AE activity in very long time series [68] or in different loading and unloading phases [69].

In more detail, for the case of foreshock seismicity, the evolution of specific NT analysis parameters versus time is studied by progressively including new events in the analysis until the time of occurrence of the main EQ event. These parameters are the already presented κ_1, S_{nt} and S_{nt-}, as well as the "average" distance $\langle D \rangle = \langle |\Pi(\omega) - \Pi_{critical}(\omega)| \rangle$ ($\omega = 2\pi\varphi$, with φ standing for the frequency in NT) between the curves of normalized power spectra $\Pi(\omega) = |\sum_{k=1}^{N} p_k exp(j\omega \chi_k)|^2$ of the evolving seismicity and the normalized power spectra at critical state calculated as $\Pi_{critical}(\omega) \approx 1 - \kappa_1 \omega^2$, for $\kappa_1 = 0.070$ [38]. Specifically, κ_1, S_{nt}, S_{nt-} and $\langle D \rangle$ are re-calculated in each step, based on the rescaled time series (χ_k, Q_k), as the total number N of the already included successive events is progressively increasing. In the resultant time evolution of κ_1, S_{nt}, S_{nt-} and $\langle D \rangle$, criticality is considered to be truly achieved when, at the same time [38,59], (i) κ_1 approaches $\kappa_1 = 0.070$ "by descending from above", (ii) $S_{nt}, S_{nt-} < S_u$, (iii) $\langle D \rangle < 10^{-2}$, and (iv) since the underlying process is expected to be self-similar, the time of criticality does not significantly change by varying the "magnitude" threshold.

In the application of seismicity NT analysis to other quantities, usually more than 20 threshold values equispaced between zero and 50% of the maximum value of the examined quantity are considered.

5. Analysis Results of Acoustic Emissions and Electrical Resistance Time Series

The application of NT analysis to AE acquired during laboratory experiments has already been addressed in [39], where NT analysis has been applied in direct analogy to the analysis of seismicity, as described in Section 4, while the quantity Q_k, the "energy" of each event, has been considered to be equal to the corresponding squared amplitude of each AE event, provided that this exceeds a certain threshold. In this work, for the case of the NT analysis of AE time series, we follow the same reasoning as [39]; specifically, we consider $Q_k = A_{max,k}^2$, provided that this is higher than a certain threshold

$\left(A_{max}^2\right)_{Th}$. However, NT analysis of electrical resistance appears for the first time here. As shown in Section 3, the amount of energy dissipated over a cracking event, W_i, is directly related to the measured resistance values (see Equation (11)).

Although the quantity Q_k in NT analysis corresponds to different physical quantities for various time series [38], an energy-related physical quantity is used for Q_k where possible. Therefore, for the case of electrical resistance time series, we consider $Q_k = |W_k|$, provided that this is higher than a certain threshold W_{Th}, while NT analysis is also applied in direct analogy to the analysis of seismicity (Section 4).

One typical example of the results obtained for each type of the recorded time series (AE and electrical resistance) is presented in Figures 5 and 6, respectively. These show the temporal evolution of the NT parameters κ_1, S_{nt}, S_{nt-} and $\langle D \rangle$ for four threshold values of AE (Figure 5) and electrical resistance (Figure 6) of the Luserna stone specimen presented in Figures 3c and 4c.

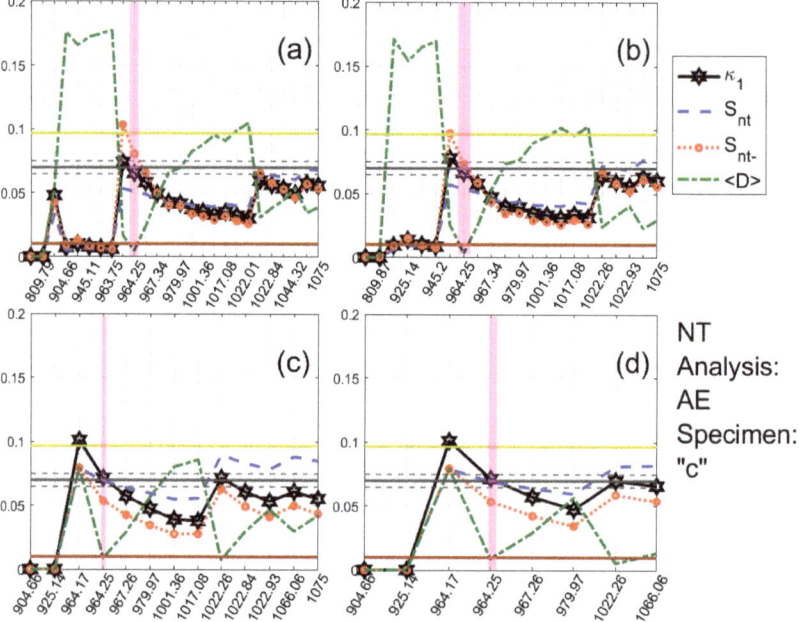

Figure 5. NT analysis of the AE time series acquired during the experiment involving specimen (c) of Figure 3. The different panels correspond to different threshold values $\left(A_{max}^2\right)_{Th}$: (a) 0.1, (b) 0.2, (c) 0.7 and (d) 2. Each panel shows, on a common vertical axis, the variation of the values of all parameters of the applied NT analysis vs. time in seconds (0 s corresponds to the initiation of the experiment). The vertical magenta patches highlight the time when criticality is approached for each threshold, i.e., when criticality conditions (cf. Section 4) are satisfied. For the ease of interpreting the results, the $\kappa_1 = 0.070$ value is shown as a solid grey horizontal line, while the following limit values have also been depicted by horizontal lines: $\langle D \rangle$ limit (10^{-2}), solid brown line; entropy limit $S_u (\approx 0.0966)$, solid light green. The horizontal grey dashed lines at 0.070 ± 0.005 define the limits of the zone within which a calculated κ_1 value is considered to be ≈ 0.070. Note that the employed events are presented equispaced in the horizontal axis following the NT representation, but the time values presented are conventional time values for easier identification of the conventional time of approach to criticality; therefore, the horizontal axis is not linear in terms of the conventional time.

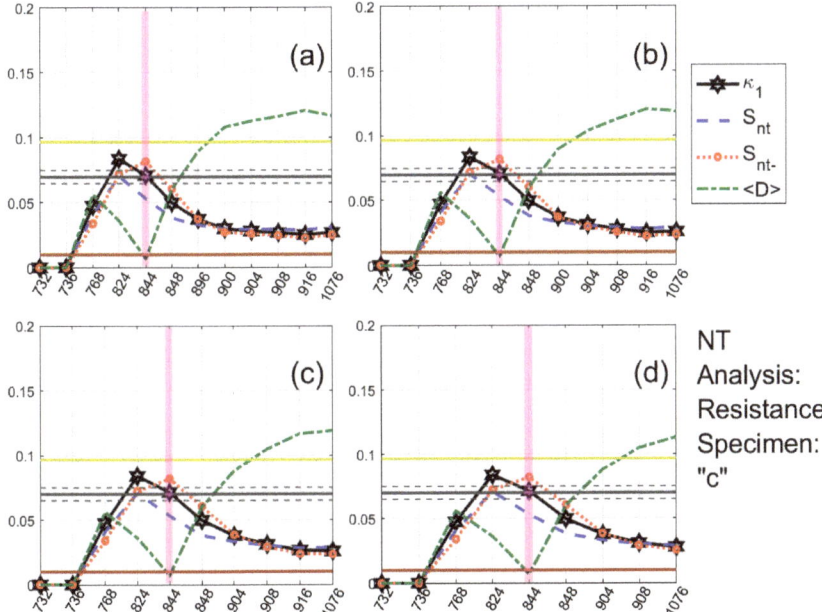

Figure 6. NT analysis of the electrical resistance time series acquired during the experiment involving specimen (c) of Figure 3. Variations of the NT analysis parameter values vs. time in seconds (0 s corresponds to the initiation of the experiment) for four different thresholds W_{Th}: (**a**) 0.00825, (**b**) 0.00975, (**c**) 0.01 and (**d**) 0.0105. The format of this figure follows the format of Figure 5.

For both of them, the same NT analysis procedure has been followed: for each threshold value, the values of the time series under analysis (i.e., $A_{max}^2(t)$ or $|W(t)|$, calculated from the original AE or electrical resistance time series, respectively, with t denoting the conventional time) are sequentially compared to the corresponding threshold (($A_{max}^2)_{Th}$ or W_{Th}, respectively) and as soon as a value of the time series under analysis exceeds the threshold, a new event is included in the NT analysis, leading to a rescaling of the (χ_k, Q_k) time series and recalculation of κ_1, S_{nt}, S_{nt-} and $\langle D \rangle$. In Figures 5 and 6, the magenta patches highlight when NT analysis criticality conditions are satisfied for each threshold value. As apparent from Figures 5 and 6, during these highlighted periods, the criteria (i)–(iii) (see the application of NT analysis to seismicity in Section 4) for the approach to criticality are simultaneously satisfied, since κ_1 approaches the value $\kappa_1 = 0.070$ "by descending from above", $S_{nt}, S_{nt-} < S_u (\approx 0.0966)$, and $\langle D \rangle < 10^{-2}$. For the AE case (Figure 5), the critical state is truly achieved in 964.25 s, the time that the highlighted periods for the presented four thresholds are overlapping, since, for this time, the criterion (iv) (see the application of NT analysis to seismicity in Section 4) is also satisfied. Correspondingly, for the electrical resistance case (Figure 6), the critical state is, according to the criterion (iv), truly achieved in 844 s, since, in that time, there is an overlap of the criticality periods for different thresholds.

By applying the same analysis procedure for the acquired AE and electrical resistance time series of all four specimens (Luserna stone and cement mortar; see Figure 3), the results presented in Table 2 were obtained. The obtained times of approach to criticality are also depicted in Figure 7 relative to the evolution of normalized applied stress and damage in terms of AE and electrical resistance.

Table 2. NT analysis results for the AE and electrical resistance time series of mortar enriched with iron oxide and Luserna stone specimens tested (cf. Figure 3).

Specimen (cf. Figure 3)	Material [†]	Time of Approach to Criticality (s)	
		NT Analysis of AE ($A_{max,k}^2$)	NT Analysis of Resistance ($|W_k|$)
(a)	CM	655–665	232–272
(b)	CM	934.83	328–580 [††]
(c)	LS	964.25	844
(d)	LS	790.32 & 900.65 [‡]	884 [‡‡]
(e)	LS	866.54	508

[†] CM: Cement mortar/LS: Luserna stone. [††] No approach to criticality was found for the dissipated energy (W) of specimen (b). However, the reported approach to criticality was identified after the NT analysis of power ($\frac{dW}{dt}$). [‡] NT analysis of specimen's (d) AE yielded two times of approach to criticality: 790.32 s and 900.65 s, for low and high threshold, $(A_{max}^2)_{Th}$, values, respectively. Note that the later approach to criticality (on 900.65 s) is more important as a possible precursor. [‡‡] The same time of approach to criticality was also identified after the NT analysis of power ($\frac{dW}{dt}$).

As a general remark, criticality in terms of electrical resistance changes systematically precedes AE criticality for all investigated specimens (Table 2, Figure 7). This result could be somehow expected, since the recorded AE activity lies in a frequency range—from a few hertz to 10 kHz—related to large cracks. As a matter of fact, previous studies [45] showed that such low-frequency AEs take place later than high-frequency AEs due to microcracking and are to be regarded as late failure precursors.

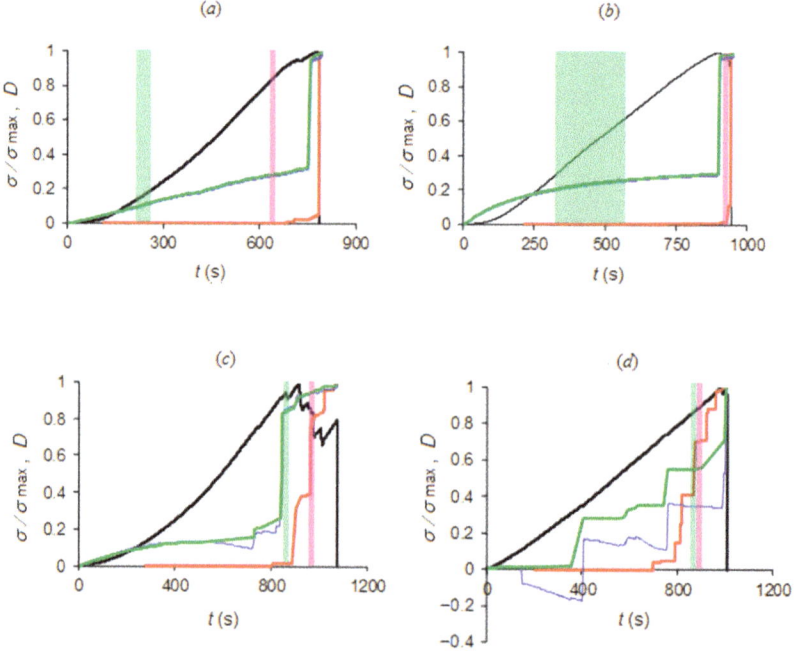

Figure 7. Cont.

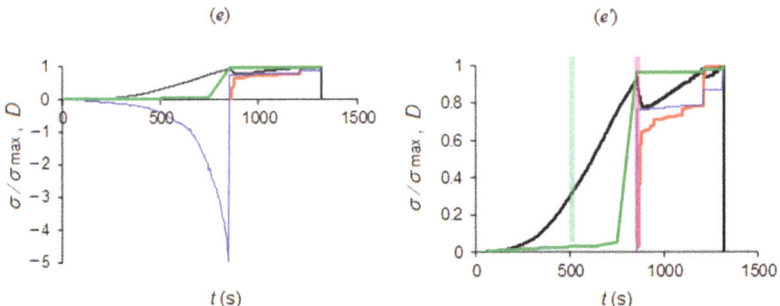

Figure 7. Times of approach to criticality by NT analysis (identified by magenta patch for AE and light green patch for electrical resistance) correlated to load history and damage evolution: the diagrams refer to CM specimens (**a,b**) and LS specimens (**c–e**) according to the inset table of Figure 3; last picture (**e'**) magnifies the positive damage range of (**e**).

Moreover, the position of the time of criticality relative to damage evolution (see Figure 7c), the two kinds of materials that can be summarized as follows:

(i) For cement mortar specimens, AE criticality appears close to the point where the damage evolution in terms of AE (red curves) starts to quickly rise, while electrical resistance criticality appears before damage evolution in terms of electrical resistance (blue curves) starts to quickly rise, at a time period where damage evolution in terms of electrical resistance has still a mild increase rate.

(ii) For Luserna stone specimens, AE criticality appears when the damage evolution in terms of AE (red curves) has already entered quick increase. For specimens (c) and (d), this occurs at or below 50% of the y-axis, during or just before the last-but-one "jump", while for specimen (e), this occurs during the negative values section, at an earlier point of damage evolution. On the other hand, electrical resistance criticality appears before damage evolution in terms of electrical resistance (blue curves) and has already entered quick increase, above 50% of the y-axis, during or just before the last-but-one "jump".

Then, some light is shed on a controversial result concerning Luserna stone, focusing on two specific examples, (c) and (d). In the case of Luserna specimen (d), the NT critical condition for the AE time series seems to be threshold-dependent, since criticality is reached at ~900.65 s for high threshold values and earlier, i.e., at ~790.32 s, for low threshold values. Such a threshold-dependent behavior can be interpreted as a deviation from self-similarity whereby a phenomenon, reproducing itself on different time, space and magnitude scales, is substantially threshold-independent. On the other hand, the uniqueness of the NT criticality condition for Luserna specimen (c)—964.25 s for all thresholds—shows evidence of self-similarity in the AE dynamics.

These observations seem to be confirmed by the Gutenberg–Richter (GR) power-law distribution of AE amplitudes, illustrating self-similarity: $N(\geq A) \propto A^{-b}$, where N is the number of AEs with peak amplitude greater than A. However, it is well known that the GR distribution has to be modified for larger events due to finite size effects, by introducing either an exponential cut-off or a second power-law with a larger b-value beyond a cross-over magnitude [70].

As shown in Figure 8, the deviation from self-similarity at larger magnitudes is more pronounced in the GR distribution of specimen (d), whereby the reduced specimen diameter intervenes, introducing finite size effects.

Considerations about self-similarity could be done also in terms of the evolving b-values (Figure 9), where different trend lines are obtained using as many threshold values: for specimen (c), trends are very similar and, then, threshold-independent, with a common minimum at 967.6 s. These findings are to be regarded as signatures of self-similar behavior. As regards specimen (d), the trend lines of b-values exhibit a stronger dependence upon the threshold magnitude, with different positions of the minimum (903.91 s for M_{Th} = 4.7, 5 and 5.3, and 1006.13 s for M_{Th} = 5.6) reflecting the twofold NT criticality condition.

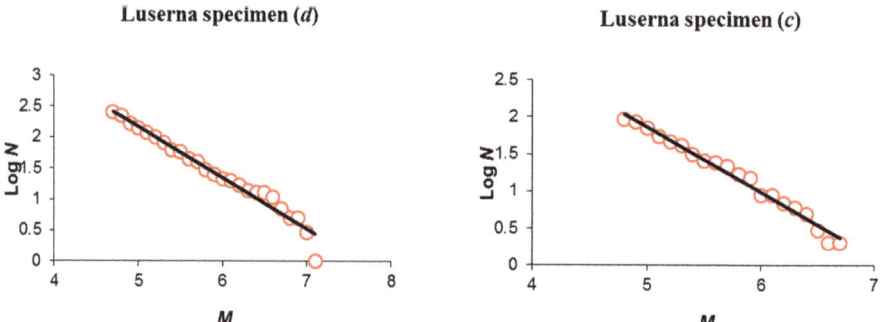

Figure 8. AE number N with magnitude greater than M as a function of M ($M = \log_{10} A$) represented by red circles; the negative slope of the fitting line is the b-value. Deviations from linearity occurring at larger magnitudes ($M \geq 7$) can be observed for the narrower specimen (*d*) (left panel), while the linear fit matches with experimental data of specimen (*c*) throughout the magnitude range (right panel).

In both cases, it is observed an upward shift of the curves (i.e., larger b-values) by increasing the threshold, consistently with the "second power-law rule" of the modified GR distribution.

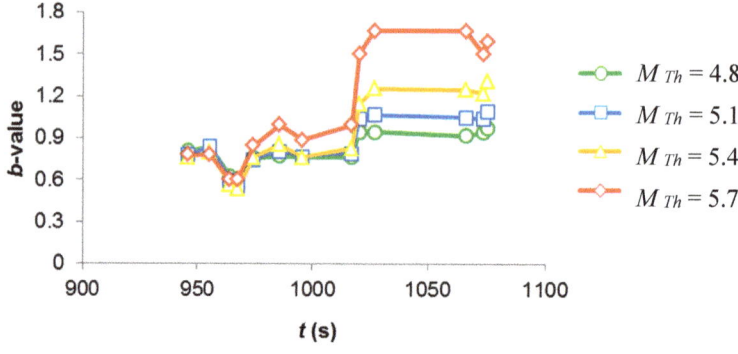

Figure 9. *Cont*.

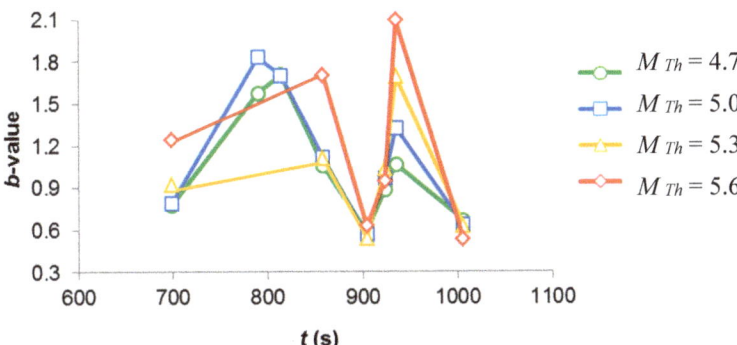

Figure 9. Trend of b-values for different threshold magnitudes. Trend lines of specimen (*c*) (upper panel) appear more self-similar than specimen (*d*) (lower panel).

6. Conclusions

The present work focused on simultaneously acquired AE and electrical resistance time series, recorded during loading tests conducted on cement mortar (CM) and Luserna stone (LS) specimens.

It was observed a brittle mechanical response from CM specimens wherein, despite the absence of significant deviations from linear elasticity, the increase in electrical resistance and AE activity during the approach to failure revealed accumulation damage. On the contrary, except for the case (*d*), LS specimens were characterized by a more ductile behavior, with numerous intermediate load drops correlated to clusters of AE events and changes in electrical resistance.

These differences are reflected by the NT analysis of the recorded AE and electrical resistance time series, where electrical properties systematically reach criticality earlier than AE: the LS–AE criticality is reached in correspondence with significantly increased AE activity, and in correspondence with abrupt stress drops, whereas LS–electrical criticality can precede the major increments in the electrical resistance. The CM–AE criticality is reached at the onset of the damage rate acceleration before the specimen collapse—case (*b*)—or slightly earlier—case (*a*)—whereas the CM–electrical criticality is less easy to understand, as it is reached when the curve of the electrical resistance has not accelerated yet.

The apparently greater unpredictability of the CM fracture behavior could be ascribed to the different material properties: LS is a metamorphic rock with a heterogeneous structure, offering more abundant and more easily identifiable fracture precursors than an artificial material, compacted during the manufacturing process, and thus more homogeneous such as CM.

Author Contributions: Conceptualization, G.N. and G.L.; methodology, G.N., O.B. and S.M.P.; software, S.M.P.; investigation, O.B. and G.N.; resources, G.L.; data curation, O.B.; writing—original draft preparation, G.N. and S.M.P.; writing—review and editing, G.N., S.M.P. and G.L. All authors have read and agreed to the published version of the manuscript.

Funding: This research received no external funding.

Conflicts of Interest: The authors declare no conflict of interest.

References

1. Yoshida, S.; Ogawa, T. Electromagnetic emissions from dry and wet granite associated with acoustic emissions. *J. Geophys. Res.* **2004**, *109*. [CrossRef]
2. Triantis, D.; Vallianatos, F.; Stavrakas, I.; Hloupis, G. Relaxation phenomena of electric signal emissions from rocks following to abrupt mechanical stress application. *Ann. Geophys.* **2012**, *55*, 207–212.

3. Sun, B.; Guo, Y. High-cycle fatigue damage measurement based on electrical resistance change considering variable electrical resistivity and uneven damage. *Int. J. Fatigue* **2004**, *26*, 457–462. [CrossRef]
4. Chen, B.; Liu, J. Damage in carbon fiber-reinforced concrete, monitored by both electrical resistance measurement and acoustic emission analysis. *Constr. Build. Mater.* **2008**, *22*, 2196–2201. [CrossRef]
5. Stavrakas, I.; Anastasiadis, C.; Triantis, D.; Vallianatos, F. Piezo stimulated currents in marble samples: Precursory and concurrent-with-failure signals. *Nat. Hazards Earth Syst. Sci.* **2003**, *3*, 243–247. [CrossRef]
6. Triantis, D.; Anastasiadis, C.; Stavrakas, I. The correlation of electrical charge with strain on stressed rock samples. *Nat. Hazards Earth Syst. Sci.* **2008**, *8*, 1243–1248. [CrossRef]
7. Kyriazopoulos, A.; Anastasiadis, C.; Triantis, D.; Brown, C. Non-destructive evaluation of cement-based materials from pressure-stimulated electrical emission. Preliminary results. *Constr. Build. Mater.* **2011**, *25*, 1980–1990. [CrossRef]
8. Potirakis, S.M.; Contoyiannis, Y.; Eftaxias, K.; Koulouras, G.; Nomicos, C. Recent Field Observations Indicating an Earth System in Critical Condition Before the Occurrence of a Significant Earthquake. *IEEE Geosci. Remote. Sens. Lett.* **2014**, *12*, 631–635. [CrossRef]
9. Potirakis, S.M.; Contoyiannis, Y.; Melis, N.S.; Kopanas, J.; Antonopoulos, G.; Balasis, G.; Kontoes, C.; Nomicos, C.; Eftaxias, K. Recent seismic activity at Cephalonia (Greece): A study through candidate electromagnetic precursors in terms of non-linear dynamics. *Nonlinear Process. Geophys.* **2016**, *23*, 223–240. [CrossRef]
10. Hadjicontis, V.; Mavromatou, C. Electric signals recorded during uniaxial compression of rock samples: Their possible correlation with preseismic electric signals. *Acta Geophys.* **1995**, *43*, 49–61. [CrossRef]
11. Hadjicontis, V.; Mavromatou, C.; Ninos, D. Stress induced polarization currents and electromagnetic emission from rocks and ionic crystals, accompanying their deformation. *Nat. Hazards Earth Syst. Sci.* **2004**, *4*, 633–639. [CrossRef]
12. Bridgnman, P.W. The effect of homogeneous mechanical stress on the electrical resistance of crystals. *Phys. Rev.* **1932**, *42*, 858. [CrossRef]
13. Russell, J.E.; Hoskins, E.R. Correlation of electrical resistivity of dry rock with cumulative damage. In Proceedings of the 11th U.S. Symposium on Rock Mechanics, Berkeley, CA, USA, 16–19 June 1969.
14. Wen, S.H.; Chung, D.D.L. Damage Monitoring of Cement Paste by Electrical Resistance Measurement. *Cem. Concr. Res.* **2000**, *30*, 1979–1982. [CrossRef]
15. Chung, D.D.L. Damage in Cement-Based Materials, Studied by Electrical Resistance Measurement. *Mater. Sci. Eng. R Rep.* **2003**, *42*, 1–40. [CrossRef]
16. Chen, G.; Lin, Y. Stress-strain-electrical resistance effects and associated state equations for uniaxial rock compression. *Int. J. Rock Mech. Min. Sci.* **2004**, *41*, 223–236. [CrossRef]
17. Olhoeft, G.R. Electrical Properties of Granite with Implications for the Lower Crust. *J. Geophys. Res.* **1981**, *80*, 931–936. [CrossRef]
18. Laštovičková, M.; Parchomenko, E.I. The electric properties of eclogites from the Bohemian Massif under high temperatures and pressures. *Pure Appl. Geophys.* **1976**, *114*, 451–460. [CrossRef]
19. Borla, O.; Lacidogna, G.; Di Battista, E.; Niccolini, G.; Carpinteri, A. Electromagnetic Emission as Failure Precursor Phenomenon for Seismic Activity Monitoring. *Soc. Exp. Mech. Ser.* **2015**, *66*, 221–229.
20. Carpinteri, A.; Lacidogna, G.; Borla, O.; Manuello, A.; Niccolini, G. Electromagnetic and neutron emissions from brittle rocks failure: Experimental evidence and geological implications. *Sadhana Acad. Proc. Eng. Sci.* **2012**, *37*, 59–78. [CrossRef]
21. Cox, S.; Meredith, P. Microcrack formation and material softening in rock measured by monitoring acoustic emissions. *Int. J. Rock Mech. Min. Sci. Géoméch. Abstr.* **1993**, *30*, 11–24. [CrossRef]
22. Lockner, D. The role of acoustic emission in the study of rock fracture. *Int. J. Rock Mech. Min. Sci. Géoméch. Abstr.* **1993**, *30*, 883–899. [CrossRef]
23. Tudik, A. Electromagnetic emission during the fracture of metals. *Sov. Tech. Phys. Lett.* **1980**, *6*, 37–38.
24. Finkel', V.M.; Golovin, Y.I.; Sereda, V.E.; Kulikova, G.P.; Zuev, L.B. Electric effects in fracture of LiF crystals in connection with the problem of the control of cracking. *Fizika Tverdogo Tela* **1975**, *17*, 770–776.
25. O'Keefe, S.G.; Thiel, D.V. A mechanism for the production of electromagnetic radiation during fracture of brittle materials. *Phys. Earth Planet. Inter.* **1995**, *89*, 127–135. [CrossRef]
26. Rabinovitch, A.; Frid, V.; Bahat, D. Surface oscillations—A possible source of fracture induced electromagnetic radiation. *Tectonophysics* **2007**, *431*, 15–21. [CrossRef]

27. Vallianatos, F.; Tzanis, A. Electric current generation associated with the deformation rate of a solid: Preseismic and coseismic signals. *Phys. Chem. Earth* **1998**, *23*, 933–938. [CrossRef]
28. Mori, Y.; Obata, Y.; Pavelka, J.; Sikula, J.; Lolajicek, T. AE Kaiser effect and electromagnetic emission in the deformation of rock sample. *J. Acoust. Emiss.* **2004**, *22*, 91–101.
29. Mori, Y.; Obata, Y.; Sikula, J. Acoustic and electromagnetic emission from crack created in rock sample under deformation. *J. Acoust. Emiss.* **2009**, *27*, 157–166.
30. Fukui, K.; Okubo, S.; Terashima, T. Electromagnetic Radiation from Rock during Uniaxial Compression Testing: The Effects of Rock Characteristics and Test Conditions. *Rock Mech. Rock Eng.* **2005**, *38*, 411–423. [CrossRef]
31. Sun, M.; Liu, Q.; Li, Z.; Wang, E. Electrical emission in mortar under low compressive loading. *Cem. Concr. Res.* **2002**, *32*, 47–50. [CrossRef]
32. Triantis, D.; Stavrakas, I.; Kyriazopoulos, A.; Hloupis, G.; Agioutantis, Z. Pressure stimulated electrical emissions from cement mortar used as failure predictors. *Int. J. Fract.* **2012**, *175*, 53–61. [CrossRef]
33. Li, J.F.; Ai, H.; Viehland, D. Anomalous electromechanical behavior of portland cement: Electro-osmotically-induced shape changes. *J. Am. Ceram. Soc.* **1995**, *78*, 416–420. [CrossRef]
34. Cao, S.; Song, W. Medium-Term Strength and Electromagnetic Radiation Characteristics of Cemented Tailings Backfill Under Uniaxial Compression. *Geotech. Geol. Eng.* **2018**, *36*, 3979–3986. [CrossRef]
35. Yoshida, S. Convection current generated prior to rupture in saturated rocks. *J. Geophys. Res.* **2001**, *106*, 2103–2120. [CrossRef]
36. Kachanov, L.M. *Introduction to Continuum Damage Mechanics*; Martinus Nijhoff: Dordrecht, The Netherlands, 1986.
37. Varotsos, P.A.; Sarlis, N.V.; Skordas, E.S.; Uyeda, S.; Kamogawa, M. Natural time analysis of critical phenomena. *Proc. Natl. Acad. Sci. USA* **2011**, *108*, 11361–11364. [CrossRef]
38. Varotsos, P.A.; Sarlis, N.V.; Skordas, E.S. *Natural Time Analysis: The New View of Time*; Springer Science and Business Media LLC: Berlin, Germany, 2011.
39. Potirakis, S.; Mastrogiannis, D. Critical features revealed in acoustic and electromagnetic emissions during fracture experiments on LiF. *Phys. A Stat. Mech. Appl.* **2017**, *485*, 11–22. [CrossRef]
40. Bak, P.; Christensen, K.; Danon, L.; Scanlon, T. Unified Scaling Law for Earthquakes. *Phys. Rev. Lett.* **2002**, *88*, 178501. [CrossRef]
41. Diodati, P.; Piazza, S.; Marchesoni, F. Acoustic emission from volcanic rocks: An example of self-organized criticality. *Phys. Rev. Lett.* **1991**, *67*, 2239–2243. [CrossRef]
42. Corral, A. Modelling critical and catastrophic phenomena. In *Geoscience: A Statistical Physics Approach, in Lecture Notes in Physics*; Bhattacharyya, P., Chakrabarti, B.K., Eds.; Springer: Berlin, Germany, 2006; Volume 705, pp. 191–221.
43. Niccolini, G.; Durin, G.; Carpinteri, A.; Lacidogna, G.; Manuello, A. Crackling noise and universality in fracture systems. *J. Stat. Mech. Theory Exp.* **2009**, *1*, 1–11. [CrossRef]
44. Niccolini, G.; Borla, O.; Lacidogna, G.; Carpinteri, A. Correlated Fracture Precursors in Rocks and Cement-Based Materials Under Stress. In *Acoustic, Electromagnetic, Neutron Emissions from Fracture and Earthquakes*; Springer: Cham, Switzerland, 2015; Volume 16, pp. 237–248. [CrossRef]
45. Schiavi, A.; Niccolini, G.; Tarizzo, P.; Carpinteri, A.; Lacidogna, G.; Manuello, A. Acoustic emissions at high and low frequencies during compression tests of brittle materials. *Strain* **2011**, *47*, 105–110. [CrossRef]
46. Niccolini, G.; Borla, O.; Accornero, F.; Lacidogna, G.; Carpinteri, A. Scaling in damage by electrical resistance measurements: An application to the terracotta statues of the Sacred Mountain of Varallo Renaissance Complex (Italy). *Rend. Lincei. Sci. Fis. Nat.* **2015**, *26*, 203–209. [CrossRef]
47. Colombo, S.; Main, I.G.; Forde, M.C. Assessing damage of reinforced concrete beam using "b-value" analysis of acoustic emission signals. *J. Mater. Civ. Eng. ASCE* **2003**, *15*, 280–286. [CrossRef]
48. Lemaitre, J.; Chaboche, J.L. *Mechanics of Solid Material*; Cambridge University Press: Cambridge, UK, 1990.
49. Krajcinovic, D. *Damage Mechanics*; Elsevier: Amsterdam, The Netherlands, 1996.
50. Lemaitre, J.; Dufailly, J. Damage measurements. *Eng. Fract. Mech.* **1987**, *28*, 643–661. [CrossRef]
51. Archie, G.E. The Electrical Resistivity Log as an Aid in Determining Some Reservoir Characteristics. *Trans. AIME* **1942**, *146*, 54–62. [CrossRef]
52. Varotsos, P.A.; Sarlis, N.V.; Skordas, E.S. Spatio-temporal complexity aspects on the interrelation between seismic electric signals and seismicity. *Pract. Athens Acad.* **2001**, *76*, 294–321.

53. Varotsos, P.A.; Sarlis, N.V.; Skordas, E.S. Long-range correlations in the electric signals that precede rupture: Further investigations. *Phys. Rev. E* **2003**, *67*, 021109. [CrossRef]
54. Varotsos, P.A.; Sarlis, N.V.; Tanaka, H.K.; Skordas, E.S. Similarity of fluctuations in correlated systems: The case of seismicity. *Phys. Rev. E* **2005**, *72*, 041103. [CrossRef]
55. Abe, S.; Sarlis, N.V.; Skordas, E.S.; Tanaka, H.K.; Varotsos, P.A. Origin of the Usefulness of the Natural-Time Representation of Complex Time Series. *Phys. Rev. Lett.* **2005**, *94*, 170601. [CrossRef]
56. Potirakis, S.M.; Schekotov, A.; Asano, T.; Hayakawa, M. Natural time analysis on the ultra-low frequency magnetic field variations prior to the 2016 Kumamoto (Japan) earthquakes. *J. Asian Earth Sci.* **2018**, *154*, 419–427. [CrossRef]
57. Varotsos, P.A.; Sarlis, N.V.; Skordas, E.S.; Tanaka, H.K.; Lazaridou, M.S. Entropy of seismic electric signals: Analysis in natural time under time reversal. *Phys. Rev. E* **2006**, *73*, 031114. [CrossRef]
58. Varotsos, P.A.; Sarlis, N.V.; Skordas, E.S. Scale-specific order parameter fluctuations of seismicity in natural time before mainshocks. *EPL Europhys. Lett.* **2011**, *96*. [CrossRef]
59. Varotsos, P.A. *The Physics of Seismic Electric Signals*; TERRAPUB: Tokyo, Japan, 2005.
60. Potirakis, S.M.; Karadimitrakis, A.; Eftaxias, K. Natural time analysis of critical phenomena: The case of pre-fracture electromagnetic emissions. *Chaos: Interdiscip. J. Nonlinear Sci.* **2013**, *23*, 023117. [CrossRef] [PubMed]
61. Potirakis, S.M.; Schekotov, A.; Contoyiannis, Y.; Balasis, G.; Koulouras, G.E.; Melis, N.S.; Boutsi, A.Z.; Hayakawa, M.; Eftaxias, K.; Nomicos, C. On Possible Electromagnetic Precursors to a Significant Earthquake (Mw = 6.3) Occurred in Lesvos (Greece) on 12 June 2017. *Entropy* **2019**, *21*, 241. [CrossRef] [PubMed]
62. Hayakawa, M.; Schekotov, A.; Potirakis, S.M.; Eftaxias, K. Criticality features in ULF magnetic fields prior to the 2011 Tohoku earthquake. *Proc. Jpn. Acad. Ser. B* **2015**, *91*, 25–30. [CrossRef] [PubMed]
63. Hayakawa, M.; Schekotov, A.; Potirakis, S.M.; Eftaxias, K.; Li, Q.; Asano, T. An Integrated Study of ULF Magnetic Field Variations in Association with the 2008 Sichuan Earthquake, on the Basis of Statistical and Critical Analyses. *Open J. Earthq. Res.* **2015**, *4*, 85–93. [CrossRef]
64. Potirakis, S.M.; Eftaxias, K.; Schekotov, A.; Yamaguchi, H.; Hayakawa, M. Criticality features in ultra-low frequency magnetic fields prior to the 2013 M6.3 Kobe earthquake. *Ann. Geophys.* **2016**, *59*, S0317. [CrossRef]
65. Potirakis, S.M.; Asano, T.; Hayakawa, M. Criticality Analysis of the Lower Ionosphere Perturbations Prior to the 2016 Kumamoto (Japan) Earthquakes as Based on VLF Electromagnetic Wave Propagation Data Observed at Multiple Stations. *Entropy* **2018**, *20*, 199. [CrossRef]
66. Sarlis, N.V.; Skordas, E.S.; Lazaridou, M.S.; Varotsos, P.A. Investigation of seismicity after the initiation of a Seismic Electric Signal activity until the main shock. *Proc. Jpn. Acad. Ser. B* **2008**, *84*, 331–343. [CrossRef]
67. Varotsos, P.A.; Sarlis, N.V.; Skordas, E.S.; Lazaridou, M.S. Identifying sudden cardiac death risk and specifying its occurrence time by analyzing electrocardiograms in natural time. *Appl. Phys. Lett.* **2007**, *91*, 064106. [CrossRef]
68. Varotsos, P.A.; Sarlis, N.V.; Skordas, E.S. Tsallis Entropy Index q and the Complexity Measure of Seismicity in Natural Time under Time Reversal before the M9 Tohoku Earthquake in 2011. *Entropy* **2018**, *20*, 757. [CrossRef]
69. Greco, A.; Tsallis, C.; Rapisarda, A.; Pluchino, A.; Fichera, G.; Contrafatto, L. Acoustic emissions in compression of building materials: Q-statistics enables the anticipation of the breakdown point. *Eur. Phys. J. Spéc. Top.* **2020**, *229*, 841–849. [CrossRef]
70. Sornette, D.; Sornette, A. General theory of the modified Gutenberg-Richter law for large seismic moments. *Bull. Seismol. Soc. Am.* **1999**, *89*, 1121–1130.

Publisher's Note: MDPI stays neutral with regard to jurisdictional claims in published maps and institutional affiliations.

© 2020 by the authors. Licensee MDPI, Basel, Switzerland. This article is an open access article distributed under the terms and conditions of the Creative Commons Attribution (CC BY) license (http://creativecommons.org/licenses/by/4.0/).

Article

Stress Dependence on Relaxation of Deformation Induced by Laser Spot Heating

Yuma Murata [1], Tomohiro Sasaki [1,*] and Sanichiro Yoshida [2]

[1] Graduate School of Science and Technology, Niigata University, Niigata 9502181, Japan
[2] Department of Chemistry and Physics, Southeastern Louisiana University, Hammond, LA 70401, USA
* Correspondence: tomodx@eng.niigata-u.ac.jp; Tel.: +81-262-6710

Abstract: This paper deals with a non-destructive analysis of residual stress through the visualization of deformation behaviors induced by a local spot heating. Deformation was applied to the surface of an aluminum alloy with an infrared spot laser. The heating process is non-contact, and the applied strain is reversible in the range of room temperature to approximately +10 °C. The specimen was initially pulled up to elastic tensile stress using a tensile test machine under the assumption that the material was subject to the tensile residual stress. The relaxation behaviors of the applied strain under tensile stress conditions were evaluated using contact and non-contact methods, i.e., two strain gauges (the contact method) and a two-dimensional electronic speckle pattern interferometer (non-contact method). The results are discussed based on the stress dependencies of the thermal expansion coefficient and the elasticity of the materials.

Keywords: non-destructive testing; optical interferometry; residual stress; thermal expansion; elasticity

1. Introduction

Residual stress is induced by various manufacturing processes such as machining, welding processes, and heat treatment. In particular, tensile residual stress causes a reduction in fatigue strength, and stress corrosion cracking. Residual stress analysis is still a subject of investigation. There are several commonly used techniques to measure residual stress destructively. The principle of these techniques is the measurement of strain released by machining processes including hole drilling and cutting with electric discharge machining. The released strain can be measured using strain sensors in contact or non-contact ways. Optical measurement techniques such as electronic speckle pattern interferometry (ESPI) [1,2], holographic interferometry [3,4], moiré interferometry [5,6], and digital image correlation (DIC) [7,8] are known as non-contact and full-field strain measurement techniques. These techniques allow the full-field measurement of residual stress in combination with the destructive way, while they involve an irreversible process. Recently, a method using microfabrication with a pulsed spot laser and digital holography has been proposed [9]. This method applies minimal destruction in a hole diameter of 20 μm. In addition, an attempt has been made to measure the released strain by annealing with laser spot heating [10]. On the other hand, diffractometry using neutrons [11] and X-rays [12] is known as the most common way of conducting non-destructive residual stress analysis. This method measures the stress through the change in the lattice constant. In addition, measurements of electromagnetic waves [13] and acoustic waves [14] using contact probes are often included in the non-destructive techniques. These non-destructive techniques have a limitation in the measurement area of several hundred mm² per measurement, leading to a time-consuming process to measure a wide range of residual stresses. A technique equipped with non-destructive and non-contact measurements is not available.

Previously, we made attempts to evaluate residual stress through the visualization of reversible deformation with optical techniques. The study focused on the stress dependence

of elasticity using the same principle as the acoustic elasticity and showed that the elastic deformation behavior applied with a tensile machine depends on the applied stress [15,16]. More recently, a study [17] investigated thermal deformation behavior under the stressed condition, focusing on the stress dependence of thermal characteristics. Both properties of elasticity and thermal expansion have stress dependency derived from the non-linearity and anharmonicity of the interatomic potential [18–20]. These studies suggested that residual stress can be estimated through the visualization of reversible thermal deformation in the temperature range of ±10 °C. However, the above methods need devices to apply reversible deformation to the material, and the possibility of a non-contact method still remains an issue.

The present study investigated the local thermal deformation behavior induced by spot heating. A heating laser beam was applied to a local area of a residually stressed specimen for a short period of time. This heating process is non-contact, and the induced temperature change is reversible in the range of room temperature to approximately +10 °C. The resultant thermal strain was measured with contact and non-contact methods. For the contact method, strain gauges were used, and for the non-contact method, an ESPI setup was used. It was expected that the thermal deformation behavior would be affected by the stress field due to the stress dependence of thermal expansion. At the same time, the local heating under the stressed condition was expected to cause strain relaxation in the surrounding area, resulting in elastic relaxation. The possibility for non-destructive residual stress analysis is discussed through the visualization of the reversible deformation behavior in the heated and surrounding areas.

2. Materials and Methods

2.1. Specimen

An industrial aluminum alloy (ISO AA5083) sheet with a thickness of 2.0 mm was used in this study. Table 1 shows the standard chemical composition of the AA5083 alloy. This alloy is a solid-solution-hardened Al-Mg-based alloy. Although the temperature change applied in this study is small enough in the range of approximately 10 °C, the alloy is microstructurally less sensitive to the heating process. The modulus of longitudinal elasticity of this material is 70 GPa, and the 0.2% proof stress is approximately 145 MPa. The sheet was cut into specimens with a length of 100 mm and a width of 10 mm by electrical discharge machining, as shown in Figure 1. The specimens were annealed at 300 °C in air for three hours to remove the as-received residual stress. The specimens used for the visualization of the strain field using speckle pattern interferometry were coated with a white lacquer spray, and those used for strain gauges were polished with #800 emery abrasive paper.

Table 1. Chemical composition of AA5083 (mass %).

Mg	Si	Fe	Cu	Mn	Cr	Zn	Ti	Al
4.0–4.9	<0.4	<0.4	<0.1	0.4–4.0	0.05–0.25	<0.25	<0.15	Bal.

Figure 1. Test specimen.

2.2. Local Spot Heating

Figure 2 shows a setup for local spot heating under tensile stress conditions, and a two-dimensional electronic speckle pattern interferometer (2D-ESPI) used for the visualization of thermal deformation, as described in the previous section. A tensile test machine was used to apply an external load to the specimen under the assumption that the material was subject to the tensile residual stress. The specimen was initially pulled up to tensile stress levels (initial stress, σ_i) at a crosshead speed of 0.002 mm/s, measuring the external load with a load cell attached to the tensile machine (KYOWA ELECTRONIC INSTRUMENTS CO., LTD, LUA-A). The initial stress condition ranged from 0 to 120 MPa, which is approximately 80% of the proof stress at maximum. The condition of initial stress $\sigma_i = 0$ MPa indicates that one end of the specimen was fixed with a clamp and the other end was free. Under each initial stress condition, the specimen was subjected to spot heating using an infrared laser (SPI Lasers red ENERGY G4, wavelength of 1064 nm). The infrared laser was irradiated perpendicularly to the surface of the specimen at an angle of 45° to the tensile axis through a half mirror, as shown in Figure 2. The spot diameter was adjusted to 1 mm by a focusing lens via a half mirror.

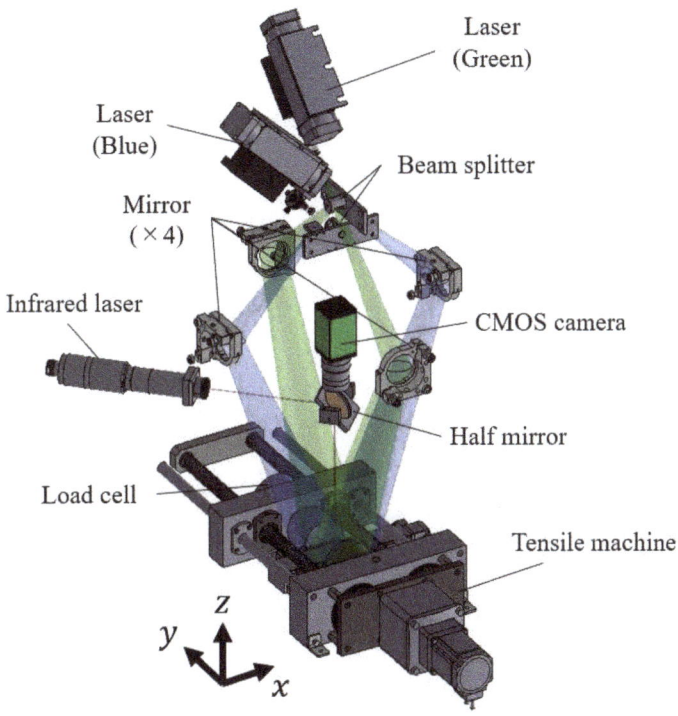

Figure 2. Optical setup and devices for local spot heating test under the stressed condition.

Figure 3 shows a typical example of a thermal image taken using an infrared camera (Optris GmbH, PI80). The temperature distribution around the heat spot is shown in Figure 4. The temperature values are plotted in the x- and y-directions as T_x and T_y, respectively. The initial profile at t = 0 exhibited a Gaussian distribution, and it was symmetric for the other directions. The maximum temperature increase at the heating point was about 10 °C from room temperature.

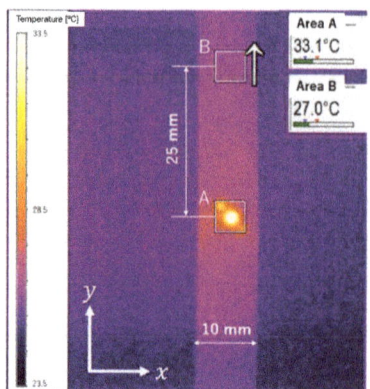

Figure 3. Thermograph of a test specimen during local heating.

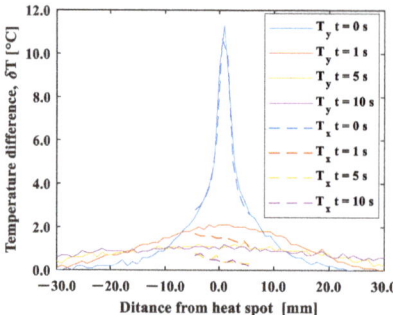

Figure 4. Temperature distribution around the heat spot.

Figure 5 shows the time variation of the average temperature in the 5 × 5 mm² areas shown by squares A and B in Figure 3. Heat diffusion occurred simultaneously with the local spot heating for 1 s. The result shows that the average temperature in area A during heating reached up to 38 °C in about one second after the start of laser irradiation, and then it decreased rapidly after the heating stopped. Meanwhile, in area B, which was 25 mm away from the center of the spot, the temperature difference was less than 1 °C, indicating that the effect of heating was negligible in this area.

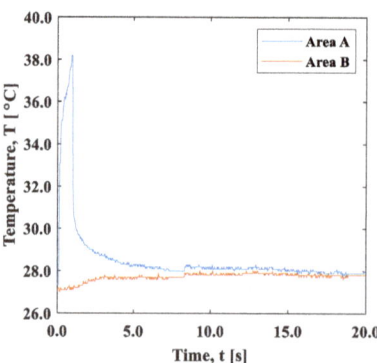

Figure 5. Time variation of temperature of areas A and B shown in Figure 3.

2.3. Strain Measurement

The thermal deformation behavior during the local spot heating was evaluated using two devices; namely, a strain gauge and a 2D-ESPI. The strain gauge measurement aimed to accurately measure the thermal strain to support the following 2D-ESPI. The capability of an in-plane and non-contact evaluation of residual stress will be discussed in the ESPI section.

(i) Strain gauge

Strain values around the local spot-heated area (area A in Figure 3) were measured using a rosette strain gauge with a central hole diameter of 2 mm and a gauge length of 1.5 mm, as shown in Figure 6. At the same time, the strain in area B, which was 25 mm away from the heat spot, was measured using a single-axis strain gauge with a gauge length of 5.0 mm. The resistance of both strain gauges was 120 ohms.

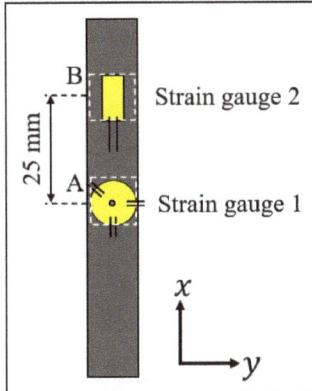

Figure 6. Strain gauge mounting position.

(ii) Two-dimensional electronic speckle pattern interferometry (ESPI)

Two optical systems sensitive to a single-axis displacement were arranged in the tensile direction (y-direction) and transverse to the tensile direction (x-direction), as shown in Figure 2. Each system was based on the "Dual-beam ESPI" sensitive to in-plane displacement [15–17]. A blue laser with a wavelength of 472.9 nm and a green laser with a wavelength of 532.3 nm (LASOS Lasertechnik GmbH, DPSS laser) were used for the light sources in the y- and x-directions, respectively. The laser beam expanded by the expander was split into two paths by a splitter. The split laser beams irradiated the surface of the specimen at the same incident angle of 18.83° through mirrors. The speckle intensity reflecting the surface changed depending on the displacement in the sensitive direction due to the optical path difference between the two interferometric arms. A prism spectroscopic 3CMOS camera (JAI Corporation, AP-1600T) was used to capture the speckles generated by the two light sources. The frame rate and resolution were 60 fps and 1280 × 960 pixels, respectively. The speckle patterns that originated from the two-color beams were split using a prism inside the camera [21]. The change in speckle intensity during the local spot heating test was calculated numerically by subtracting each frame of the image from a later frame taken after thermal expansion. The interferometric fringe patterns representing the displacement field can be obtained for each light source. Displacement u, v in the x, y-directions when the number of fringes increases by n_x, n_y can be given by

$$\begin{bmatrix} u \\ v \end{bmatrix} = \begin{bmatrix} \frac{\lambda_x n_x}{2 \sin \theta} \\ \frac{\lambda_y n_y}{2 \sin \theta} \end{bmatrix} \quad (1)$$

where λ_x and λ_y are the wavelengths of the light sources for the two optical configurations, and θ is the incident angle. The angle of θ was 18.83°. By calculating the displacement in the x- and y-directions independently, it is possible to measure the two-dimensional in-plane displacement distribution. In addition, carrier fringes were introduced by rotating the mirror, as employed in the previous study [17]. Figure 7 shows an example of a carrier fringe image. A fringe pattern with a constant interval was artificially introduced by aligning the mirror on one optical arm. This makes the fringe analysis based on the spatial frequency easier. The actual displacement field was obtained by subtracting the displacement from the carrier from the displacement of a given image. Approximately four carrier fringes per 10 mm were introduced in the tensile direction. The strain field can be obtained through the following process.

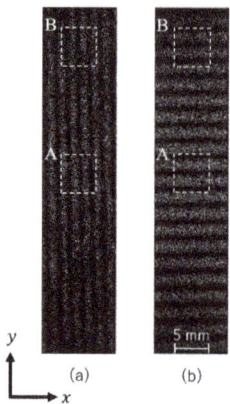

Figure 7. Carrier fringe image: (**a**) x-direction; (**b**) y-direction.

Figure 8a shows a cropped image of a 5 × 5 area (101 × 101 pixels) in Figure 7b. An intensity distribution, as shown in Figure 8b, is obtained by averaging intensity values in the x-direction. Since the number of fringes increases with the difference in the displacement, the spatial frequency of the speckle intensity represents the strain level. The Fourier spectrum computed from the intensity distribution is shown in Figure 8c. To increase the resolution of the frequency analysis, the sampling points of the intensity profile were interpolated to 100 times the original plot data. The peak frequency near the frequency of the carrier, from +200 to −100 around 200 n/mm, was searched in the spectrum. The peak frequency indicates the number of fringes n per unit length. The mean strain at the center coordinate of the cropped area was determined from the number of fringes using Equation (2).

$$\varepsilon = \begin{bmatrix} \varepsilon_{xx} & \varepsilon_{xy} \\ \varepsilon_{yx} & \varepsilon_{yy} \end{bmatrix} = \begin{bmatrix} \frac{\partial u}{\partial x} & \frac{1}{2}\left(\frac{\partial u}{\partial y} + \frac{\partial v}{\partial x}\right) \\ \frac{1}{2}\left(\frac{\partial v}{\partial x} + \frac{\partial u}{\partial y}\right) & \frac{\partial v}{\partial y} \end{bmatrix} \cong \begin{bmatrix} \frac{u}{l_x} & \frac{1}{2}\left(\frac{u}{l_y} + \frac{v}{l_x}\right) \\ \frac{1}{2}\left(\frac{v}{l_x} + \frac{u}{l_y}\right) & \frac{v}{l_y} \end{bmatrix} \quad (2)$$

where l_x, l_y are the length of the cropped area, and u, v are the displacements in the x- and y-directions in the cropped area, respectively. The strain distribution was obtained by shifting the cropped area by one pixel in the sensitive directions.

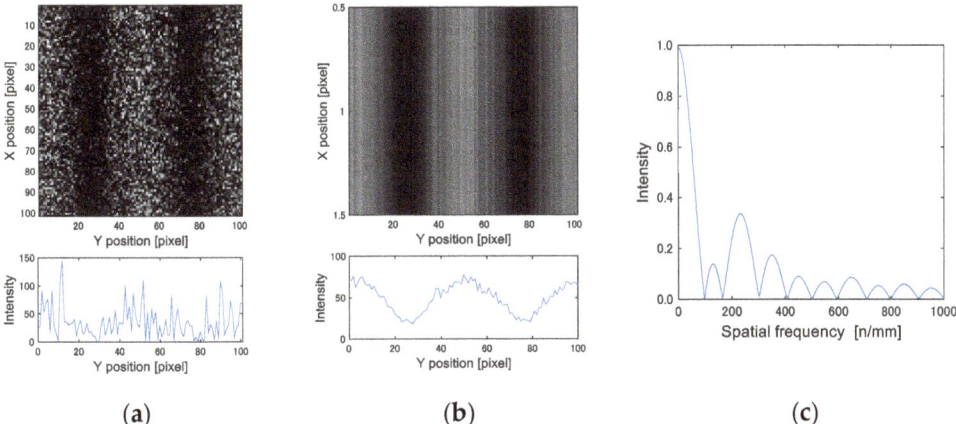

Figure 8. Fringe image analysis by FFT: (**a**) fringe image and intensity profile; (**b**) averaged intensity profile; and (**c**) FFT spectrum.

3. Results and Discussion

3.1. Relaxation Behavior of Strain and Temperature

It is difficult to measure temperature and strain changes during heating because of the rapid and unstable temperature change. Thus, we focused on the cooling process after the laser spot heating. The thermal diffusion in the cooling process can be discussed based on an instantaneous heat source applied to the surface. The equation of heat conduction is generally expressed by

$$\frac{\partial T(x,y,z,t)}{\partial t}\frac{1}{K} = \left(\frac{\partial^2 T}{\partial x^2} + \frac{\partial^2 T}{\partial y^2} + \frac{\partial^2 T}{\partial z^2}\right) \quad (3)$$

where $T(x,y,z,t)$ is the temperature, $K = k/\rho c$ is the thermal diffusivity, and k, ρ, and c are the thermal conductivity, the density, and the thermal capacity, respectively. When an initial temperature distribution at $t = 0$ is given by a function of coordinates as $T_0(x,y,z)$, Equation (1) is satisfied by the following solution [22]:

$$T(x,y,z,t) = \frac{1}{\left(\sqrt{4\pi Kt}\right)^3} \int_{-\infty}^{\infty}\int_{-\infty}^{\infty}\int_{-\infty}^{\infty} T_0(x',y',z') \exp\left\{-\frac{(x'-x)^2 + (y'-y)^2 + (z'-z)^2}{4Kt}\right\} dx'dy'dz' \quad (4)$$

When a point heat source is instantaneously applied at point $(0,0,0)$, at $t = 0$, the initial temperature is expressed by

$$T_0(x,y,z,0) = \delta(x)\delta(y)\delta(z) \quad (5)$$

where $\delta(x)$ is the Dirac delta function. For heat conduction on the surface of a semi-infinite body, where the point heat source is applied to the surface ($z = 0$) in the $x - y$ plane, the solution to Equation (3) is given by

$$T(x,y,z,t) = \frac{2Q_0}{\rho c\left(\sqrt{4\pi Kt}\right)^n} \exp\left\{-\frac{x^2 + y^2 + z^2}{4Kt}\right\} \quad (6)$$

where Q_0 is the volumetric heat source applied to the initial point, and n represents the spatial dimension of heat diffusion. We set the time t in Equation (6) as follows. In the experiment, we set the origin of time ($t_{experiment} = 0$) to the moment when the heating source was removed. To evaluate Equation (6), we used a finite time step to express the instantaneous heating. These operations required us to adjust the origin of t so that the spatial temperature profile resulting from Equation (6) was consistent with the experimental counterpart. We calibrated t so that Equation (6) yielded the temperature profile closest to the experimental counterpart.

Figure 9 shows the temperature obtained by Equation (6) for the position $(0, y, 0)$ plotted on the experimental value measured by the infrared camera. The calculation was conducted for the spatial dimensions $n = 2$ and 3. The spatial and temporal changes are shown in Figure 9a,b, respectively. The calculated values shown by the dashed and chained lines ("Cal.") are roughly consistent with the measured value "Ex." in the shorter time within 0.1s, while larger temperature drops are exhibited. This is attributed to the fact that the thermal model assumes the heat diffusion in the semi-infinite body. The actual heat flow may be slower in the longer time due to the heat conduction reaching the end of the specimen. In addition, the spatial dimension $n = 2$ is in better agreement, implying a dominant heat flow in the two-dimensional diffusion in the x-y plane.

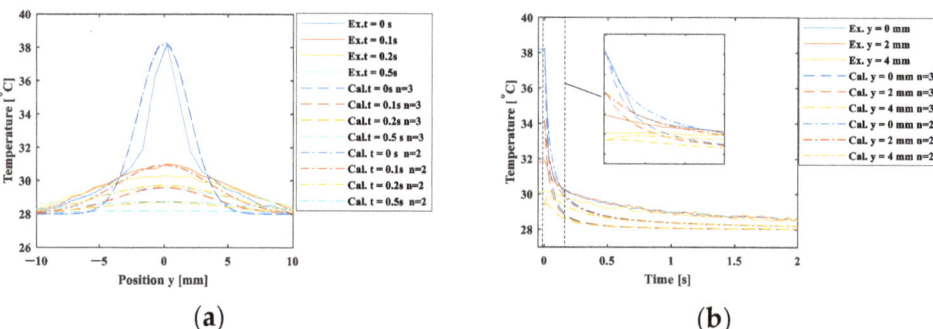

Figure 9. Comparison of the cooling process in the experiment and the point heat source model. (a) spatial changes; (b) temporal changes.

On the other hand, Figure 10 shows the strain change around the heating point (area A). In the x-direction (Figure 10a), the maximum strain of approximately 2.5×10^{-4} was roughly consistent with the linear expansion per 10 °C, and it showed no significant difference in the initial tensile stress condition of 0 to 120 MPa. The maximum strain ε_{yy} in the y-direction (Figure 10b) was almost equal to ε_{xx} at the initial stress of 0 MPa, while under the initial tensile stress condition, it obviously became smaller. This result indicates that the thermal expansion was hindered due to the constraint in the y-direction. The thermal expansion coefficient α can be regarded as a constant in the temperature change within 10 °C. According to the data of the aluminum alloy obtained by Takeuchi et al., the change in the thermal expansion coefficient α in this temperature range is on the order of 10^{-8} [K^{-1}] [23]. The thermal strain in area A was estimated as $\epsilon_T = \alpha \Delta T$, using the temperature change measured with the infrared camera shown in Figure 9a. Here, the temperature difference between the positions $x = 0$ mm and $x = 2.5$ mm was used for ΔT and 2.38×10^{-5} was used for the thermal expansion coefficient, α. The estimated thermal strain is plotted as a function of time in Figure 10a by a dotted line. The estimated strain value deviated from the value measured with the strain gauge; the estimated thermal strain exhibited a more rapid decrease than the measured strain. The measured strain includes the thermal strain and the elastic strain due to the constraint of the surrounding area. When the thermal strain was instantaneously removed by heat diffusion in the cooling process, elastic strain relaxation subsequently occurred. The rapid drop in the thermal strain ϵ_T shown in Figure 10a implies that the elastic strain relaxation was independent of the heat diffusion process discussed in Figure 9. The disagreement between the temperature measurement and strain measurement rules out the role of thermal strain in the observed strain measurement.

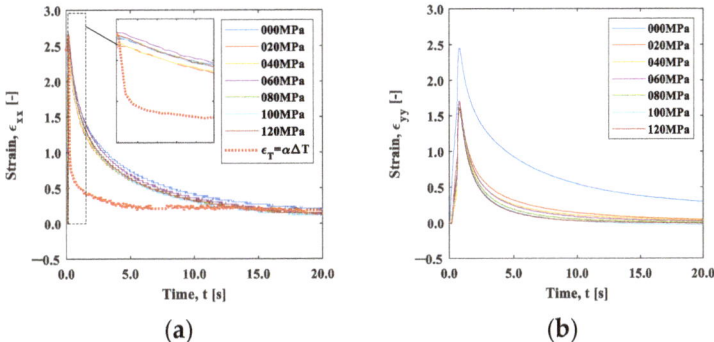

Figure 10. Strain–time curve measured by a strain gauge: (**a**) x-direction; and (**b**) y-direction.

We now discuss the observed strain relaxation behavior. It is well known that viscoelastic materials exhibit exponential strain relaxation due to their viscosity. Although the origin of viscosity in this experiment is unclear, the strain relaxations shown in Figure 10 indicate exponential-like decay. Figure 11a,b are natural logarithm plots of the strain change during the cooling after the maximum temperature in Figure 10. It can be seen that the slope is not completely linear, indicating that the exponential-like decay cannot be represented by a single decay constant. However, a close look at the initial part of the graph indicates that the trend between t = 0 s and t = 0.4 s is approximately linear, as the inserts in these figures exhibit.

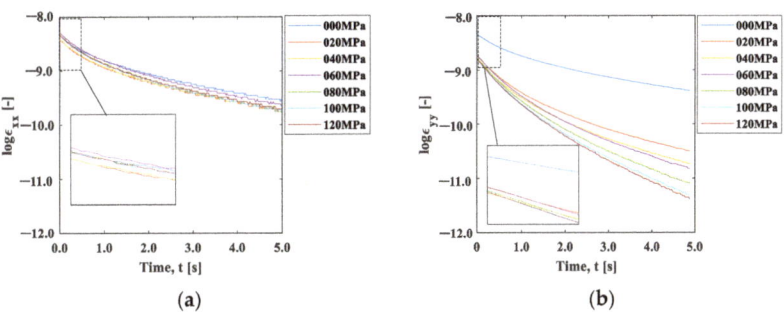

Figure 11. Log (strain)–time curve: (**a**) x-direction; and (**b**) y-direction.

We evaluated this initial slope, called the coefficient of strain relaxation (CSR), by least squares fitting the natural logarithmic curve within 0.4 s in Figure 11. Figure 12a shows the results plotted as a function of the initial stress condition. The CSR_y decreased significantly with the initial stress, indicating that the relaxation occurred faster with the increase in the initial stress. In contrast, the CSR_x showed a slight increase with the initial strain, indicating that the strain relaxation behavior was opposite to CSR_y. The elastic modulus has stress dependency originating from the non-linearity and anharmonicity of the interatomic potential [18]. It is known that the elasticity decreases under tensile stress due to the fact that the slope of the potential curve decreases as the interatomic distance increases from its equilibrium position. It should be noted that the condition $\sigma_i = 0$ is essentially different from the other initial stress conditions because one end of the specimen is under no constraint. The difference in the stress dependence between CSR_x

and CSR_x observed in Figure 12a can be ascribed to Poisson's effect. The tensile stress in the y-direction leads to the compressive stress in the x-direction.

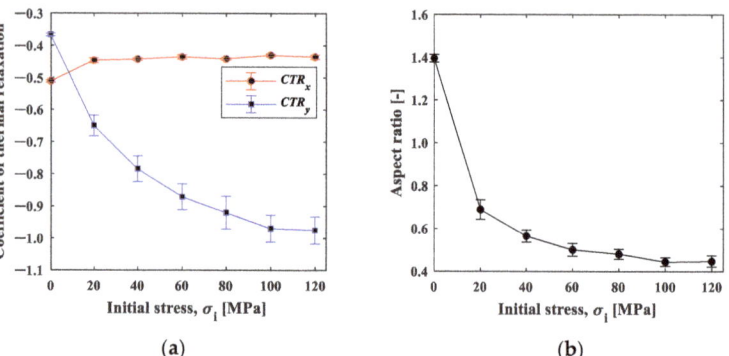

Figure 12. (a) CSR–initial stress curve plot; and (b) aspect ratio of CSR_x/CSR_y.

Figure 12b shows the aspect ratio, CSR_x/CSR_y. The aspect ratio under the initial stress $\sigma_i = 0$ MPa was about 1.4, indicating that larger shrinkage occurred in the x-direction. This may be due to the fact that the deformation was constrained in the x-direction with the clamp. The aspect ratio decreased with the increase in the initial stress, and then it approached Poisson's ratio in elasticity. This fact supports our proposition that the influence of elasticity, not the thermal effect, is dominant in the cooling process.

Figure 13a shows the strain change in area B, 25 mm away from the heat spot. The strain changed in the negative direction (compressive direction) in the initial tensile stress conditions, except for the condition of 0 MPa. As shown in Figure 5, the temperature difference in area B was within +1 °C, and the thermal effect was small enough to neglect. This result indicates that the elastic strain was recovered by the thermal expansion of the spot-heated area. The change in strain may follow the exponential function, and this also applies to area A. Figure 13b shows the minimum strain in the strain–time curve plotted against the initial stress. The minimum strain decreased depending on the initial stress, demonstrating the stress dependence of the compressive strain relaxation.

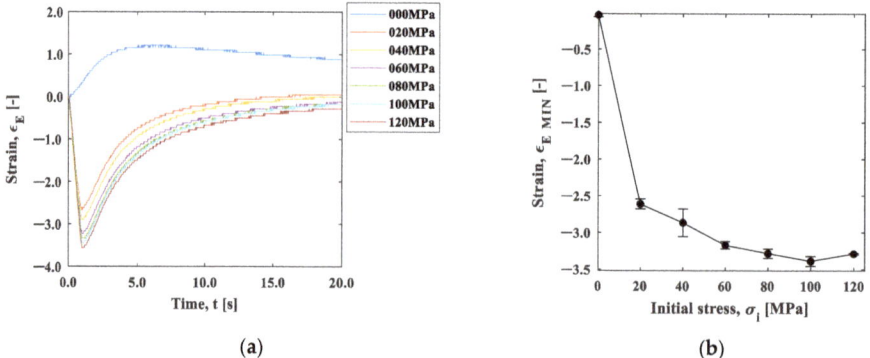

Figure 13. (a) Strain–time curve; and (b) strain–initial stress curve.

Figure 14a shows the natural logarithm of the strain after the minimum strain at the initial stress of 20–120 MPa. All stresses are shown linearly; therefore, Figure 15b shows the slope of the line up to 1 s in Figure 14a as the coefficient of elastic relaxation (CER). Figure 14b shows that the CER increased with the increase in the initial stress. In other words, the rate of elastic relaxation strain became slower due to the decrease in the elastic modulus with increasing stress. As shown in Figure 12, the rate of thermal shrinkage of the heated area (area A) was faster in the y-direction. In contrast, the elastic relaxation of the less heat-affected area (area B) shown in Figure 14 increased as the initial stress increased. The rate of elastic relaxation became slower as the initial stress increased due to the stress dependence of the elastic modulus, as shown in Figure 14. These results show that the initial stress can be estimated from the above factors.

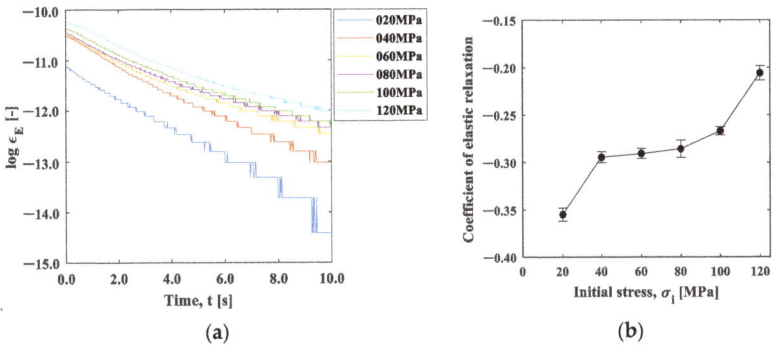

Figure 14. (a) Log (strain)–time curve; and (b) CER–initial stress curve.

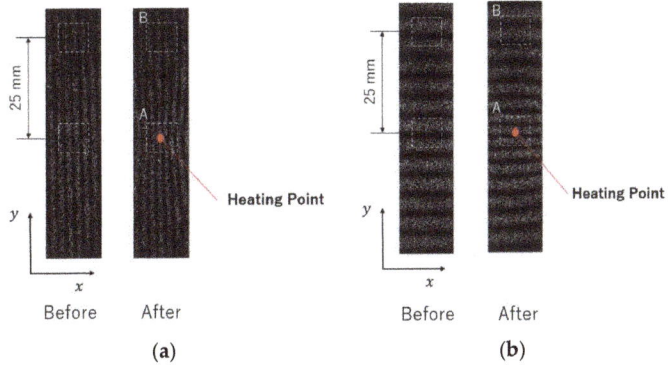

Figure 15. Fringe images before and after the spot heating, in the (a) x-direction and (b) y-direction.

3.2. Strain Change Measured by ESPI

Figure 15a,b show typical fringe patterns measured by the 2D-ESPI. The initial stress was 120 MPa. The left images show the carrier fringes introduced before the heating. The fringe intervals became narrower toward the center of the heating point, as shown in the right images, indicating that thermal expansion occurred.

Figure 16 shows the time variation of the mean strains in area A. The strain–time curve exhibits almost the same pattern as that obtained in the strain gauge measurement; the thermal strain increased instantaneously, followed by an exponential decrease in the strain. On the other hand, the strain value was slightly larger than that measured with the strain gauge. The ESPI can directly measure the strain field of the heating point, while the strain gauge shows the averaged strain value in the gauge length that is applied around the heating point. This may result in a larger value being returned by the ESPI. In addition, the strain value measured with the ESPI includes higher noise. This temporal noise may depend on the interval of the carrier fringe initially introduced. In this experiment, the strain was calculated from the spatial frequency (number of fringes in the region of interest) of the carrier fringes. The ratio of the fringe number before/after the deformation became larger as the number of fringes decreased, leading to high sensitivity for displacement, while the analysis was affected by thermal fluctuation and mechanical vibration. In addition, the resolution of the analysis had a limitation due to the pixel resolution of the CMOS camera. The resultant data contain strain noise on the order of 10^{-5}. Although this noise level is greater than that of the strain gauge, Figure 16 indicates that the S/N (signal-to-noise) ratio was satisfactorily small.

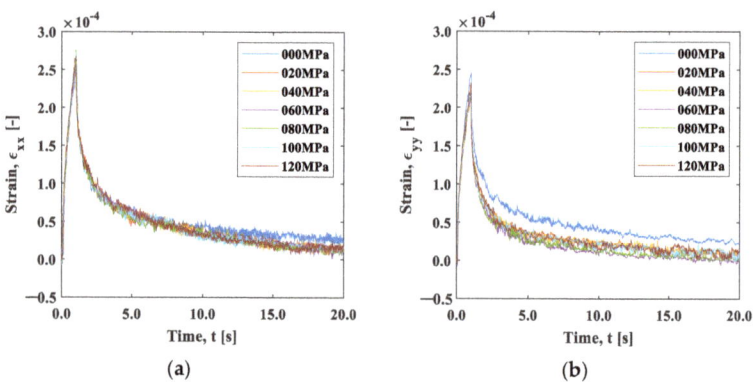

Figure 16. Strain–time curve obtained using the ESPI: (**a**) x-direction; and (**b**) y-direction.

Here, the stress dependences can be discussed in the same manner as the strain gauge measurement. Figure 17 shows the natural logarithm of the strain. The first-order approximation of the curve for 0.4 s after heating and the slope for each stress are shown in Figure 18. The strain relaxation coefficient obtained by the ESPI measurement showed stress dependency, similar to the strain gauge measurement; the CSR in the y-direction decreased with the initial stress, and the CSR in the x-direction and its aspect ratio also showed the same tendency as the strain gauge measurement. Furthermore, the maximum strain in the unheated area (area B) during the local heating test was plotted against the initial stress, as shown in Figure 19. It is confirmed that the nonthermal strain relaxation increased with increasing stress, although there was some scatter. The above results show that the ESPI allows the same residual stress analysis as the strain gauge measurement.

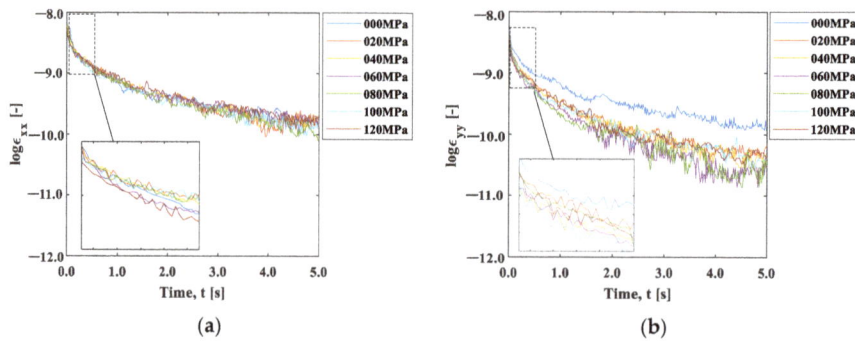

Figure 17. Time–log (strain) curve obtained using the ESPI: (**a**) x-direction; and (**b**) y-direction.

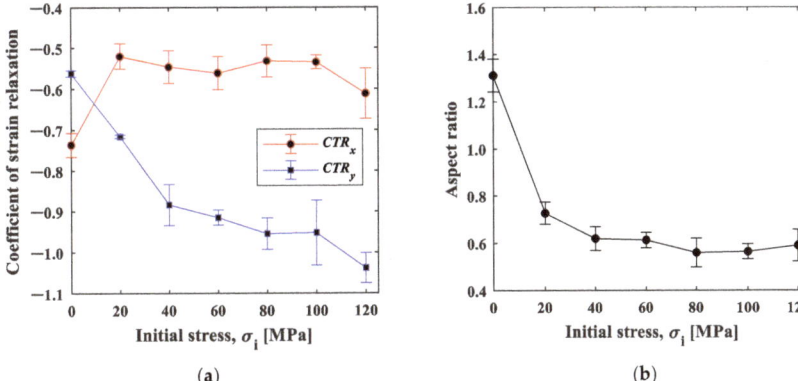

Figure 18. (a) CSR–initial stress curve; and (b) aspect ratio.

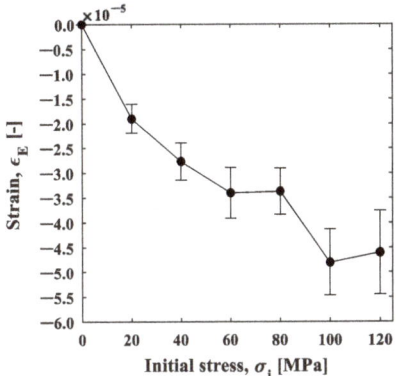

Figure 19. Strain–initial stress plot for area B obtained by the ESPI measurement.

The full-field measurement can be performed using an ESPI. Figure 20 shows the spatial strain distribution in the y-direction measured during the cooling process. The initial stress σ_i was 120 MPa. The strain exhibited a Gaussian distribution at $t = 0$ s (immediately after the heating stopped), while the width of the function was larger than the temperature distribution shown in Figure 4. This also indicates that the strain relaxation occurred in a different manner to the thermal diffusion. As the distance from the heat spot increased, the strain decreased towards the negative range, indicating compressive deformation. In this study, the strain distribution was obtained through the frequency analysis of the carrier fringe introduced before the heating. The strain fluctuation observed in the region where the strain was close to zero was due to the fact that, in this region, the high fringe density compromises the spatial resolution of the image. As revealed in the strain gauge measurements, the relaxation of tensile strain in the heated area and the compressive strain in the non-heated area occurred simultaneously during the experiment. The boundary of tension/compression in the strain profile may provide an indication of the position where the strain was balanced. At the cooling time of 1s, the strain exhibited tensile strain at a distance of about 10 mm from the heated area, above which it became compressive. The time variation of the boundary position between ε_T and ε_C is shown in Figure 21a. With the elapse of time, the boundary position moved away from the heated area, converging to -22.0 mm at $t = 3.0$. The strain relaxation induced by the removal of thermal expansion propagated in this time. Furthermore, area B, which was 25 mm away from the heat spot, can be regarded as the area that was not affected by the heat conduction. The relationship between $|\varepsilon_T|$ and $|\varepsilon_C|$ showed a linear relation, implying that $\Delta\varepsilon_T/\Delta\varepsilon_C \simeq const$. This fact directly indicates that the strain in the heat spot was constantly balanced with the nonthermal strain in the less heat-affected area, regardless of the thermal diffusion.

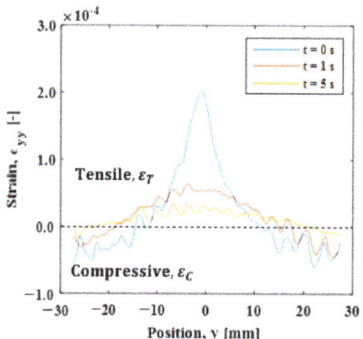

Figure 20. Strain distribution in the y-direction.

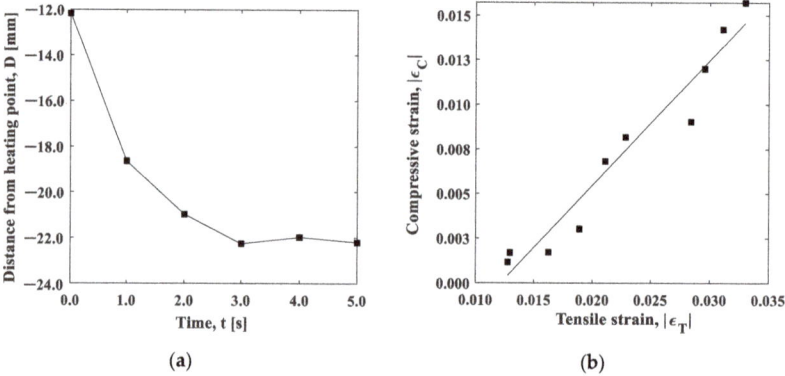

Figure 21. (a) Boundary position between compression and tension; and (b) relationship between $|\varepsilon_T|$ and $|\varepsilon_C|$.

4. Conclusions

The present study investigated the effect of the stress condition on the thermal deformation behavior through local spot heating in the temperature range of room temperature to approximately +10 °C. The strain was measured using a strain gauge in a contact way, and a two-dimensional ESPI in a non-contact way. The following results were obtained:

(1) Relaxation of the positive (tensile) strain in the local spot-heated area occurred more slowly than the thermal relaxation due to the heat diffusion, and it showed an exponential decay behavior.

(2) The coefficient of strain relaxation obtained by the strain relaxation curve in the shorter time depended on the tensile stress initially applied, which was attributed to the stress dependency of elasticity.

(3) In the less heat-affected area (the area far from the heated area), the compressive strain was induced by the thermal expansion of the heated area. The compressive strain in the cooling process also showed stress dependency.

(4) The two-dimensional ESPI allowed the visualization of the above strain relaxation behavior in a non-contact way. These results indicate the feasibility of non-destructive

and non-contact residual stress estimation through the evaluation of the above relaxation coefficients.

Author Contributions: Conceptualization, T.S. and S.Y.; investigation, Y.M. and T.S.; project administration, T.S. and S.Y.; writing—original draft preparation, Y.M.; writing—review and editing, S.Y. and T.S. All authors have read and agreed to the published version of the manuscript.

Funding: This work was supported by The Ministry of Education, Culture, Sports, Science and Technology (MEXT) KAKENHI No. 18H01722.

Institutional Review Board Statement: Not applicable.

Informed Consent Statement: Not applicable.

Data Availability Statement: Not applicable.

Conflicts of Interest: The authors declare no conflict of interest.

References

1. Barile, C.; Casavola, C.; Pappalettera, G.; Pappalettere, C. Analysis of the effects of process parameters in residual stress measurements on Titanium plates by HDM/ESPI. *Measurement* **2014**, *48*, 220–227. [CrossRef]
2. Wu, S.Y.; Qin, Y.W. Determination of residual stresses using large shearing speckle interferometry and the hole drilling method. *Opt. Lasers Eng.* **1995**, *23*, 233–244. [CrossRef]
3. Nelson, D.V.; Makino, A.; Fuchs, E.A. The holographic-hole drilling method for residual stress determination. *Opt. Lasers Eng.* **1997**, *27*, 3–23. [CrossRef]
4. Pisarev, V.S.; Bondarenko, M.M.; Chernov, A.V.; Vinogradova, A.N. General approach to residual stresses determination in thin-walled structures by combining the hole drilling method and reflection hologram interferometry. *Int. J. Mech. Sci.* **2005**, *47*, 1350–1376. [CrossRef]
5. Ya, M.; Miao, H.X.; Lu, J. Determination of residual stress by use of phase shifting moiré interferometry and hole-drilling method. *Opt. Lasers Eng.* **2006**, *44*, 68–79.
6. Jiang, Y.; Xu, B.-s.; Wang, H.-d.; Liu, M.; Lu, Y.-h. Determination of residual stresses within plasma spray coating using Moiré interferometry method. *Appl. Surf. Sci.* **2011**, *257*, 2332–2336.
7. Peng, Y.; Zhao, J.; Chen, L.S.; Dong, J. Residual stress measurement combining blind-hole drilling and digital image correlation approach. *J. Constr. Steel Res.* **2021**, *176*, 106346. [CrossRef]
8. Xu, Y.; Bao, R. Residual stress determination in friction stir butt welded joints using a digital image correlation-aided slitting technique. *Chin. J. Aeronaut.* **2017**, *30*, 1258–1269.
9. Martínez-García, V.; Pedrini, G.; Weidmann, P.; Killinger, A.; Gadow, R.; Osten, W.; Schmauder, S. Non-contact residual stress analysis method with displacement measurements in the nanometric range by laser made material removal and SLM based beam conditioning on ceramic coatings. *Surf. Coat. Technol.* **2019**, *371*, 14–19. [CrossRef]
10. Pechersky, M.J.; Miller, R.F.; Vikram, C.S. Residual stress measurements with laser speckle correlation interferometry and local heat treating. *Opt. Eng.* **1995**, *10*, 34. [CrossRef]
11. Jiang, W.; Wan, Y.; Tu, S.T.; Wang, H.; Huang, Y.; Xie, X.; Li, J.; Sun, G.; Woo, W. Determination of the through-thickness residual stress in thick duplex stainless steel welded plate by wavelength-dependent neutron diffraction method. *Int. J. Press. Vessel. Pip.* **2022**, *196*, 104603. [CrossRef]
12. Ao, S.; Li, C.; Huang, Y.; Luo, Z. Determination of residual stress in resistance spot-welded joint by a novel X-ray diffraction. *Measurement* **2020**, *161*, 107892. [CrossRef]
13. Yelbay, H.I.; Cam, I.; Gür, H.C. Non-destructive determination of residual stress state in steel weldments by Magnetic Barkhausen Noise technique. *NDT E Int.* **2010**, *43*, 29–33. [CrossRef]
14. Kurashkin, K.; Mishakin, V.; Rudenko, A. Ultrasonic Evaluation of Residual Stresses in Welded Joints of Hydroelectric Unit Rotor Frame. *Mater. Today Proc.* **2019**, *11*, 163–168. [CrossRef]
15. Yoshida, S.; Sasaki, T.; Craft, S.; Usui, M.; Haase, J.; Becker, T.; Park, I.K. Stress Analysis on Welded Specimen with Multiple Methods. *Adv. Opt. Methods Exp. Mech.* **2014**, *3*, 143–152.
16. Yoshida, S.; Sasaki, T.; Usui, S.; Sakamoto, S.; Girney, D.; Park, I.K. Residual Stress Analysis Based on Acoustic and Optical Methods. *Materials* **2016**, *9*, 112. [CrossRef]
17. Sasaki, T.; Yoshida, S.; Ogawa, T.; Shitaka, J.; McGibboney, C. Effect of Residual Stress on Thermal Deformation Behavior. *Materials* **2019**, *12*, 4141. [CrossRef]
18. Hughes, D.S.; Kelly, J.L. Second-order elastic deformation of solids. *Phys. Rev.* **1953**, *92*, 1145–1149. [CrossRef]
19. Rosenfield, A.R.; Averbach, B.L. Effect of stress on the thermal expansion coefficient. *J. Appl. Phys.* **1956**, *27*, 154–156. [CrossRef]
20. Bert, C.W.; Fu, C. Implications of stress dependency of the thermal expansion coefficient on thermal buckling. *J. Press. Vessel Technol.* **1992**, *114*, 189–192. [CrossRef]

21. Li, X.; Ye, Y.; Zhang, D. Simultaneous 3D ESPI using a 3CCD Camera. In Proceedings of the International Conference on Advanced Technology in Experimental Mechanics, Niigata, Japan, 7–11 October 2019; p. 10.
22. Carslaw, H.S.; Jaeger, J.C. *Conduction of Heat in Solids*; Oxford University Press: London, UK, 1959.
23. Takeuti, Y.; Nyuko, H.; Noda, N.; Zaima, S. A consideration on thermal stresses in a composite cylinder made of aluminum alloy and steel with temperature dependence of material constants. *Light Met.* **1978**, *28*, 247–252. [CrossRef]

Article

Dynamic ESPI Evaluation of Deformation and Fracture Mechanism of 7075 Aluminum Alloy

Shun Takahashi [1], Sanichiro Yoshida [2], Tomohiro Sasaki [1,*] and Tyler Hughes [2]

1. Graduate School of Science and Technology, Niigata University, Niigata 9502181, Japan; f19b107f@mail.cc.niigata-u.ac.jp
2. Department of Chemistry and Physics, Southeastern Louisiana University, Hammond, LA 70402, USA; sanichiro.yoshida@selu.edu (S.Y.); Joshua.hughes-2@selu.edu (T.H.)
* Correspondence: tomodx@eng.niigata-u.ac.jp; Tel.: +81-25-262-6710

Citation: Takahashi, S.; Yoshida, S.; Sasaki, T.; Hughes, T. Dynamic ESPI Evaluation of Deformation and Fracture Mechanism of 7075 Aluminum Alloy. *Materials* **2021**, *14*, 1530. https://doi.org/10.3390/ma14061530

Academic Editor: Alessandro Pirondi

Received: 9 February 2021
Accepted: 17 March 2021
Published: 20 March 2021

Publisher's Note: MDPI stays neutral with regard to jurisdictional claims in published maps and institutional affiliations.

Copyright: © 2021 by the authors. Licensee MDPI, Basel, Switzerland. This article is an open access article distributed under the terms and conditions of the Creative Commons Attribution (CC BY) license (https://creativecommons.org/licenses/by/4.0/).

Abstract: The deformation and fracture mechanism in 7075 aluminum alloy is discussed based on a field theoretical approach. A pair of peak-aged and overaged plate specimens are prepared under the respective precipitation conditions, and their plastic deformation behaviors are visualized with two-dimensional electronic speckle pattern interferometry (ESPI). The in-plane velocity field caused by monotonic tensile loading is monitored continuously via the contour analysis method of ESPI. In the plastic regime, the peak-aged specimen exhibits a macroscopically uniform deformation behavior, while the annealed specimen exhibits non-uniform deformation characterized by a localized shear band. The occurrence of the shear band is explained by the transition of the material's elastic resistive mechanism from the longitudinal force dominant to shear force dominant mode. The shear force is interpreted as the frictional force that drives mobile dislocations along the shear band. The dynamic behavior of the shear band is explained as representing the motion of a solitary wave. The observed decrease in the solitary wave's velocity is accounted for by the change in the acoustic impedance with the advancement of plastic deformation.

Keywords: field theory; deformation and fracture; 7075 aluminum alloy; electronic speckle pattern interferometry

1. Introduction

Micro-fractures in structural materials can rapidly grow in their scale and directly lead to serious accidents. Even in aluminum alloys, known to be highly ductile materials, failure occurs in an uncontrollable fashion. Because of their high specific strength, corrosion resistance, weldability, and inexpensiveness, aluminum alloys are widely used for the parts of various structures. Thus, understanding of the fracture mechanism in aluminum alloys has been subjects of intensive study for a long time. However, much remains unexplained.

A challenging aspect in this regard is the difficulty in the description of the transition from plastic deformation to fracture. It is generally known that the fracture of ductile metals is initiated by non-uniformity in the microstructure, such as anisotropy of crystal grains, inhomogeneity of precipitates and solid solutions and develop to macroscopic non-uniformity. It is also known that the progression of localized yielding and plastic deformation develop to necking and eventually lead to fracture. However, the dynamics of the transitional stage is not well understood.

As an indicator of the progression from the micro- to macro-scale non-uniformity in deformation, the phenomenon known as the Portevin-Le Chatelier (PLC) effect [1] is of great interest. The PLC effect is typically observed in tensile experiments of aluminum alloy specimens and exhibits the following two characteristics. The first is the zigzag characteristics of the stress–strain curve known as the serration. The second is the local shear deformation called the "shear band". The former represents the instability in the shear stress and the latter the local strain concentration associated with the stress instability.

The most intriguing aspect of this effect is that it qualitatively relates the micro- and macro-dynamics of deformation and its progression to fracture. In the microscopic view, it is generally accepted that the origin of plastic instability is dynamic strain aging responsible for the dynamic interactions between mobile dislocations and solute atoms in alloys [2]. The effect of strain rate and the microstructural factors such as the amount of solute atom and grain size still has been theoretically, as well as experimentally investigated by a considerable number of researchers e.g., [3].

In order to connect those microscopic and macroscopic dynamics, and thereby deformation and fracture mechanisms quantitatively, it is essential to take a comprehensive approach that describes both microscopic deformation mechanisms, such as interatomic bonding and dislocation motion in crystals, and fracture mechanics, such as necking and cracking on the same theoretical foundation. A recent field theory of deformation and fracture [4,5] has the ability to meet the requirement. Based on the fundamental physical principle known as symmetry in physics, this theory describes all stages of deformation including elasticity, plasticity and fracture mechanism on the same basis without relying on phenomenology and scale-independently. According to this theory, deformation can be comprehensively described as the wave dynamics that govern the displacement field of deforming objects and the transition to fracture can be characterized by the change in the form of the wave. More specifically, as the deformation transitions from the deformation to fracturing stage, the wave changes its form from sinusoidal type waves to a solitary waves.

On the experimental end, a number of authors use various optical techniques to visualize the above-mentioned phenomena. Hassan et al. [6] presented the first comprehensive literature review on remote deformation measurement in the presence of discontinuities using the digital image correlation technique. Ranc et.al [7] use infrared pyrometry to visualize the heat generation during the shear band formation. These studies have correlated the temporal and spatial behavior of the shear band with the nature of serrations. However, they do not clarify the detailed mechanism underlying in the transition from deformation to fracture.

From the viewpoint of implementation of the above-mentioned field theory in a diagnostic algorithm, the optical interferometric technique known as the electronic speckle-pattern interferometry (ESPI) is useful. Yoshida et al. [8] discusses the specific wave types that differentiate stages of deformation in association with the corresponding optical interferometric fringe patterns. In [9], Yoshida et al. extend the discussion to the transition to the fracture and identify the shear band as manifestation of the solitary wave dynamics in association with micro-fracture and corresponding dislocation dynamics.

While ESPI is a powerful tool to implement the field theory, it is necessary to analyze a large number of interferometric fringe patters for diagnosis. It is desirable to use a computerized method to automate the fringe analysis, at least partially. Due to the non-uniformity nature of the deformation, the interference fringes are non-uniform spatially. In addition, the deformation in the late stage of plastic regime does not transient at a constant rate. These situations make it difficult to apply the conventional methods such as the phase shift method [10]. Recently, we have devised an algorithm to facilitate the process of fringe analysis under the given situation. Thus, the aim of this paper is twofold. In the first part, we present the computerized fringe analysis method. In the rest of the paper, we report our recent study in which we used this fringe analysis algorithm to visualize the non-uniform deformation and fracturing behavior in Al–Zn–Mg–Cu alloy specimens. We will discuss the physics behind the observed deformation behaviors and clarify the interconnection to the dislocation dynamics. The effect of heat treatment on the specimen is discussed as well.

2. ESPI Optical Measurement System and Analysis Algorithm

2.1. Optical Arrangement of Electronic Speckle Pattern Interferometry

The deformation behaviors of Al–Zn–Mg–Cu alloys were measured with the two-dimensional ESPI described in the previous work [9]. The experimental set-up is shown in Figure 1. The two-dimensional displacement fields in the directions x and y on the measured surface can be obtained by computing the intensity difference as a contour pattern, called the "fringe pattern". The relation of the displacement ζ and the phase difference between the interferometric arms $\Delta\phi$ is expressed by (see Appendix A):

$$\Delta\phi = 4\pi \frac{\zeta \sin \theta}{\lambda} \quad (1)$$

In this study, the tensile test was conducted at a constant strain rate of 8.33×10^{-3} [s^{-1}]. The fringe pattern was calculated for several time intervals, in order to maintain the accuracy of the analysis. The velocity $v = \partial \zeta / \partial t$ was calculated by dividing the displacement value by the time interval. The tensile load was measured with a strain gauge type load cell. The tensile strain was calculated from the displacement of crosshead and the gauge length of specimen.

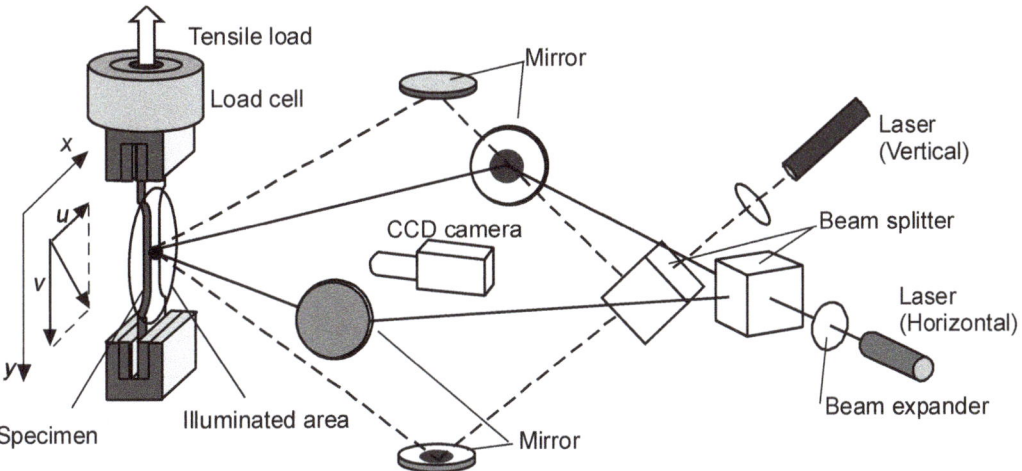

Figure 1. Optical setup of 2-D electronic speckle pattern interferometer [9].

2.2. Fringe Pattern Analysis

The displacement field can be obtained by extracting the fringe pattern. The intensity of superimposed speckles on the measurement point (x, y) is generally expressed by the following equation [11]:

$$I(x, y) = 2A^2 + 2A^2 \cos \phi(x, y) \quad (2)$$

where, A is amplitude of light source, ϕ is the initial phase difference between the speckles resulting from the two interferometric arms. Since each speckle has phase, the phase ϕ can be approximated as taking a random distribution in the measurement field like white noise. The superimposed speckle intensities before and after the deformation are expressed as follows:

$$I_{Before}(x, y) = 2A^2 + 2A^2 \cos \phi \quad (3)$$

$$I_{After}(x, y) = 2A^2 + 2A^2 \cos(\phi + \Delta\phi) \quad (4)$$

where, $\Delta\phi$ is the phase difference resulting from the deformation of material. The $\Delta\phi$ can be estimated through the subtraction of the speckle intensities before and after. The absolute difference of intensity before and after deformation is calculated as follows:

$$\left|I_{After} - I_{Before}\right| = 4A^2 \left|\sin\frac{2\phi + \Delta\phi}{2} \sin\frac{\Delta\phi}{2}\right| \quad (5)$$

Here, the first term of right hand $\sin\frac{2\phi+\Delta\phi}{2}$ contains the initial phase of the superimposed speckle, while ϕ randomly changes depending on the material surface. Since the second term $\Delta\phi/2$ becomes zero when $\Delta\phi$ is equal to a multiple of 2π regardless the ϕ, we can determine the phase $\Delta\phi$ where the speckle appears as "dark fringes". To determine the $\Delta\phi$ reducing the speckle noise caused by the first term, image processing was conducted. The process consists of five steps of (i) speckle noise reduction using Gaussian filter, (ii) phase determination with partial differential image, (iii) binarization, (iv) morphological processing and detection fringes and (v) interpolation as shown in Figure 2. Details of the steps are described as follows:

- Step (i): Speckle noise reduction using Gaussian filter.

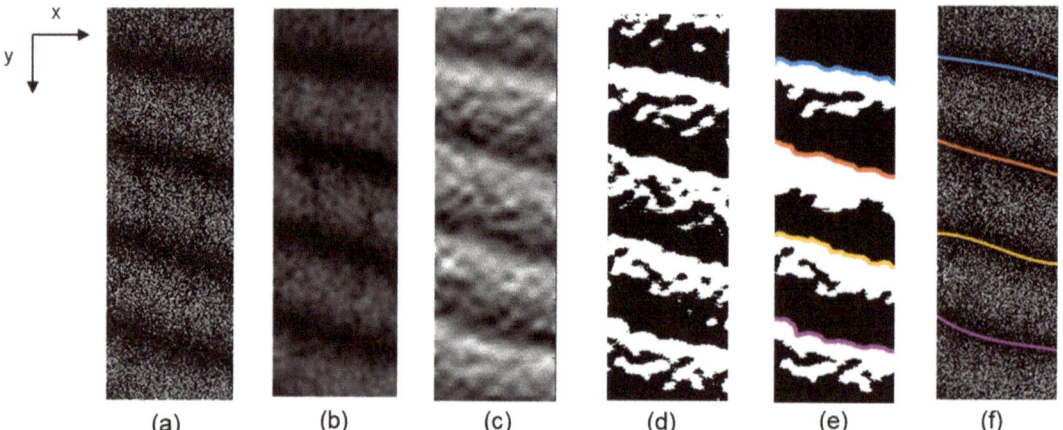

Figure 2. (a) original image (b) Gaussian filtered image (c) partial differential image (d) binarized image (e) morphology processed image (f) detected fringe line.

Figure 2a shows a specific example of fringe pattern obtained in the tensile test of aluminum alloy. The fringe pattern represents in-plane displacement component v, in the direction y. The pattern includes four fringes due to the displacement on the measurement surface and the noise resulting from the speckle noise. To remove the high frequency component of noise, we applied a weighted Gaussian filter as follows:

$$I_{filtered}(x,y) = \sum_i \sum_j h(i,j) I(i-x, j-y) \quad (6)$$

$$h_g(i,j) = e^{-\frac{i^2+j^2}{2\sigma^2}} \quad (7)$$

$$h(i,j) = \frac{h_g(i,j)}{\sum_i \sum_j h_g(i,j)} \quad (8)$$

where, σ is standard deviation of the Gaussian distribution, indexes i and j represent distances from center position of filter. The filtered image is obtained by a convolution between the intensity of the original image and the Gaussian filter. The filter is a square and its size is calculated by $2[2\sigma + 1]$. It is possible to remove high-frequency components of the speckle noise by the convolution operation with a fringe image. The frequency characteristic of filter is determined by standard deviation σ. Figure 3 shows frequency characteristics for each sigma value. Mean interval of fringe appeared in Figure 2a is roughly 70 pixels. The smallest fringe interval obtained in this study is 20 pixels as discussed in the later sections. In other words, the number of fringes obtained in this study is around 17 per image length. According to Figure 3, the standard deviation at $\sigma = 4$ is low enough to cover the measured fringe resulting from the deformation. Figure 2b shows the filtered image at the $\sigma = 4$. The resultant image shows that Gaussian filter can remove the high frequency noises without losing the main fringe information.

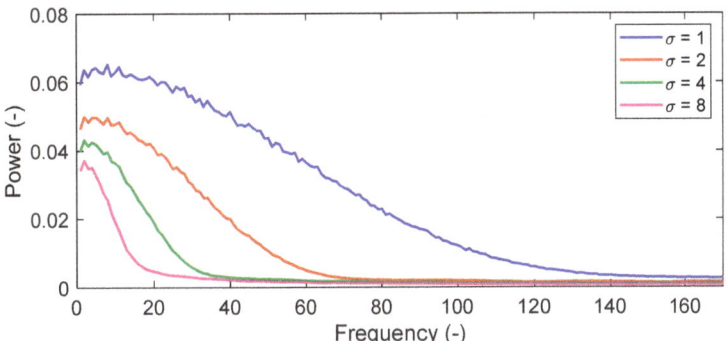

Figure 3. Frequency characteristics of the Gaussian filter.

- Step (ii): Phase determination with partial differential image

To determine the phase at positions of "dark fringe", we devised a new algorithm using partial differentiation as follows. The filtered speckle intensity can be calculated as follows:

$$I_{filtered} = 4A^2 \left| \sin \frac{\Delta \phi}{2} \right| \tag{9}$$

Then, the partial differentiation in the direction y of Equation (9) can be written as follows:

$$\frac{\partial I_{filtered}}{\partial y} = \begin{cases} 4A^2 \frac{\Delta \phi'}{2} \cos \frac{\Delta \phi}{2} & 2n\pi < \frac{\Delta \phi}{2} < (2n+1)\pi \\ -4A^2 \frac{\Delta \phi'}{2} \cos \frac{\Delta \phi}{2} & (2n-1)\pi < \frac{\Delta \phi}{2} < 2n\pi \end{cases} \tag{10}$$

It should be noted that $\partial I / \partial y$ switches discontinuously positive to negative at the phase difference, $\Delta \phi = 2n\pi$. Figure 4 shows the intensity profile of Figure 2b in the direction y and its partial derivative obtained by Equation (9). The intensity is centerline of the direction x. Since the intensity change in the derivative curve becomes discontinuous, the phase can be easily discriminated. Figure 2c shows the image obtained by applying the Equation (10) to Figure 2b. The result is scaled in 256 gradation.

- Step (iii): Binarization

The partially differentiated intensity is normalized in integer value from -1 to 1 as shown in Figure 4. In the differentiated profile, the position where the value is zero means the valley of original intensity profile, "the dark fringe". The differentiated image is binarized with a threshold of 0, in other words, all color values under 0 become 0, while all others become 1. This process detects the dark fringe as a boundary of the region without depending on the bias of intensity of the fringe pattern. The binarized image is obtained as shown in Figure 2d.

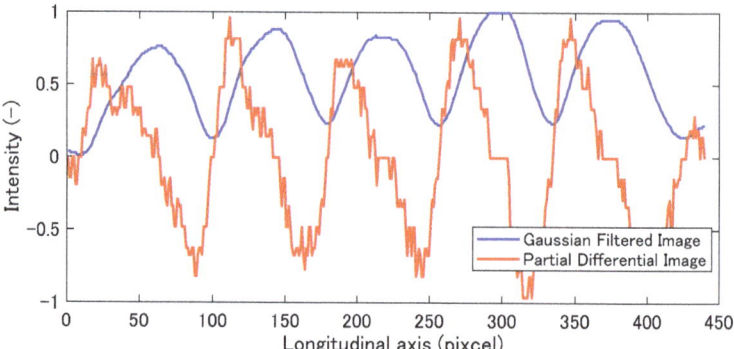

Figure 4. Intensity distribution in the direction x.

- Step (iv): Morphological processing

Figure 2d still has tears and minute regions. In order to extract only the main fringe region, the binarized image was subjected to morphological process of erosion and dilation. Figure 2e shows the morphological processed image. By detecting the boundary of the black and white area of the processed image, the valley part of the fringe pattern can be extracted. The colored line in Figure 2e is a plot of the border where the area changes from black to white for each area.

- Step (v): Detection fringes and interpolation

The border line of the area from Figure 2e is looks discontinuity, however, the actual deformation of the specimen should occur continuously. In the previous works [1–4], we demonstrated that the elastic/plastic deformation can be expressed based on the field theory. The deformation process will be discussed based on the second order derivative function of the displacement. Thus, it is accurate enough to fit the curve with a quadratic function. Figure 2f shows fitted lines overlaid in the original speckle image. Regarding the displacement in the tensile direction, the phase at the fixed end of specimen was set to 0, and the order of fringe, n was indexed forward to the crosshead side. Then, the phase values between the dark fringes were interpolated by piecewise cubic hermite interpolating polynomial (PCHIP) algorithm [12]. Regarding the displacement in the widthwise direction u, the fringe order n was indexed from the left end to the right end, followed by the interpolation in the same manner of tensile direction. Assuming that the displacement in the direction x is bilaterally symmetric, all phase values were shifted so that the phase at the center coordinates of the specimen is zero. Consequently, two-dimensional displacement vector $\zeta(u,v)$ is obtained. Strain and rotation tensors are defined as follows:

$$E = \begin{bmatrix} \varepsilon_{xx} & \varepsilon_{xy} \\ \varepsilon_{yx} & \varepsilon_{yy} \end{bmatrix} = \begin{bmatrix} \frac{\partial u}{\partial x} & \frac{1}{2}\left(\frac{\partial u}{\partial y}+\frac{\partial v}{\partial x}\right) \\ \frac{1}{2}\left(\frac{\partial v}{\partial x}+\frac{\partial u}{\partial y}\right) & \frac{\partial v}{\partial y} \end{bmatrix} \quad (11)$$

$$\Omega = \frac{1}{2}(\Delta \times \zeta) = \begin{bmatrix} 0 & \frac{1}{2}\left(\frac{\partial u}{\partial y}-\frac{\partial v}{\partial x}\right) \\ \frac{1}{2}\left(\frac{\partial v}{\partial x}-\frac{\partial u}{\partial y}\right) & 0 \end{bmatrix} \quad (12)$$

Here, assuming a two-dimensional plane strain condition, the displacement along z axis, $w = 0$, $\partial u/\partial z = \partial v/\partial z = 0$. The rotation vector ω in the plane strain condition is given as follows:

$$\omega = \begin{bmatrix} \omega_x \\ \omega_y \\ \omega_z \end{bmatrix} = \begin{bmatrix} 0 \\ 0 \\ \frac{1}{2}\left(\frac{\partial v}{\partial x}-\frac{\partial u}{\partial y}\right) \end{bmatrix} \quad (13)$$

3. Deformation Process Observed in the Tensile Test

3.1. Material and Specimen

The material used in this study was a sheet of industrial Al–Zn–Mg–Cu alloy (AA7075) with 5 mm thick. The chemical composition of alloy is shown in Table 1. To investigate the effect of matrix hardness on the deformation process, the following two types of heat treatments were employed. First, the material was solution treated at 480 °C for 2 h and hardened at 120 °C for 24 h up to peak hardness (7075-T6 alloy). In the microstructural view, the solute atoms form precipitates, resulting in high yield strength. The T6 treated alloy was subsequently over aged at 400 °C for 30 min. The matrix is softened by coarsening the precipitates (7075-annealed alloy). The optical micrographs of specimens are shown in Figure 5. Mean grain sizes of two specimens are roughly 150 µm. As shown in Figure 6, the materials were cut into tensile specimens of 10 mm in gauge width and 25 mm in gauge length.

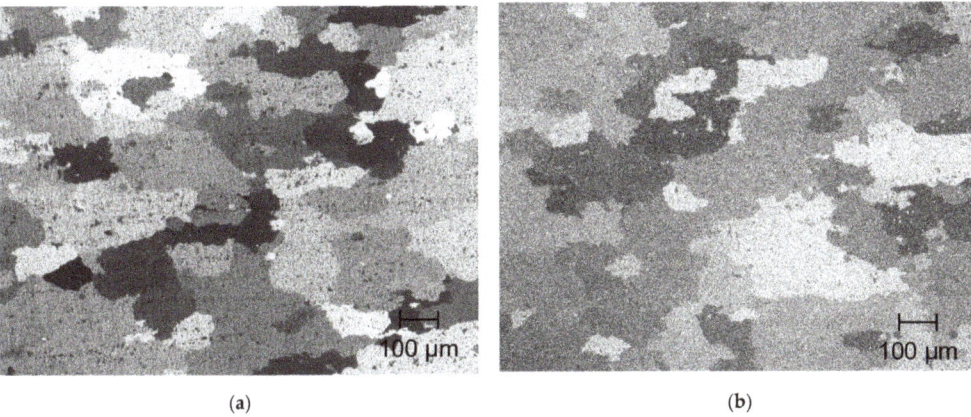

Figure 5. Optical micrographs of (a) 7075-T6 specimen and (b) 7075-annealed specimen.

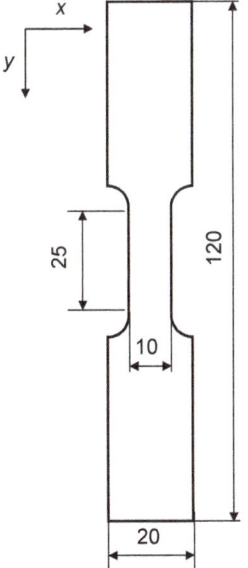

Figure 6. Shape of the Specimen.

Table 1. Chemical composition of AA7075 (mass %).

Alloy	Si	Fe	Cu	Mn	Mg	Cr	Zn	Ti
7075	~0.40	~0.50	1.2~2.0	~0.30	2.1~2.9	0.18~0.28	5.1~6.1	~0.20

3.2. Experimental Results

Figures 7–9 show stress–strain curves (S-S curve) of 7075-T6 alloy and 7075- annealed alloy, and the fringe patterns observed at several strain levels. The fringe images shown in the figure represent respectively.

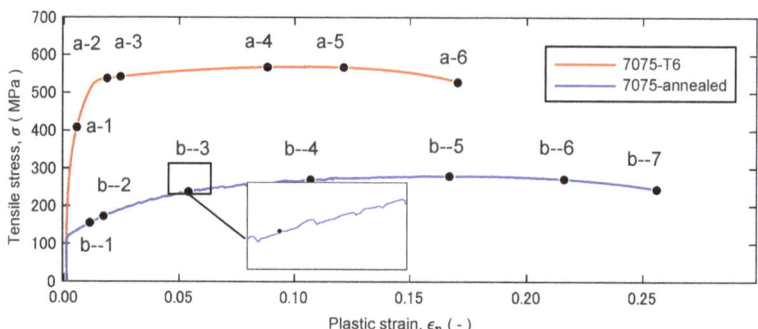

Figure 7. Stress–strain curves of 7075-T6 and 7075-annealed alloys.

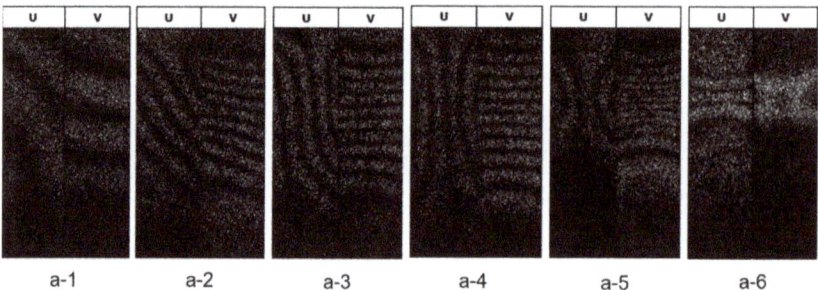

Figure 8. Evolution of fringe patterns observed in tensile test of 7075-T6 alloy. Images (a-1 to a-6) show the patterns observed at the strains marked in Figure 7.

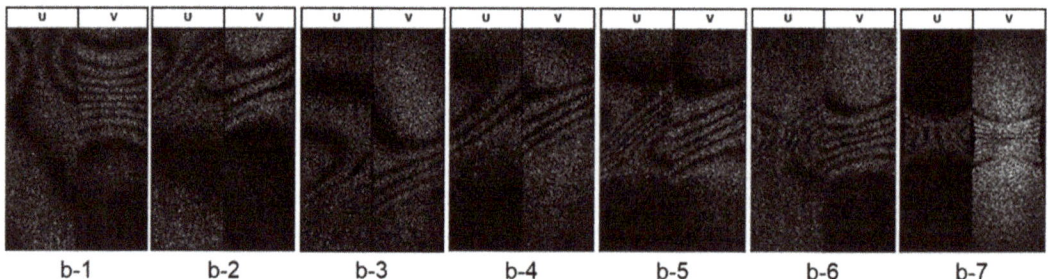

Figure 9. Evolution of fringe patterns observed in tensile test of 7075-annealed alloy. Images (b-1 to b-7) show the patterns observed at the strains marked in Figure 7.

3.2.1. 7075-T6 Alloy

7075-T6 alloy shows a homogenous plastic deformation. Figure 8a-1–a-6 respectively show the fringe patterns observed at the strains marked in Figure 7. At the stress under 500 MPa, the fringe pattern of longitudinal displacement v (v-fringe) is almost vertical to the tensile axis and an equal interval in all area of specimen. The v-fringes start to curve and concentrate to a neck of specimen (Figure 8a-1). Some number of vertical fringes to the tensile axis appears in the displacement u (u-fringes), indicating that the yielding initiated with shear and rigid body rotation. At the yielding the number of fringes increases, while the fringes with equal interval shows homogenous deformation. The deformation shows macroscopically homogeneous (Figure 8a-2–a-4). At the maximum load of 550 MPa in the S-S curve, the v-fringes concentrated to center of specimen with curving of u-fringes and necking of specimen subsequently begin (Figure 8a-5). The fringe concentration forms a bright pattern, resulting in a fracture at the center of bright pattern.

3.2.2. 7075-Annealed Alloy

7075-annealed alloy shows inhomogeneous plastic deformation characterized by a serrated curve as shown in Figure 7 in blue. The deformation process is completely different from that observed in the 7075-T6 alloy. The serration phenomenon is closely related to instability of the plastic deformation, and it is well known as the Portevin-Le Chatelier (PLC) effect. The fringes in the elastic range show similar patterns to 7075-T6 alloy. The fringe starts to concentrate just after the yielding (Figure 9b-2). The concentrated fringes exhibit a drift motion along the parallel part of specimen. The v-fringes and u-fringe subsequently tilt to an angle of 45°, forming a shear band as shown in Figure 9b-2–b-3. The shear band drifts along the parallel part of specimen and disappears at the neck part Figure 9b-4–b-6. The serrated curve is clearly seen in the S-S curve in the plastic strain $\varepsilon_p > 0.05$. The plastic deformation progresses with the drift motion of the shear band. The activity of shear band repeats until near the maximum stress. At the maximum stress in the 7075-annealed alloy (Figure 9b-7), the shear band activity becomes stationary at the center of specimen, then the material fails.

4. Discussion

In this section, we discuss the deformation behaviors observed in the present experiment based on the field theoretical interpretation of deformation and fracture. The field theoretical behavior of the deformation in the transition from the pre-yield stage to the post-yield stage, the plastic deformation dynamics in association with the mobile dislocation dynamics, and the transition from the late plastic stage to the fracturing stage in association with the corresponding wave dynamics are addressed. Through the present study we have found an interesting contrast between T6 and annealed in their deformation behaviors. The former exhibit more brittle-material like behavior and the latter more ductile-material like behavior in the transition to the fracturing stage. We will discuss the contrast from the field theoretical viewpoint.

4.1. General Arguments of Deformation and Fracture Based on Field Theory
4.1.1. Equation of Motion

According to the field theory, the material force acting on a unit volume of solid under deformation is given as follows:

$$\rho \frac{\partial v}{\partial t} = (\lambda + 2G)\nabla^2 \xi - G(\nabla \times \omega) - \sigma_0 \rho (\nabla \cdot v) v - (E \delta x \Delta x)(\partial_x^4 \xi) \qquad (14)$$

Here ρ is the density of the deforming material, v is the particle velocity vector, ξ is the particle displacement vector, G is the shear modulus, ω is the rotation vector, σ_0 is the material constant representing the degree of viscosity, E is the Young's modulus, Δx is the width of a concentrated strain and δx is the small width that borders the concentrated strain from the rest of the specimen. (See below under "Solitary wave" for more descriptions about

Δx and δx). On the right-hand side of this equation, the first term represents the continuum mechanical longitudinal elastic force, the second term the shear elastic force associated with volumetric rotation, the third term the plastic energy dissipative longitudinal (velocity damping) force, and the last term the elastic force associated with nonuniform normal strain distribution.

4.1.2. Rotational Nature of Plasticity

It is generally true that in the post-yield regime the material still possesses elasticity. This is evidenced by the fact that a material about to fracture due to a tensile load shrinks back to some extent if the load is removed. The field theory characterizes the yield phenomenon as the shear instability. When an elastic material yields, the elastic-deformation dynamics shifts from the one dominated by the longitudinal elasticity to the one dominated by the shear elasticity. In terms of Equation (14), this transition is represented by the shift that the second term dominates over the first term on the right-hand side, i.e., $(\nabla \times w)$ becomes visible in the differential displacement field.

4.1.3. Irreversibility of Plasticity

The irreversibility of plasticity is represented by the longitudinal energy dissipative mechanism represented by the third term on the right-hand side of Equation (14). This energy dissipative mechanism and the fracture mechanism as its final stage can be argued with the analogy to the electrical breakdown of gas media. Formulaically, the term $(\nabla \cdot v)$ resembles the electric charge density $\rho_e = \varepsilon(\nabla \cdot E)$ if we replace the velocity vector v with the electric field vector E. Here ε is the electric permittivity. The electric conduction current density is given as $j = \rho_e W_d = \varepsilon(\nabla \cdot E) W_d$ where W_d is the drift velocity proportional to the electric field via the electric mobility μ_e as $W_d = \mu_e E$. Thus, $j = \varepsilon(\nabla \cdot E) \mu_e E$. According to Maxwell electrodynamics, the conduction current density is proportional to the electric field via the material constant known as the conductivity as $j = \sigma E$. Equating these two expressions, we find $j = \sigma E = \varepsilon(\nabla \cdot E) \mu_e E$, hence, $\sigma = \mu_e \varepsilon(\nabla \cdot E)$. We can view the term $\sigma_0 \rho (\nabla \cdot v)$ as corresponding to $\mu_e \varepsilon(\nabla \cdot E)$, and the entire term $\sigma_0 \rho (\nabla \cdot v) v$ as resembling the conduction current density $j = \varepsilon(\nabla \cdot E) \mu_e E$. When conduction current flows under the influence of the electric field, the energy is dissipated as heat. This energy loss is known as Ohmic loss. We can interpret the plastic energy loss mechanism as resembling the Ohmic loss mechanism. Based on this analogy, we call the quantity $\nabla \cdot v$ as the deformation charge density.

4.2. Experimental Observations

Increase in $\nabla \times w$ and Localization of $\nabla \cdot v$ with Development of Deformation.

Figure 10 shows the changes in the $\nabla \times w$ field and the deformation charge density $\nabla \cdot v$ observed with the development of plastic deformation. Here, ε_p denotes the plastic strain, and for each plastic strain the quiver plot displays the two-dimensional $\nabla \times w$ vector and the surface plot displays the charge density $\nabla \cdot v$. These fields are evaluated on the -plane based on the fringe analysis described in Section 2.2. With the increase in the plastic strain, the $\nabla \times w$ field becomes less uniform. At the same time, the charge distribution is localized in a certain region of the specimen where the charge density increases.

While commonly showing the above-mentioned change in the $\nabla \times w$ field and the charge density $\nabla \cdot v$, the T6 and annealed specimens exhibit certain differences. The T6 specimen exhibits a more horizontally symmetric feature, both in the $\nabla \times w$ field, and in the charge distribution as compared with the annealed specimen. Figure 10a shows that when the charge density is concentrated near the vertical center, the T6 specimen still shows horizontally symmetric distribution in both the $\nabla \times w$ field and the charge density $\nabla \cdot v$. The concentrated charge distribution shows an X-like shape. Contrastively, the annealed specimen exhibits a horizontally asymmetric feature in the $\nabla \times w$ and $\nabla \cdot v$ patterns, as the $\nabla \times w$ field starts to be less uniform. The concentrated charge distribution runs across the specimen width forming a slanted band structure.

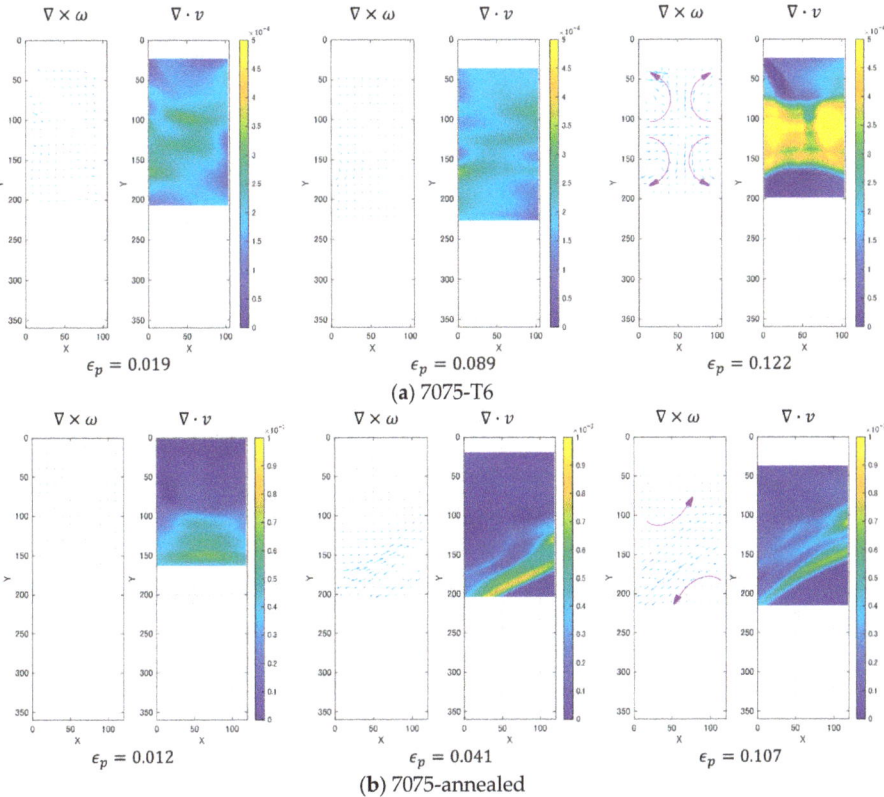

Figure 10. Typical $\nabla \times \omega$ map in the plastic deformation processes.

The $\nabla \times \omega$ field shows an increase with the development of plastic deformation. Figure 11 plots the magnitude of $\nabla \times \omega$ vector averaged over the approximately 8 × 8 mm area where the $\nabla \cdot v$ pattern is drawn in Figure 10. The average $|\nabla \times \omega|$ is plotted as a function of plastic strain ε_p. From the yield point (corresponding to $\varepsilon_p = 0$ in Figure 11) onward, the average $|\nabla \times \omega|$ increases sharply. After the initial sharp rise, in both specimens commonly, the average $|\nabla \times \omega|$ decreases at a slower rate to a minimum and resumes increasing to the highest peak at approximately the same rate as it decreases from the first peak.

The two specimens show difference in the temporal behavior of the average as well. In the case of the T6 specimen, the average $|\nabla \times \omega|$ shows the above-mentioned minimum only once as seen in Figure 11a. On the other hand, the annealed specimen shows multiple minima, exhibiting the oscillatory behavior observed in Figure 11b.

Careful analysis on the $\nabla \times \omega$ field and the $\nabla \cdot v$ distribution in Figure 10 reveals that the concentrated charge density $\nabla \cdot v$ appears next to the region where the $\nabla \times \omega$ field grows. In the case of the T6 specimen, a symmetric pattern of the concentrated charge density appears between a pair of grown $\nabla \times \omega$. The quiver plot for $\varepsilon_p = 0.122$ in Figure 10a shows an example of such a pattern. The arrows inserted in the quiver plot illustrate the approximate direction of the $\nabla \times \omega$ vectors, which represents the pattern of shear force in the area. We can interpret that the X-shaped pattern of $\nabla \cdot v$ is formed at the boundary of the vertical pair of the curly $\nabla \times \omega$ field. On the right half of the specimen, the material on the upper side of the boundary rotates clockwise and the lower side counterclockwise. On the left half of the specimen, the pattern is opposite to the right half. The upper side of the boundary rotates counterclockwise and the lower side clockwise.

This combination of four rotations forms a horizontally inward velocity. It is possible to interpret this as the indication that the specimen starts to undergo necking. In the stress–strain curve shown in Figure 7, $\varepsilon_p = 0.122$ is marked a-5. From this point onward, the stress decreases monotonically. In this process, the symmetric $\nabla \cdot v$ pattern keeps increasing the degree of localization, turning an X-shaped pattern (a-6). Macroscopically, this indicates an increase in the degree of necking.

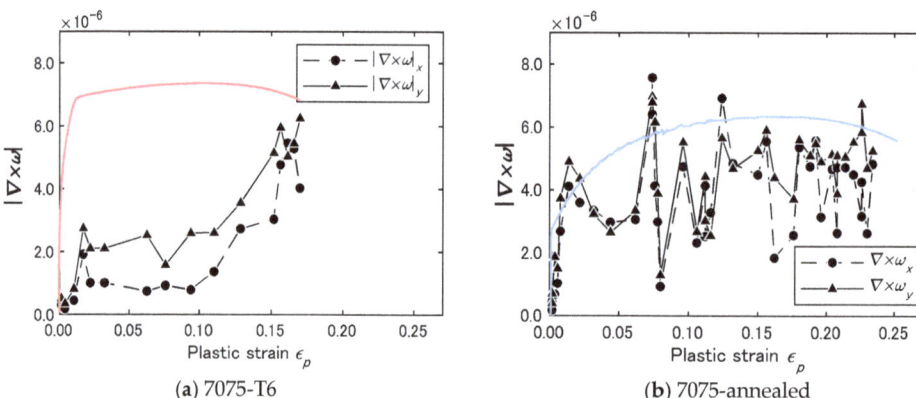

Figure 11. Variation of $|\nabla \times \omega|$ (average over shear band zone) during the tensile test.

The annealed specimen displays a different behavior. The slant band structured pattern of $\nabla \cdot v$ appears along the boundary of the same rotations of $\nabla \times \omega$ field. Figure 10b $\varepsilon_p = 0.107$ indicates that on the upper and lower sides of the boundary, the rotations are counterclockwise. It is interesting to compare this pattern with the four rotations observed in the T6 specimen in Figure 10a $\varepsilon_p = 0.122$. We can interpret that the two rotations observed in the annealed specimen correspond to the upper left and lower right rotations observed in the annealed specimen. In an earlier stage, the annealed specimen exhibits the same symmetric type fringe pattern as the T6 specimen (see Figure 9b-1). It is likely that when the plastic strain develops to the level where the slant band structure appears, the clockwise rotations observed in the early stage on the upper right and lower left of the boundary disappear. As a result, the pair of the counterclockwise rotations exert strong shear force along the boundary.

We can interpret the above-mentioned transition from the "X"-shaped to the slant band structure as being associated with the transition from the normal force dominating regime to the shear force dominating regime. In the early stage of deformation when the deformation is in the linear elastic regime, the longitudinal elastic force represented by the first term on the right-hand side of Equation (14) is dominant. With the development of plastic deformation, the longitudinal resistant force becomes less effective, and instead, the shear resistant force represented by the second term on the right-hand side of Equation (14) becomes more active. It is naturally understood that at a certain point, the shear force dominates over the longitudinal force. This transition of dominant resistant force mechanism accompanies two events. The first is a stress drop. This is because the shear modulus is lower than the longitudinal (Young's) modulus. The second is the formation of shear force across the boundary over the entire width of the specimen.

4.3. Activity of Shear Band

4.3.1. Shear Elastic Dynamics in Shear Band

Figures 9–11 indicate that the increase in $\nabla \times \omega$ is accompanied by a concentration of $\nabla \cdot v$. When $\nabla \times \omega$ grows to a certain level, the charge $\nabla \cdot v$ takes the form of band structure. In the case of T6 the band is in an "X"-shape and its appearance coincides with the necking. In the case of annealed it is more a shear band like (running from one side to the other side of the specimen as a slant line) until the final fracture. The slant band structures run at the boundary of two developed $\nabla \times \omega$ fields.

The fringe pattern observed with the ESPI setup represents the spatial dependence of the differential displacement occurring during the time step Δt used for the image subtraction. Thus, each fringe can be interpreted as the contour of the velocity measured in the unit of Δt, and thus proportional to the velocity in the unit of m/s, $d\xi/dt$. When the fringes are linear parallel, we can express the fringe pattern as follows:

$$\frac{d\xi}{dt}(u,v) = \begin{bmatrix} c_{11}x + c_{12}y \\ c_{21}x + c_{22}y \end{bmatrix} \tag{15}$$

Here the linearity allows us to express u and v as a linear function of x and y where $c_{11} \ldots c_{22}$ are all constant, and the constant of proportionality between the velocity in the unit of Δt and m/s is absorbed in $c_{11} \ldots c_{22}$. The fact that the linear fringes are parallel allows us to use the same constants $c_{11} \ldots c_{22}$ for the entire boundary region.

$$\omega_z = \frac{\partial v}{\partial x} - \frac{\partial u}{\partial y} = \int (c_{21} - c_{12}) dt \tag{16}$$

Equation (16) indicates that ω_z is independent of the space coordinates, which in turn indicate that $\nabla \times \omega = 0$ as follows:

$$\nabla \times \omega = \left(\frac{\partial \omega_z}{\partial y} - \frac{\partial \omega_y}{\partial z}\right)\hat{x} + \left(\frac{\partial \omega_x}{\partial z} - \frac{\partial \omega_z}{\partial x}\right)\hat{y} + \left(\frac{\partial \omega_y}{\partial x} - \frac{\partial \omega_x}{\partial y}\right)\hat{y} = \frac{\partial \omega_z}{\partial y}\hat{x} - \frac{\partial \omega_y}{\partial x}\hat{y} \tag{17}$$

The condition $\nabla \times \omega = 0$ indicates that one of the pre-fracture criteria holds in the shear band region. Notice that the quiver plots in Figure 10 indicate that the regions on both sides of the shear band undergo differential rotations exerting the horizontal resistive force represented by $\partial \omega_z / \partial y \hat{x}$ and the vertical resistive force represented by $\partial \omega_z / \partial x \hat{y}$. The formation of a shear band can be interpreted as that these differential rotational forces cause strong, localized shear force along the boundary and that consequently the shear banded area exhibits the pre-fracture condition.

While developed shear forces are present above and below a shear band, the above observation that $\nabla \times \omega = 0$ in the shear band region indicates that this region does not exert the resistive force. This is the basis of the claim that a pre-fracture criterion is met in a shear band. Figure 11b indicates that this type of pre-fracture condition is recoverable. The field theory interprets this recovery as a phenomenon analogous to the spark discharge. Although the recovery mechanism from this pre-fracture condition has not been clarified, it is possible that some sort of atomic rearrangement in association with motions of mobile dislocations underlies the recovery (see a section below).

4.3.2. Shear Band as a Constant Charge $\nabla \cdot v$

The approximately parallel and linear fringe pattern inside a shear band indicates that the differential displacement inside the band represents normal strain in a direction perpendicular to the fringes. From Equation (15), we can express the shear band fringe pattern as follows, and thereby interpret it as a constant deformation charge $\nabla \cdot v$.

$$\nabla \cdot v = \frac{\partial}{\partial t}\left(\frac{\partial u}{\partial x} + \frac{\partial v}{\partial y}\right) = c_{11} + c_{22} = const(damping\ coefficient) \tag{18}$$

Under the condition that c_{11} and c_{22} are constant, the shear band drifts keeping its shape. This is a constant movement of a deformation charge. Figure 12 shows fringe patterns in the annealed specimen that we can interpret as representing deformation charges. The patterns shown in Figure 12a are observed in a stage earlier than (b). The plots above these fringe images are the stress–strain characteristics for the respective stages.

The locations on the stress–strain curve at which each image is observed are indicated by markers and the lines connecting with the fringe patterns. Careful observation of the strain-stress curve reveals the following aspects: (1) The three images in Figure 12a are not associated with a sharp stress drop; (2) The two left images in (b) are observed after a sharp stress drop and the rightmost images is observed after the next stress drop. Of these six images, the ones observed after a stress drop exhibit a shear band like structure. The one after the stress rise does not show a band structure. These aspects indicate that the formation and motion of the shear-band like fringe pattern (or the deformation charge) are associated with a stress drop.

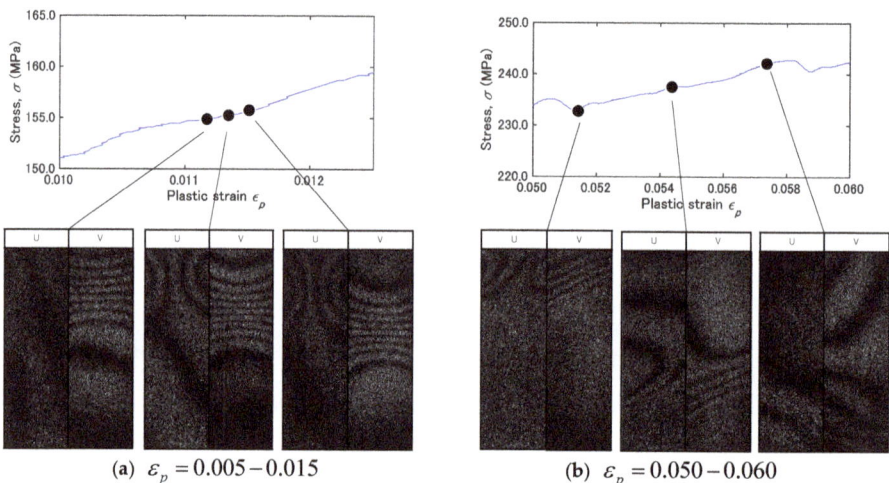

Figure 12. Shear band activity in 7075-annealed alloy observed in the strain range of (a) $\varepsilon_p = 0.005 - 0.015$ and (b) $\varepsilon_p = 0.050 - 0.060$.

This observation is consistent with the above-argued stress drop resulting from the transition of longitudinal-force dominant to shear-force dominant resistive mechanism. Macroscopically, this sequence of stress drop and resumption can be interpreted as representing the serration.

4.3.3. Connection to Mobile Dislocations and Serrations

Figure 10b indicates that the shear force $\nabla \times \omega$ vectors on the opposite sides of a shear band curls oppositely as indicated by curved arrows inserted in the rightmost quiver plot for each specimen. The mutually opposite curly vector fields exert strong differential shear force along the boundary. This differential shear force can be interpreted as the frictional force that drives mobile dislocations.

Based on the above interpretation that the differential shear force drives mobile dislocations, we can make the following arguments. Mobile dislocations are stopped by obstacles such as the grain boundary. This interpretation is consistent with the experimental observation that the speed of shear band exhibits the same time dependence as that of mobile dislocations [4]. Being driven by a frictional force, the mobile dislocations cause energy dissipation as they move. This is consistent with the field theoretical interpretation that the resistive force represented by $\sigma_0 \rho (\nabla \cdot v) v$ term in Equation (14) is velocity damping force. The energy dissipative dynamics of mobile dislocations accounts for the plastic irreversibility. Once mobile dislocations reach a side of the specimen, the left side in the case of $\varepsilon_p = 0.107$ in Figure 10b, the dislocations do not move any longer. This explains the end of shear band drift and resumption of stress rise as the specimen recovers from the partial pre-fracture. The rightmost image and corresponding stress–strain characteristics exhibit that the disappearance of shear band like fringe pattern and appearance of curved fringe patterns accompanied by a fringe pattern.

4.3.4. Solitary Wave

The field theory explains that the movement of a shear band corresponds to a motion of a solitary wave [4]. The condition $\nabla \times \omega = 0$ makes to rewrite Equation (14) in the following form. Note that at this stage the linear elastic force term $(\lambda + 2G)\nabla^2 \zeta$ is inactive as well.

$$\rho(\partial_t v) = \sigma_0 \rho v (\partial_x v) - \frac{(E\delta x \Delta x)}{c_s}(\partial_x^3 v) \tag{19}$$

Here, the following replacement is used to rewrite Equation (14) into Equation (17)

$$\partial_x v = \frac{\partial \zeta}{\partial x} = \frac{\partial \zeta}{\partial t}\frac{dt}{dx} = \frac{1}{c_s}\frac{\partial \zeta}{\partial t} \tag{20}$$

Equation (17) is known as Korteweg-de Vries (KdV) equation and has a solution in the following form:

$$v = a\,\text{sech}^2(b(x - c_w t)) \tag{21}$$

$$c_w = \frac{\sigma_0 a}{3} \tag{22}$$

$$b = c_w \sqrt{\frac{\rho}{4E\delta x_s \Delta x_s}} \tag{23}$$

Here, a is the amplitude of the solitary wave, b is a constant decided by the material's property and the shape of the shear band, and c_w is the propagation velocity of the solitary wave.

It is interesting to discuss the behavior of shear band observed in the experiment in conjunction with the above expressions of the solitary wave. Figure 13a shows the change of velocity of shear band observed during the tensile test. The band velocity shows an exponential decrease similar to the decay in the amplitude of the $|\nabla \times \omega|$ observed in Figure 11b. Equation (22) indicates that the solitary wave velocity is proportional to the amplitude of the solitary wave. As the deformation develops toward the final fracture, the level of energy dissipation increases, damping the particle movement. This makes the amplitude of the velocity solitary wave decrease. The exponential decrease in c_w is understandable because velocity damping causes exponential decrease in the displacement and velocity fields. The same damping mechanism explains the exponential decay of the $|\nabla \times \omega|$ amplitude in Figure 11b.

Now let us discuss the behavior of the solitary wave in terms of the motion of the deformation charge $\nabla \cdot v$. From the analogy to the electromagnetic field, we can argue that the motion of the deformation charge corresponds to the conduction current. For a given electric potential, conduction current flows depending on the resistance (impedance). In the electric field, when the electric charge density is concentrated in a local region, the increase in the charge density lowers the impedance, which further increases the local charge density. Once this positive feedback mechanism dominates, the gas medium electrically breaks down quickly. It is expected that in the deformation field the impedance decreases as well by a similar positive feedback mechanism [4]. Since we are dealing with displacement wave, it is reasonable to discuss the acoustic impedance given in the following form.

$$z = \frac{\kappa}{v_p} = \frac{d\sigma/d\varepsilon}{c_w} \tag{24}$$

Here, κ is the stiffness of the medium and v_p is the wave velocity. In the present context, the stiffness can be evaluated as $d\sigma/d\varepsilon$ and the wave velocity is c_w.

Figure 13b plots the impedance evaluated by substituting $d\sigma/d\varepsilon$ into κ and c_w into v_p in Equation (24). As expected, impedance decreases monotonically with the plastic strain ε_p. Figure 13a indicates that c_w also decreases with ε_p, which tends to increase impedance z in Equation (24). The fact that the impedance decreases despite the decrease in c_w on the denominator of Equation (24) indicates that the decrease in the stiffness is greater than the increase in c_w. As discussed above, the decreases in the wave velocity c_w can be attributed to the increase in the damping characteristics of the material. The similar damping has been observed in acoustic measurement [13]. The velocity acoustic wave in solid metals is dependent on the dislocation structure. The progress of plastic deformation with the increase of dislocation density results in the damping of acoustic velocity. The observed monotonic decrease in impedance indicates that the stiffness decreases faster than the increase in the damping effect.

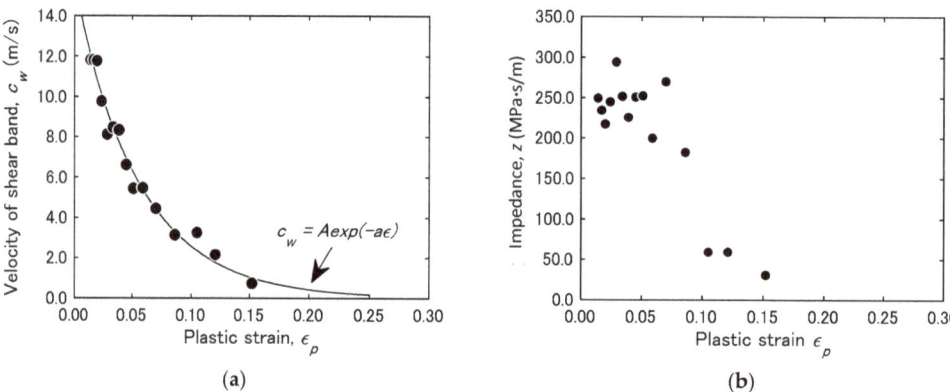

Figure 13. Velocity and impedance of shear band in 7075-annealed alloy.

4.3.5. Comparison of T6 and Annealed Specimens

It is interesting to compare the T6 and annealed specimens for various behaviors. Figure 6 clearly shows that the T6 specimen exhibits more brittle behavior in the stress–strain characteristics and the annealed specimen exhibits more ductile behavior. This contrast can be observed in other features of the two specimens.

First, look at the difference in $(\nabla \times w)_x$ and $(\nabla \times w)_y$. These are the x- and y-component of shear displacement. Figure 11a indicates that in the T6 specimen is factor of two to three greater than $(\nabla \times w)_x$. When $\nabla \times w$ reaches the maximum value, this relation of $(\nabla \times w)_x$ to $(\nabla \times w)_y$ becomes unclear. The ratio of these two quantities essentially represents Poisson's effect. It makes sense that as $\nabla \times w$ reaches the maximum value, the deformation is characterized more by plasticity than elasticity, hence the Poisson's effect becomes less prominent. The argument that the annealed exhibits more ductile behavior is consistent with the fact that the annealed specimen shows more serrations in the stress–strain characteristics in Figure 7. Figure 11b indicates that the annealed specimen shows that $(\nabla \times w)_y$ is somewhat greater than $(\nabla \times w)_x$ but the difference is not as clear as the T6 case. Being more ductile material, the annealed specimen does not show Poisson's effect at the same level as the T6 specimen. Figure 11 also shows that the T6 specimen exhibits the increase in $\nabla \times w$ once. On the other hand, the annealed specimen shows the increase and decrease in $\nabla \times w$ several times repetitively. The discussions made in the above sections indicate that each cycle of increase in $\nabla \times w$ corresponds to the motion of mobile dislocations to a side of the specimen. The observed contrast in the pattern of $\nabla \times w$ increase/decrease characterizes that the annealed specimen is more ductile.

Another interesting observation is that the concentrated charge density pattern of the T-7 specimen regains horizontal symmetry toward the stage of final fracture as presented by Figure 9b-7 fringe image. It is likely that the symmetric feature in the charge density is the natural behavior of the specimen as the applied load and the specimen geometry is symmetric. It is the dislocation's motion that causes the observed asymmetric behaviors of the deformation field. Towards the end of the ductile deformation, the dislocations stop their activity, hence the deformation field regains the symmetric feature. Being a more brittle material, the T6 specimen does not exhibit this dislocation-induced asymmetric behavior in the deformation field.

5. Conclusions

The plastic deformation behavior of 7075-T6 and 7075-annealed has been visualized using ESPI and discussed based on the field theory. The following conclusions are obtained:

1. The proposed method for the fringe analysis using image processing can detect the fringe contours without a phase-stepping method. The dynamic deformation behavior of 7075-alloys can be evaluated using this method.
2. With the increasing of the plastic strain, the $\nabla \times \omega$ field becomes less uniform. At the same time, the charge distribution is localized in a certain region of the specimen where the charge density increases. These transitions from symmetric to asymmetric deformation field are caused by transitions from longitudinal-force dominant resistive mechanisms to shear-force dominant mechanisms.
3. The deformation localization is closely related to the shear elastic dynamics. In microscopic view, the plastic deformation is initiated via driving the mobile dislocation. The shear force ($\nabla \times \omega$) vectors on the opposite sides of a shear band curls oppositely. The mutually opposite curly vector fields exert strong differential shear force along the boundary. This differential shear force can be interpreted as the frictional force that drives mobile dislocations.
4. The T6 specimen exhibits more brittle behavior in the stress–strain characteristics and the annealed specimen exhibits more ductile behavior. The observed contrast in the pattern of $\nabla \times \omega$ increase/decrease characterizes that the T6 specimen is more brittle and annealed is more ductile.
5. The fringe patterns of the T6 and annealed specimens show similar characteristics in the initial stage, differ in the progressive stage of deformation, and show similarity again in the final stage. This is because the dislocations are not active in the initial and final stages.

Author Contributions: Conceptualization, S.T. and S.Y.; investigation, S.T., T.S. and T.H.; project administration, T.S. and S.Y.; writing—original draft preparation, S.T.; writing—review and editing, S.Y., T.S. and T.H. All authors have read and agreed to the published version of the manuscript.

Funding: This work was supported by MEXT-KAKENHI No.18H01722. The Ministry of Trade, Industry and Energy (MOTIE) and Korea Institute for Advancement of Technology (KIAT), Korea through the International Cooperative R&D program. (Project No. P0006842).

Institutional Review Board Statement: Not applicable.

Informed Consent Statement: Not applicable.

Data Availability Statement: The data presented in this study are available on request from the corresponding author.

Conflicts of Interest: The authors declare no conflict of interest.

Appendix A. Principle of ESPI

ESPI is an optical measurement method that observes the displacement distribution on the measurement surface by photographing the speckle pattern that produced by the superposition of laser beams passing through two or more optical paths on the measurement surface with a Charge Coupled Device (CCD) camera and applying image processing. The displacement distribution obtained by ESPI is displayed as contour interference fringes, and the strain distribution can be obtained by analyzing the fringe pattern. ESPI advantages by its full-field of view, non-contact measurement, and large measurement area. In addition, ESPI can measure in real-time because the CCD camera and image processing can be performed simultaneously by a computer. ESPI can detect small displacements on the order of several hundred [nm] over a large measurement area.

The principle of ESPI is described below. When a laser beam illuminates the surface of a material, the laser beam is diffused reflected due to the roughness of the surface. Due to the interference of the diffused reflected laser, a speckled pattern is observed on the material surface (Figure A2). When the material is deformed, the interference conditions of the laser beam on the surface also change, and the intensity of the speckle pattern also change. Figure A2 shows a schematic diagram of the optical system for in-plane displacement measurement by ESPI. The laser beam is emitted from the semiconductor laser is expanded by the beam expander and then split into two directions by the beam splitter. The split laser beams are reflected by mirrors and illuminate from two directions symmetrical to the normal of the measurement surface. At this time, the two lights interfere each and form a speckle pattern on the specimen surface. The interference speckle pattern is recorded by a CCD camera, and the absolute value of the difference in intensity between the image before deformation (Figure A3a) and the image after deformation (Figure A3b) yields the stripe image as Figure A3c.

Figure A4 shows the optical path difference when the measurement surface illuminated by the laser beam moves u in the in-plane direction. The figure shows the case where two laser beams, Laser1 and Laser2, split by a half-mirror, are incident on point A and the surface moves u in the in-plane direction from point A to point B due to deformation of the material. In this case, assuming that Laser1' and Laser2' are incident on point B, the optical path difference between Laser1 and Laser1' is $L_1 = U \sin \theta$, and that between Laser2 and Laser2' is $L_2 = U \sin \theta$, and the optical path difference between the two directions is $L = U \sin \theta$. Next, we consider the change in the intensity of the laser beam when point A is displaced to point B due to the deformation of the material. The state equation E of the laser beam moving in the forward direction on the x-axis at time $t(s)$ is expressed as follows:

$$E(x,t) = A \cos \frac{2\pi}{\lambda}(x - c_a t) = A \cos(kx - \omega t) \tag{A1}$$

where A is the amplitude of the laser beam, λ is the wavelength, c_a is the speed of light in the atmosphere, $k = \frac{2\pi}{\lambda}$, and $\omega = k c_a$. Expressing this using the complex function is as follows:

$$E(x,t) = A e^{i(kx - \omega t)} \tag{A2}$$

It follows that the intensity of the light at a fixed observation time $T(s)$ is expressed as follows:

$$I(x,t) = \frac{1}{T} \int_0^T E \cdot E^* dt \tag{A3}$$

Substituting Equation (A2) into Equation (A3)

$$I(x) = \frac{1}{T} \int_0^T A e^{i(kx - \omega t)} \cdot A e^{-i(kx - \omega t)} dt = A^2 \tag{A4}$$

The square of the amplitude of the laser light is the intensity of the light. From Equation (A2), the two laser beams irradiated to point A, Laser1 and laser2, can be expressed as follows:

$$U_1 = A_1 e^{i(kx - \omega t)} \tag{A5}$$

$$U_2 = A_2 e^{i(k(x+\Delta x) - \omega t)} \tag{A6}$$

A_1 and A_2 are the amplitudes of the respective laser beams, and Δx is the small optical path difference between Laser1 and Laser2 before the material deformation. This optical path difference generates an interference speckle pattern. The equation of laser beam at point A is expressed as a superposition of U_1 and U_2 as follows:

$$E_A = U_1 + U_2 = A_1 e^{i(kx - \omega t)} + A_2 e^{i(k(x+\Delta x) - \omega t)} \tag{A7}$$

The intensity of the superposition of the two laser beams is as follows:

$$\begin{aligned} I_A &= \frac{1}{T}\int_0^T \left(A_1 e^{i(kx-\omega t)} + A_2 e^{i(k(x+\Delta x)-\omega t)} \right)\left(A_1 e^{-i(kx-\omega t)} + A_2 e^{-i(k(x+\Delta x)-\omega t)} \right) dt \\ &= A_1^2 + A_2^2 + 2A_1 A_2 \cos(k\Delta x) \end{aligned} \tag{A8}$$

The intensity of the interference speckle pattern of the two laser beams varies with $k\Delta x$. When point A is displaced to point B due to deformation of the material, the laser beams at point B, Laser1' and Laser2' express as follows:

$$U_{1'} = A_1 e^{i(k(x+L_1) - \omega t)} \tag{A9}$$

$$U_{2'} = A_2 e^{i(k(x+\Delta x + L_2) - \omega t)} \tag{A10}$$

The equation of the laser beam at point B and the Intensity of it are expressed as follows:

$$E_B = U_{1'} + U_{2'} = A_1 e^{i(k(x+L_1) - \omega t)} + A_2 e^{i(k(x+\Delta x + L_2) - \omega t)} \tag{A11}$$

$$\begin{aligned} I_B &= \frac{1}{T}\int_0^T \left(A_1 e^{i(k(x+L_1)-\omega t)} + A_2 e^{i(k(x+\Delta x+L_2)-\omega t)} \right)\left(A_1 e^{-i(k(x+L_1)-\omega t)} + A_2 e^{-i(k(x+\Delta x+L_2)-\omega t)} \right) dt \\ &= A_1^2 + A_2^2 + 2A_1 A_2 \cos k(\Delta x + L_1 + L_2) \\ &= A_1^2 + A_2^2 + 2A_1 A_2 \cos(k\Delta x + 2ku \sin\theta) \end{aligned} \tag{A12}$$

The absolute value of the intensity of the laser beam before and after the deformation is as follows:

$$\begin{aligned} I_A - I_B &= 2A_1 A_2 \{\cos k\Delta x - \cos(k\Delta x + 2ku\sin\theta)\} \\ &= 4A_1 A_2 \sin(ku\sin\theta)\sin(k\Delta x + ku\sin\theta) \end{aligned} \tag{A13}$$

From this equation, the intensity of the interference speckle pattern due to deformation changes with $ku \sin\theta$. Expressing this equation using the phase change as $\Delta\varphi$ and $k = 2\pi/\lambda$ is as follows:

$$\Delta\varphi = \frac{4\pi}{\lambda}\sin\theta \tag{A14}$$

Whenever the phase $\Delta\varphi$ becomes a multiple of 2π, the intensity of the interference speckle pattern becomes 0. Therefore, the fringe pattern (interference fringe) appears on a surface with a displacement gradient and the displacement when n interference fringes appear is expressed as follows:

$$u = \frac{n\lambda}{2\sin\theta} \tag{A15}$$

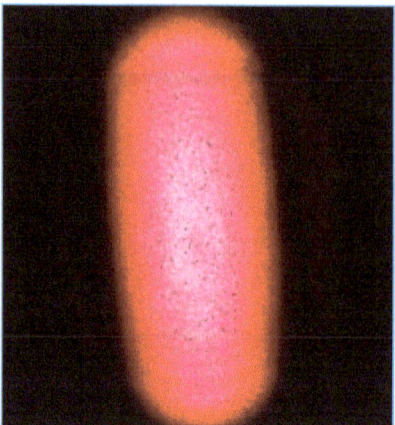

Figure A1. Laser speckle pattern.

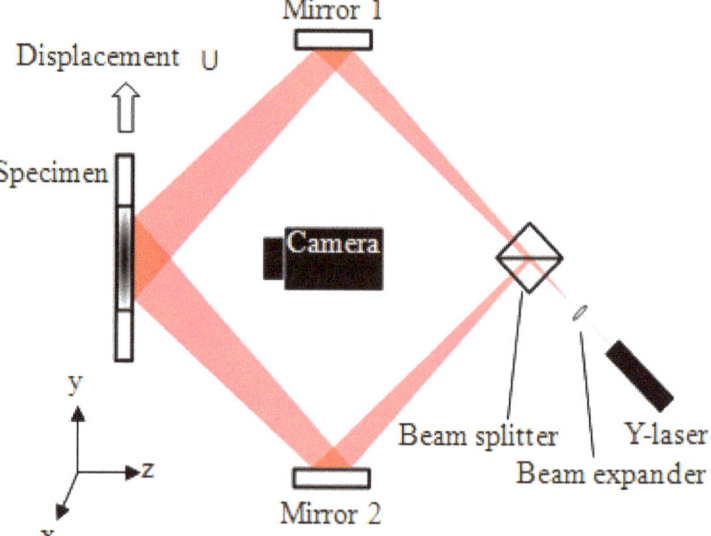

Figure A2. Optical system of 2D-ESPI.

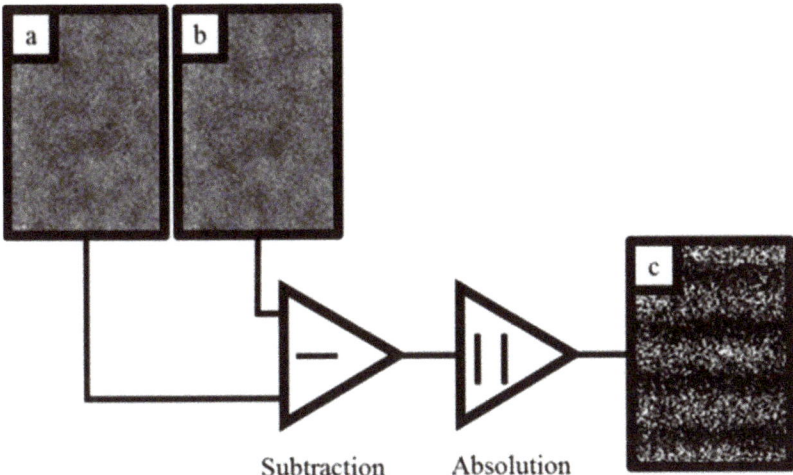

Figure A3. Speckle pattern of before deformation. (**b**) Speckle pattern of after deformation. (**c**) Fringe pattern image calculated from (**a**) and (**b**).

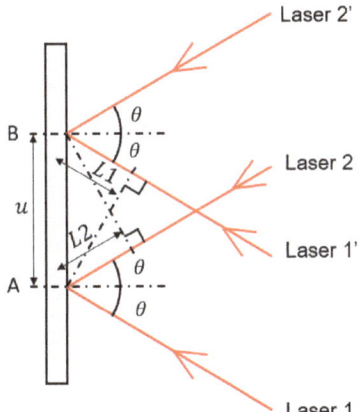

Figure A4. Schematic illustration of difference in optical path.

References

1. Yilmaz, A. Portevin–Le Chatelier effect: A review of experimental findings. *Sci. Technol. Adv. Mater.* **2011**, *11*, 11–16. [CrossRef] [PubMed]
2. McCormick, P.G. The Portevin-Le Chatelier effect in an Al-Mg-Si alloy. *Acta Metal.* **1971**, *19*, 463–471. [CrossRef]
3. Sarmah, R.; Anathakrishna, G. Correlation between band propagation property and the nature of serrations in the Portevin–Le Chatelier effect. *Acta Metall.* **2015**, *91*, 192–201. [CrossRef]
4. Yoshida, S. Comprehensive Description of Deformation of Solids as Wave Dynamics. *Math. Mech. Complex Syst.* **2015**, *3*, 243–272. [CrossRef]
5. Yoshida, S. *Deformation and Fracture of Solid-State Materials*; Springer: New York, NY, USA, 2015.
6. Hassan, G.M. Deformation measurement in the presence of discontinuities with digital image correlation: A review. *Opt. Lasers Eng.* **2021**, *137*, 106394. [CrossRef]
7. Ranc, N.; Du, W.; Ranc, I.; Wagner, D. Experimental studies of Portevin-Le Chatelier plastic instabilities in carbon-manganese steels by infrared pyrometry. *Mater. Sci. Eng. A* **2016**, *663*, 166–173. [CrossRef]
8. Yoshida, S.; Sasaki, T. Deformation Wave Theory and Application to Optical interferometry. *Materials* **2020**, *13*, 1363. [CrossRef] [PubMed]

9. Yoshida, S.; McGibboney, C.; Sasaki, T. Nondestructive Evaluation of Solids Based on Deformation Wave Theory. *Appl. Sci.* **2020**, *10*, 5524.
10. Pati, A.; Rastogi, P. Approaches in generalized phase shifting interferometry. *Opt. Lasers Eng.* **2005**, *43*, 475–490. [CrossRef]
11. Shirohi, R.S. Speckle Methods in Experimental Mechanics. In *Speckle Metrology*; Shirohi, R.S., Ed.; CRC Press: Boca Raton, FL, USA, 1993; pp. 99–154.
12. Fritsch, F.N.; Carlson, R.E. Monotone Piecewise Cubic Interpolation. *SIAM J. Numer. Anal.* **2006**, *17*, 238–246. [CrossRef]
13. Erofeev, V.I.; Malkhanov, A. Nonlinear acoustic waves in solids with dislocations. *Procedia IUTAM* **2017**, *23*, 228–235. [CrossRef]

Article

Application of Digital Image Correlation in Space and Frequency Domains to Deformation Analysis of Polymer Film

Caroline Kopfler [†], Sanichiro Yoshida [*,†] and Anup Ghimire [†]

Department of Chemistry & Physics, Southeastern Louisiana University, Hammond, LA 70402, USA; caroline.kopfler@selu.edu (C.K.); anup.ghimire@selu.edu (A.G.)
* Correspondence: syoshida@selu.edu
† These authors contributed equally to this work.

Abstract: Using speckle patterns formed by an expanded and collimated He-Ne laser beam, we apply DIC (Digital Image Correlation) methods to estimate the deformation of LLDPE (linear low-density polyethylene) film. The laser beam was transmitted through the film specimen while a tensile machine applied a load to the specimen vertically. The transmitted laser light was projected on a screen, and the resultant image was captured by a digital camera. The captured image was analyzed both in space and frequency domains. For the space-domain analysis, the random speckle pattern was used to register the local displacement due to the deformation. For the frequency-domain analysis, the diffraction-like pattern, due to the horizontally-running, periodic groove-like structure of the film was used to characterize the overall deformation along vertical columns of analysis. It has been found that when the deformation is small and uniform, the conventional space domain analysis is applicable to the entire film specimen. However, once the deformation loses the spatial uniformity, the space-domain analysis falls short if applied to the entire specimen. The application of DIC to local (windowed) regions is still useful but time consuming. In the non-uniform situation, the frequency-domain analysis is found capable of revealing average deformation along each column of analysis.

Keywords: linear low-density polyethylene; digital image correlation; optical non-destructive testing; speckle; fourier scaling theorem; gaussian filtering; optical methods

1. Introduction

Commercial shipping facilities commonly use linear low-density polyethylene (LLDPE) for packaging freight. For efficiency, wrapping machines are responsible for packaging shipments by rotating thin LLDPE film around the objects. These machines perform this action as quickly as possible to optimize both time and cost of the process. At increased wrapping rates, however, LLDPE film stretches more, ultimately leading to its failure.

The unique characteristics of this film make the material advantageous and multi-functional across industries. LLDPE is a thermoplastic polymer characterized by a predominantly linear backbone and a high proportion of short branches. Due to the material's structure, it cannot be packed tightly, which gives it a hazy, transparent appearance. In comparison to traditional low-density polyethylene, LLPDE has a crystalline structure with little to no elastic memory or recovery, resulting in increased tensile strength and elongation [1]. During manufacturing, the LLDPE molecules align with the spooling direction to create striations on the film. Up close, these striations present themselves as periodic grooves on the film's surface, seen below in Figure 1. These grooves cause non-uniform surface height variation, which complicates conventional optical methods based on speckles. In this study, we explore a way to use the structural pattern for optical-base deformation analysis.

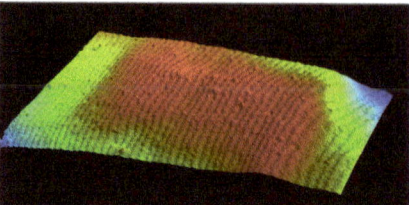

Figure 1. Surface structure of LLDPE film.

While many non-destructive optical testing methods are available to evaluate material deformation [2,3], speckle techniques are most attractive for analysis of LLDPE film. When coherent light is reflected on a rough surface or transmitted through an inhomogeneous medium, the optical field of an imaging device forms a pattern consisting of many bright and dark spots. These spots, known as speckles, result from the coherent superposition of light rays scattered by the medium. Because the light rays reaching a particular spot on the image plane follow optical paths defined specifically by the corresponding section of the medium, the speckle field reveals the specimen's fingerprints at a given time and position. By analyzing a speckle field formed on the image plane, we can probe the spatiotemporal behavior of the medium, such as the displacement or velocity of the scattering particles.

Traditionally, the subtraction method used in speckle pattern studies [2,3] retrieves displacement through phase analysis of the speckles. Digital Image Correlation [4–6] detects the displacement by evaluating the correlation between the speckle fields of the original and deformed states. By relating the spatiotemporal behavior within a dynamic model, we can characterize the material's deformation properties, such as elasticity and viscosity. These dynamic models are often used to investigate biological materials [7–9], which are generally within systems experiencing Brownian motion [10]. While the technique in this study is similar to those used to mechanically characterize dynamic physiological systems, it is important to note that we focus on a quasi-static system, where factors like speckle decorrelation time are negligible. Additionally, the film's surface height variation presents a problem when optically evaluating change in thickness. This observation motivates exploration of alternative methods for evaluating surface structure and deformation of the material.

The present study applies conventional DIC methods [4,11] using the speckle field generated by the laser beam transmitted through the film specimen, unlike painted speckles often used in similar deformation analysis [12]. The coexistence of scattering-induced speckles and periodic grooves, however, complicates analysis due to the formation of diffraction-like patterns superposed onto the scattering-induced speckles on a projected image. While this image looks like typical patterns from diffraction grating, the interval between neighboring grooves is orders of magnitude greater than the optical wavelength. Therefore, we interpret the pattern on the screen as the result of destructive interference between multiple rays and refer to it as the linear dark-fringe-like interference pattern, simple referred to as linear dark-fringes.

We find that we can use the periodicity of the grooves to estimate the film's deformation. In the frequency domain, this quasi-constant interval produces fairly sharp peaks in the Fourier spectrum. As the film stretches, the interval between linear dark-fringes increases, which shifts the peak frequency on the Fourier spectrum. From this shift, we can estimate the amount of stretch [13,14]. Here, random speckles superposed to linear dark-fringes compromises accuracy of the analysis [15]. We find that proper Gaussian pre-filtering and selection of frequency range can improve overall accuracy of the analysis [16]. The effectiveness of this method is discussed in the following sections, as well as space-domain analysis using DIC methodology for comparison [4,5,14].

2. Materials and Methods

2.1. Structural Inhomogenity in LLDPE Film

The linear backbone and short branches of the LLDPE molecules align in the direction of spooling during the extrusion manufacturing process, resulting in periodic structural grooves observable on the film's surface, as illustrated in Figure 1. When a light source passes through the transparent film, a projection of the dark fringes and speckle patterns are visible. Figure 2a shows a sample optical pattern projected on a screen.

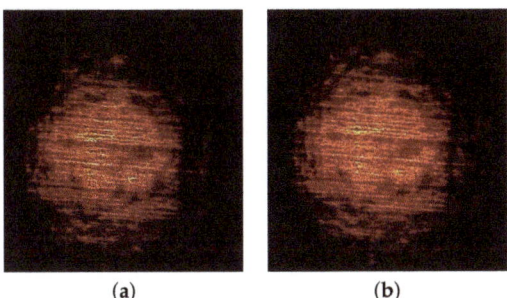

(a)   (b)

Figure 2. Sample digital images collected during the experiment. (a) Original (un-stretched) Image, (b) Stretched Image.

When an external load stretches the film, the linear dark-fringes and speckles change independently. Figure 2b shows the projected optical pattern after an external load stretches the film sample vertically. Careful examination of Figure 2b will reveal that the interval of the horizontally running linear dark-fringes increases from Figure 2a and that speckle patterns expand vertically.

Recall, Figure 1 shows the image formed by a surface profiler (Bruker Contour GT-KO, courtesy of University of New Orleans). Vision-64 Analysis Software indicates approximately 40 peaks along the y-axis over a span of 25 mm. This indicates that the pitch of the periodic structure is 25 mm/40 ≈ 600 µm, which is three orders of magnitude greater than the laser's wavelength of 632.8 nm. It is unlikely that this periodic structure forms diffraction patterns similar to those formed by diffraction grating. However, the groove-like structure causes the transmitted light to form a periodic pattern of destructive interference that resembles a diffraction pattern.

The random nature of speckle patterns on the projected image makes application of traditional optical interferometric techniques difficult based on the phase of the source-light such as holographic inteferometry [15,17]. Speckle-pattern interferometry is applicable, but linear dark-fringes compromise the accuracy because the spatial shifts of speckles and linear dark-fringes are different from each other. The speckle-pattern shift results from physical displacement of the local area, whereas the shift of the linear dark-fringes results from the deformation of structural grooves. When changes are small and uniform, the two types of shift may represent the same deformation; however, when deformation is non-uniform, the shifts are likely to behave differently.

2.2. Experimental Setup

The experimental setup used for data collection consisted of several parts. Initially, several LLDPE film samples with thicknesses of 10 µm and 20 µm were cut into 4 cm squares and attached to an ADMET tensile machine's stationary and dynamic grips. Recall that the film's surface structure contains grooves in line with its spooling direction. Thus, in separate experiments, the samples were oriented so that these striations were perpendicular (horizontal) and parallel to the pulling direction (vertical).

A 17 mW Helium-Neon laser provided the light source for experiments. The laser beam expands as it propagates, and collimation is necessary for obtaining accurate interferometric patterns. Initially, we suspected that polarization could affect the film's transmission characteristics, which was investigated by varying the polarization. We did not find a noticeable effect of polarization, which lead us to the experimental setup shown in Figure 3.

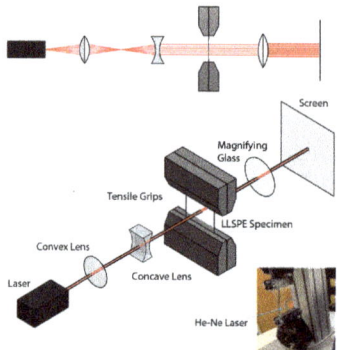

Figure 3. Experimental setup, showing the laser passing through the specimen.

Additionally, the spot size of the beam is increased by a factor of 10. The beam expander consists of two lenses with focal lengths of 25 mm and 250 mm. The radius of the laser beam on the specimen is 5 mm and covers approximately 20% of the film's surface. Additionally, the beam's waist size is 0.5 mm. Multiple studies including beam spot size measurements along the propagation axis and cross-sectional beam profile measurements using a charge-coupled device camera verified that the laser beam is in the Gaussian mode (TEM00) [18]. We adjusted the f# of the imaging lens to optimize speckle size [19,20].

Digital images were taken as the tensile machine pulled the specimen. The digital camera was set to take images as the material stretched in elongation intervals of 50 μm. The image correlation procedure, described in the next section, compares pairs of images to evaluate the corresponding deformation. The camera specifications follow in Table 1.

Table 1. Camera specifications.

Focal Length	52 mm
Aperature	$f/2$
Frame Rate	1/33 s

2.3. Principles of Operation
2.3.1. One-Dimensional DIC Method in Space-Domain

The following method finds the average stretch, or normal strain, over a column (or row) of a projected image. Here, the normal strain is defined by Equation (1), where dl is the elongation (compression) and l is the initial length [13].

$$\epsilon = \frac{dl}{l} \tag{1}$$

The four steps below describe the analytical process for 1D analysis in the space-domain.

1. Stretch the original, un-stretched image by artificially stretching the axis for a given scaling factor. Figure 4 exhibits an example of an original sample image (a) stretched numerically along the vertical axis by a factor of 8% (b).

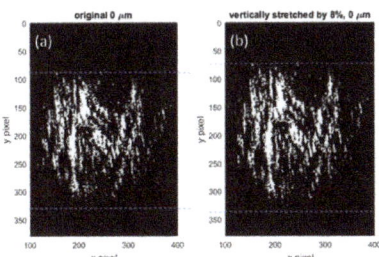

Figure 4. Original image (**a**) and artificially stretched by 8% (**b**).

2. Compare the physically stretched image with the artificially stretched image by computing the correlation coefficient C_{cor} defined by Equation (2) over a column for a vertical stretch or a row for a horizontal stretch.

$$C_{cor} = \frac{cov(a,b)}{\sqrt{cov(a,a)cov(b,b)}} \quad (2)$$

The variables a and b are the column (row) vectors containing gray-scale values of un-stretched and stretched image. In Equation (3), $cov(a,b)$ is the covariance of vectors a and b [21].

$$cov(a,b) = \frac{\sum_{i=1}^{n}(a_i - \bar{a})(b_i - \bar{b})}{n-1} \quad (3)$$

Here, $\bar{a}$ and $\bar{b}$ are the mean values of the respective vectors' elements.

3. Repeat 1 using a different stretching factor and compute the cross-correlation. Iterate this procedure to find the maximum cross-correlation. Determine the stretch of the selected column (row) as the stretch factor that maximizes the cross-correlation.
4. Repeat 1–3 for all the columns (rows) to find the average stretch (normal strain) for each column (row).

2.3.2. Two-Dimensional DIC Method in Space-Domain

This method is generally known as the convolutional DIC technique [22]. A small window called a kernel is set up in the original (un-stretched) image, as shown in Figure 5a. Kernels are groups of elements within an image. An element represents a pixel value that corresponds to the target position within the kernel. In the stretched image, the DIC algorithm moves the kernel vertically and horizontally on a pixel-by-pixel basis and computes the correlation using Equation (4) at each coordinate point [23]. Here in Equation (4), p_x represents the movement of the kernel, where the first function $f(x)$ represents the gray-scale pixel value at coordinate x before the stretch and $g(x)$ is the gray-scale pixel value at x after the stretch.

$$(f \star g)(p_x) = \int_{-\infty}^{\infty} f(\tau)g(p_x + \tau)d\tau \quad (4)$$

The left-hand side of Equation (4) indicates that the cross-correlation is a function of kernel movement p_x. In the same fashion, the kernel moves along the y-axis by p_y, and the pair of p_x and p_y that maximizes the cross-correlation is recorded. The pair (p_x, p_y) constitutes the displacement vector as the center pixel of the kernel [23]. By examining all the area of interest in this way, we can estimate the local displacement vector.

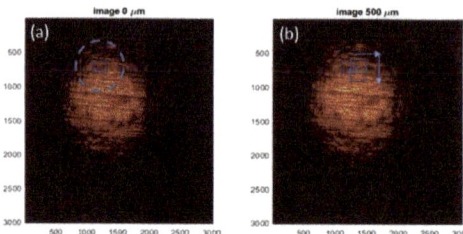

Figure 5. A windowed pattern chosen as the kernel to compute the correlation between the two images.

For clarification, the DIC process using Equation (4) in a numerical method calculates the similarity between two signals. In the context of this study, the two images serve as the *input* and *output* signals, respectively, where the input function is the image before elongation, and the output function is the image after elongation.

2.3.3. Frequency Domain Method

The one-dimensional DIC method discussed above uses the speckle pattern along the entire length of a column of interest. This method works well when the speckle patterns change in the same fashion as the artificial elongation, which assumes uniform stretch. In reality, the deformation of the film specimen becomes non-uniform at a low level of elongation. This situation makes it difficult to apply the one-dimensional DIC method to a realistic situation where deformation of the wrapping film undergoes non-uniform stretching. The convolutional DIC method is applicable in this situation; however, it is unrealistic and time-consuming.

The frequency domain method solves the problem by utilizing differentiation and scaling properties of the Fourier transform. When deformation of the film specimen is not uniform at the local level so that the one-dimensional space domain method is not applicable, it is often the case that the deformation is uniform enough at the global scale. The periodic groove-like structures can serve as a gauge to evaluate overall deformation. Under this condition, the periodicity of dark-fringes due to the groove-like structures form a quasi-single peak in the frequency domain. As the film stretches, the peak shifts on the frequency axis, which can be interpreted as scaling in the space domain. In this situation, we can estimate the scaling factor in the frequency domain as follows.

Let function $f(x)$ be representing the gray-scale variation of the speckles along the x-axis, ξ be the new axis after the compressing/stretching the original axis by a factor of α ($\xi = \alpha x$), $g(\xi)$ be the derivative of $f(\xi)$ with respect to ξ ($g(\xi) = df(\xi)/d\xi$), and ω be the angular frequency associated with the Fourier transform of $g(\xi)$. From the differential and scaling properties of Fourier transform, we can obtain the following equation that relates the Fourier transform of $g(\xi)$, $G_\alpha(\omega)$, and axis scaling factor α.

$$G_\alpha(\omega) = \left| \mathcal{F}\left\{ \frac{df(\xi)}{d\xi} \right\} \right| = \frac{\omega}{\alpha} G\left(\frac{\omega}{\alpha}\right) \tag{5}$$

Here, $\mathcal{F}$ denotes the Fourier transform operation. Appendix A describes this logic in detail, including the derivation of Equation (5).

Equation (5) indicates that $G_\alpha(\omega)$ takes the same form on the frequency axis regardless of the value of α; at the peak frequency $(\omega/\alpha)_{peak} \equiv \phi_{peak}$, $G_\alpha(\omega) = \phi_{peak} G(\phi_{peak})$, at $(\omega/\alpha) = \phi_0$, $G_\alpha(\omega) = \phi_0 G(\phi_0)$, ... If the area is enclosed by $G_\alpha(\omega)$ and the frequency axis (see the simplified illustration in Figure A1), it is proportional to $1/\alpha$. Thus, comparing this area at scaling factor α with its corresponding area in the un-stretched situation, we can find α.

2.3.4. Analytical Steps

We took the following steps to implement the above algorithm.

1. Gaussian filter the optical image projected on the screen. In this algorithm, the local speckles due to the scattering of the film material become noise while the periodic pattern of the linear dark-fringes produces the signal. Since the local speckle pattern has higher spatial frequency than the linear dark-fringes, low-pass filter the optical image to increase the signal-to-noise ratio.
2. Numericaly differentiate the gray-scale value of the original image with respect to the coordinate variable that is set parallel to the tensile axis. This step is to evaluate $df(\xi)/d\xi$ term in Equation (5).
3. Take the FFT of the differentiated gray-scale obtained in step 2. Call this resultant spectrum the original Fourier spectrum. Numerically integrate the original Fourier spectrum for a selected spatial frequency range. This step is to evaluate the area enclosed by the Fourier spectrum and the frequency axis. Call the resultant value the original spectrum-frequency area. The frequency range for this integration should contain the spectral peak and exclude the high frequency region removed by the Gaussian filter. Since the optical intensity varies at each image captured for various reasons such as the change in the background optical intensity, normalize the spectrum-frequency area for the selected frequency range by dividing it by the Fourier spectrum-area of the entire frequency range.
4. Repeat steps 1–3 after the specimen stretches to the current elongation. This process yields the Fourier spectrum for a given stretch factor and the corresponding (current) spectrum-frequency area. Call the resultant spectrum-frequency area the current normalized spectrum-frequency area. Iterate this step for other stretch factors by further elongating the specimen. This procedure yields multiple current normalized spectrum-frequency areas.
5. Compare the current normalized spectrum-frequency areas obtained in step 4 with the original normalized spectrum-frequency area. From the ratio of the current spectrum-frequency area to the original spectrum-frequency area, determine the axis compression factor α. From the axis compression factor, evaluate the stretching factor as $\epsilon = 1/\alpha$.

We present the actual process of the above steps in Section 3.3.

3. Results and Discussion
3.1. Visible Estimation of Displacement Due to Stretch

Prior to the application of the above method, we estimate the horizontal and vertical displacement of the image Figure 4a when the tensile load elongates the specimen by 50 μm. For this estimation we select four coordinate points (call the reference points) in the image where the intensity patterns are distinctive. Figure 6a shows the four reference points $(x_1, y_1) - (x_4, y_4)$ when the specimen is elongated by 100 μm from the un-stretched state. Figure 6b indicates the coordinates of these distinctive intensity patterns when the specimen is elongated by an additional 50 μm.

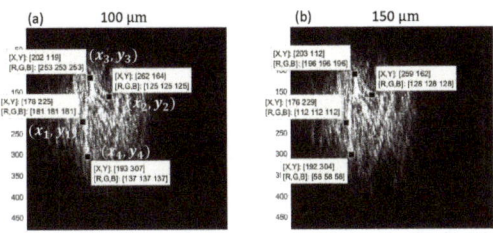

Figure 6. Four reference points used for visual estimation of displacement.

The difference in the coordinates of each reference point gives us the displacement vector. Tables 2 and 3 list the change in these coordinates before and after the additional elongation of 50 µm. Here the former table is for the horizontal displacement and the latter table is for the vertical displacement.

Table 2. Horizontal normal strain estimated from representative coordinates.

	(x_1, y_1)	(x_2, y_2)	$x_2 - x_1$
100 µm	(178.3, 225.8)	(262.2, 164.0)	262.2 − 178.3 = 83.9
150 µm	177.0, 223.0)	259.2, 162.6)	259.2 − 177.0 = 82.2
change in length normal strain	$\Delta(x_2 - x_1)$ ϵ_{xx}		82.2 − 83.9 = −1.7 −1.7/83.9 = −0.020 = −2.0%

Table 3. Vertical normal strain estimated from representative coordinates.

	(x_3, y_3)	(x_4, y_4)	$y_4 - y_3$
100 µm	(202.6, 119.1)	(193.4, 307.2)	307.2 − 119.1 = 188.1
150 µm	(203.8, 112.1)	(192.9, 304.4)	304.4 − 112.1 = 192.3
change in length normal strain	$\Delta(y_4 - y_3)$ ϵ_{yy}		192.3 − 188.1 = 4.2 4.2/188.1 = 0.022 = 2.2%

From Tables 2 and 3, we can estimate the horizontal and vertical strain caused by the additional 50 µm elongation to be 2.0% (horizontal compression) and 2.2 (vertical stretch). Notice that the magnitude of the horizontal and vertical strain are mutually similar, not reflecting Poisson's ratio of LLDPE (approximately 0.4) [1], indicating that the deformation is not elastic. In the following section, we evaluate the results of the space-domain DIC methods described above referring to the data shown in Tables 2 and 3.

3.2. Digital Image Correlation Method in Space-Domain

3.2.1. One-Dimensional DIC

Figure 7 shows the correlation data from the one-dimensional DIC method. Here, the correlation coefficient based on Equation (2) is plotted against the stretching factor ϵ for artificial elongations of 50 µm to 100 µm, and 100 µm to 150 µm (from the left to right). The upper graphs are for the horizontal strain and the lower graphs are for the vertical strain. Since the specimen stretches vertically, the horizontal strain is compressive and the vertical strain is tensile, according to Poisson's effect. The data shown in Figure 7 is along row 200 in Figure 6a for the horizontal cases and column 200 for the vertical cases.

The correlation data in Figure 7 under each condition show a peak value. The artificial stretching factor corresponding to a peak (called the peak stretching factor) indicates when the intensity pattern of the original image is stretched for this amount and the intensity profile along the column or row shows the highest correlation with the intensity profile of the same column or row in the stretched image.

According to Figure 7, the estimated horizontal compression for the elongation from 50 µm to 100 µm, and 100 µm to 150 µm are, respectively, 0.2% and 0.05%. The estimated vertical stretch is 0.8% and 0.44%. These compression and stretch are an order of magnitude lower than the values in Tables 2 and 3, which indicates that the one-dimensional DIC method does not accurately evaluate the actual deformation.

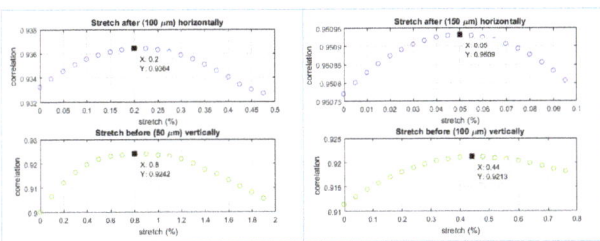

Figure 7. Results from one-dimensional image correlation method in space-domain.

Figure 8 shows the vertical stretch evaluated from the one-dimensional DIC method as a function of elongation. At elongations greater than 250 μm, the speckle fields before and after the corresponding deformation lose the correlation [24]. Consequently, the plot corresponding to Figure 7 does not show a clear peak. We speculate that when the elongation increases to 250 μm or higher, the deformation becomes significantly inhomogeneous and the stretch tends to become concentrated in a small region, reducing the overall correlation of the entire image. The nonlinear behavior observed in Figure 8 seems to result from this reduction in correlation. The increase in the error bar with the elongation supports this speculation.

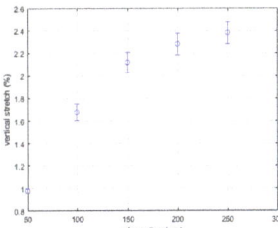

Figure 8. Vertical stretch evaluated from one-dimensional DIC.

These observations indicate that the intensity correlation in the space-domain is not a good method to evaluate the stretching factor for an entire column or row.

3.2.2. Two-Dimensional DIC

Figure 9 is a quiver plot that exhibits the vector field of the displacement experienced by the specimen when the elongation increases from 100 μm to 150 μm.

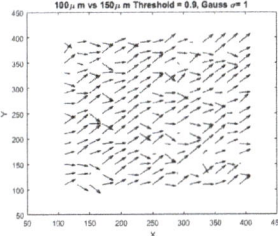

Figure 9. Results from two-dimensional digital image correlation method in space-domain.

It is seen that along column 200 the displacement vectors are oriented at approximately 45°. This orientation is consistent with the above observation that the magnitude of the horizontal and vertical strain is at the same level.

The differences in orientation between vectors can be attributed to a variety of factors, including error. However, preliminary results from our thermal imaging study, presented in Appendix B, indicate that the film specimen exhibits an alternating pattern of stretch and compression along the columns, implying that the observation in Figure 9 may have significance. Additional research, including more in-depth analyses in the frequency domain, is required to confirm this behavior, and will be the subject of our future work.

3.3. One-Dimensional Image Scaling Method in Frequency Domain

Image Scaling and Gaussian Filtering

High frequency speckles (the speckle noise) superposed on the linear dark-fringes compromises the accuracy of this method. Gaussian filtering reduces the speckle noise and, therefore, is an effective way to process the image prior to applying this technique. Figure 10 shows the effect of Gaussian filtering with two different filtering parameters σ (standard deviation).

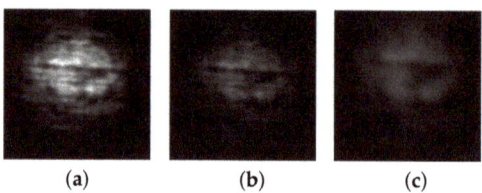

Figure 10. Comparison between unfiltered and Gaussian filtered images. (a) Unfiltered, (b) Filtered, $\sigma = 20$, (c) Filtered, $\sigma = 30$.

Choosing a good frequency range for the analysis is not straightforward. One idea is to use the frequency range that the filtering does not alter. Figure 11 shows the Fourier spectrum along a vertical line near the horizontal center of Figure 10 (a) (unfiltered) and (b) (Gaussian-filtered with standard deviation σ of 20).

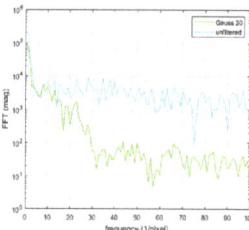

Figure 11. Fourier spectrum of Gauss filtered with $\sigma = 20$ (pixel) and unfiltered.

It is seen that the frequency range of 1–10 pixel^{-1} is unaffected by the filtering. Thus, the image shown in Figure 10b is used for the rest of the analysis.

According to the above argument, we use the frequency components in the range of 1–10 pixel^{-1} and discard all other frequency components including the DC (0 pixel^{-1}) component as it represents the uniform background intensity. We repeat the procedure for the following seven elongation data; no elongation, elongation of 50, 100, 200, 300, 400, and 500 μm. Since the specimen stretched vertically, the Fourier spectrum compresses wias elongation increases. As the spectrum compresses on the frequency axis, the area of the spectrum decreases.

Using the analytical steps described in Section 2.3.4, the spectral compression was evaluated by computing the area of the spectrum in the frequency range of 1–10 pixel^{-1}. Due to the variation of the total optical intensity between measurements for reasons mentioned in Section 2.3.4, the Fourier spectrum of the differential intensity over 1–10 pixel^{-1}

is normalized to the differential intensity's total intensity, which is the sum of gray-scale values of the numerically differentiated data.

The above procedures yield Figure 12. The vertical axis of Figure 12a is the area of the normalized Fourier spectrum of the differential intensity. The solid line is the best linear fit to the data points.

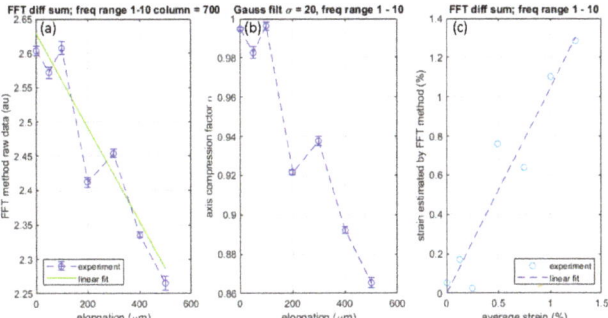

Figure 12. (a) Normalized Fourier spectrum vs. elongation; (b) Calibrated Fourier spectrum vs. elongation; (c) Comparison of stretching factor evaluated with frequency-domain image scaling method and average normal strain evaluated from elongation.

Since the y-intercept of this line represents the FFT spectrum area before the specimen stretches, the division of all other data points by the y-intercept provides us with the axis compression factor α as shown in Figure 12b. The specimen's stretching (the normal strain) ϵ is the reciprocal of α. We can find the average normal strain by dividing the elongation by the specimen's length. Thus, Figure 12c compares the specimen's stretching factor evaluated by the above-described FFT method (the experimental ϵ) with the average normal strain calculated from the elongation (the estimated ϵ). The dashed-line is the best linear fit. The experimental ϵ and estimated ϵ show reasonable agreement indicating a linear relation with the slope of unity.

4. Concluding Remarks

Summary and Findings

In summary, this study applies DIC and speckle pattern techniques to characterize unique patterns observed in LLDPE film undergoing tensile deformation. This insight is of particular interest when considering desirable financial outcomes in industries like commercial shipping that aim to optimize time, cost, and efficiency. Images of the film's surface undergoing deformation were projected onto a screen using a linearly polarized, collimated Helium-Neon laser beam. Resulting digital images contain speckle and diffraction-like fringe patterns and were used in both the spatial and frequency domain analyses as well. In the space domain, the random speckle pattern was used to register local displacement generated from the deformation. The fringes, however, exhibit periodic features consistent with the structural grooves due to the polymer arrangement on the film's surface, and compromises the registration of local displacement. These periodic groove patterns are used in the frequency domain analysis. Conventional DIC in the space domain applies when deformation is uniform over the entire specimen or analysis is limited to small, localized regions. Overall, spatial DIC is found to be unreliable and inefficient for the present application. The frequency domain analysis, however, is found to be capable of revealing average global deformation under non-uniform conditions. Ultimately, we conclude that analysis in the frequency domain using the linear groove patterns is superior to the traditional methods because it is capable of revealing the global average deformation.

More quantitatively, the study has led to the following findings.

1. The space-domain one-dimensional DIC exhibits an order of magnitude smaller (a factor of two smaller at best) strain than the expected value at the 1% or lower strain level. The speckle patterns lost correlation when the strain level becomes approximately four times higher. We suspect the reason behind this finding is as follows. The speckles are formed by the diffusive nature of the transmitted light due to the randomness of the short branches that form the polymer. When the film specimen stretches these short branches shift depending on their original orientations. Therefore, the shift of the associated speckle patterns are not necessarily in line with the direction of the stretch. Hence, at a certain point of elongation, the speckle pattern starts to change randomly.
2. Frequency domain analysis appears superior to its space-domain counterparts. Unlike the speckle patterns due to the random structure of the polymer, the frequency domain analysis uses the periodic structural grooves. Consequently, the linear dark-fringes resulting from this periodic structure correlates well with the stretch of the film specimen. The random speckle patterns compromise this correlated change in the linear dark-fringes. Proper low-pass filtering diminishes this compromising effect. In the present case, Gaussian filtering with the standard deviation of 20 is found effective. With this configuration, the estimated global strain shows reasonable agreement with the expected strain level at least up to 1.4%.
3. The two-dimensional DIC based on the convolutional algorithm is found effective to some extent. However, some of the vectors in the resultant local displacement field exhibit seemingly incorrect directions. Whether these vectors are incorrectly produced by the algorithm or possibly representing the actual material's behavior is an open question. The preliminary result of additional thermal imaging experiments indicate a similar behavior of deformation. It is the subject of our future research.

Author Contributions: S.Y. conceived of the study; C.K. performed experiments; C.K and S.Y. analyzed data and wrote the manuscript; A.G. performed two-dimensional DIC. All authors have read and agreed to the published version of the manuscript.

Funding: This research was partially supported by the Ministry of Trade, Industry, and Energy (MOTIE) and Korea Institute for Advancement of Technology (KIAT), Korea through the International Cooperative R&D program (Project No P 0006842).

Institutional Review Board Statement: Not applicable.

Data Availability Statement: Not applicable.

Conflicts of Interest: The authors declare no conflict of interest.

Abbreviations

The following abbreviations are used in this manuscript:

DIC   Digital Image Correlation
FFT   Fast Fourier Transform
LLDPE   Linear Low-Density Polyethylene

Appendix A. Verification of the Fourier Scaling Property

As the LLDPE film stretches, the space between linear dark-fringes, or the spatial frequency of the fringes, changes. Section 2.3.3 describes how this study uses the Fourier scaling property (Equation (5)) for frequency-domain analysis.

This method evaluates this change in the spatial frequency in the frequency domain as follows.

Consider that $f(x)$ represents the gray-scale of the projected image before the stretch along the x-axis. The Fourier transform of this function is given as follows.

$$F(\omega) = \int_{-\infty}^{\infty} f(x) e^{-j\omega x} dx \qquad (A1)$$

Stretching/compressing the material is equivalent to multiplying the coordinate axis by a factor α as $\xi = \alpha x$. This multiplication alters the Fourier transform as follows.

$$F_\alpha(\omega) = \int_{-\infty}^{\infty} f(\alpha x) e^{-j\omega x} dx = \int_{-\infty}^{\infty} f(\xi) e^{-j\omega \frac{\xi}{\alpha}} \frac{d\xi}{\alpha} = \frac{1}{\alpha} \int_{-\infty}^{\infty} f(\xi) e^{-j\omega \frac{\xi}{\alpha}} d\xi \quad (A2)$$

Next consider Fourier transform of $g(\xi) = df(\xi)/d\xi$, $G_\alpha(\omega)$. From Equation (A2),

$$G_\alpha(\omega) = \frac{1}{\alpha} \int_{-\infty}^{\infty} g(\xi) e^{-j\omega \frac{\xi}{\alpha}} d\xi = \frac{1}{\alpha} \int_{-\infty}^{\infty} \frac{df(\xi)}{d\xi} e^{-j\omega \frac{\xi}{\alpha}} d\xi \quad (A3)$$

Integrate the right-hand side of Equation (A3) by parts.

$$G_\alpha(\omega) = \frac{1}{\alpha}\left\{\left[f(\xi) e^{-j\omega \frac{\xi}{\alpha}}\right]_{-\infty}^{\infty} - \int_{-\infty}^{\infty} f(\xi)\left(\frac{d}{d\xi} e^{-j(\frac{\omega}{\alpha}\xi)}\right) d\xi\right\} \quad (A4)$$

We can ignore the first term on the right-hand side of Equation (A4) as it represents the highest frequency components of the complex conjugate pair of the Fourier transform. The highest frequency components correspond to the pixel-to-pixel gray-scale variation. Our purpose here is to find the change in the spatial frequency of the linear dark-fringes. The interval of the linear dark-fringes involves multiple pixels. Thus, $G_\alpha(\omega)$ becomes as follows.

$$G_\alpha(\omega) = \frac{-1}{\alpha} \int_{-\infty}^{\infty} f(\xi)\left(\frac{d}{d\xi} e^{-j(\frac{\omega}{\alpha}\xi)}\right) d\xi = j\left(\frac{\omega}{\alpha}\right) \int_{-\infty}^{\infty} f(\xi) e^{-j(\frac{\omega}{\alpha}\xi)} d\xi \quad (A5)$$

Here, we use integration by parts and drop the first term for the same reason as Equation (A4). We can interpret the right-hand side of Equation (A5) as the Fourier transform of function $g(\xi)$, $\mathcal{F}\{g(\xi)\}$, with the scaled frequency ω/α. So,

$$|G_\alpha(\omega)| = \left|\mathcal{F}\left\{\frac{df(\xi)}{d\xi}\right\}\right| = |\mathcal{F}\{g(\xi)\}| = |\mathcal{F}\{g(\alpha x)\}| = \left(\frac{\omega}{\alpha}\right)\left|F\left(\frac{\omega}{\alpha}\right)\right| \quad (A6)$$

The equality for the last two terms of Equation (A6) can be interpreted as a result of the combination of the differentiation and scaling properties of Fourier transform [17]. Figure A1 demonstrates the relation between a function's stretching in the space domain and the compression of the Fourier transform of the function in the frequency domain [15]. Here Figure A1a shows the plots of the original function, along with the same function stretched by factors of 10% and 20% in the space domain. Figure A1b plots the magnitude of the Fourier transforms corresponding to the three plots in (a). It is seen that the spatial stretch compresses the Fourier spectrum.

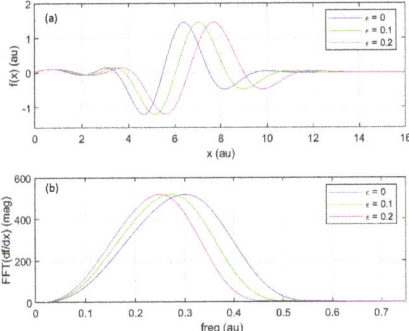

Figure A1. (a) Sample function with stretching, (b) Fourier transforms of functions in (a).

Equation (5) used in Section 2.3.3 indicates that when the specimen stretches the Fourier transform of the spatial derivative of the function that represents the spatial variation of the gray-scale level compresses in the same fashion as Figure A1. This method uses this property to estimate the stretching factor from the compression of the Fourier spectrum.

Appendix B. Thermal Imaging

A FLIR infrared camera was used to capture these images while the specimen was deforming. Preliminary data using thermal imaging are shown below in Figure A2. The upper row presents the temperature of the specimen along with the background (laboratory air) temperature. The frame number shown above each image indicates the region enclosed by a rectangle is the film specimen. Due to the phenomenon known as the thermoelastic effect, it is expected that the film specimen increases/decreases its temperature when it is stretched/compressed. Figure A2 indicates that as the deformation develops.

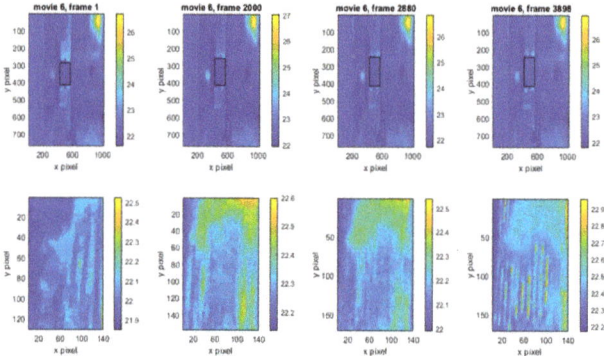

Figure A2. Preliminary analysis using thermal imaging.

The top row of images in Figure A2 shows the full image including ambient interference, while the bottom row shows the image zoomed in to the specimen (indicated above in the box). The original length of the specimen was 2.9 mm, and the final elongation was 3.6 mm. The frame numbers correspond to vertical strain of 0%, 9.95%, 14.3%, and 19.4%, respectively.

References

1. Ahmed, M. Importance of Linear Low Density Polyethylene (LLDPE). Technologies in Industry 4.0. Available online: https://www.technologiesinindustry4.com/2021/08/importance-of-linear-low-density-polyethylene-lldpe.html (accessed on 7 January 2022).
2. Sciammarella, C.A.; Sciammarella, F.M. *Experimental Mechanics of Solids*; Wiley: Hoboken, NJ, USA, 2012.
3. Leendertz, J.A. Interferometric displacement measurement on scattering surfaces utilizing speckle effect. *J. Phys. E* **1970**, *3*, 214–218. [CrossRef]
4. Hild, F.; Roux, S. Digital Image Correlation. In *Optical Methods for Solid Mechanics: A Full-Field Approach*; Hack, E., Rastogi, P.K., Eds.; Wiley-VCH: Weinheim, Germany, 2012; pp. 183–225.
5. Hild, F.; Roux, S. Comparison of Local and Global Approaches to Digital Image Correlation. *Exp. Mech.* **2012**, *52*, 1503–1519. [CrossRef]
6. Nwanoro, K.; Harrison, P.; Lennard, F. Investigating the Accuracy of Digital Image Correlation in Monitoring Strain Fields Across Historical Tapestries. *Strain* **2021**, *58*, 12401. [CrossRef]
7. Brake, J.; Jang, M.; Yang, C. Analyzing the relationship between decorrelation time and tissue thickness in acute rat brain slices using multispeckle diffusing wave spectroscopy. *J. Opt. Soc. Am.* **2016**, *33*, 270–275. [CrossRef] [PubMed]
8. Hajjarian, Z.; Nia, H.T.; Ahn, S.; Grodzinsky, A.J.; Jain, R.K.; Nadkarni, S.K. Laser Speckle Rheology for evaluating the viscoelastic properties of hydrogel scaffolds. *Sci. Rep.* **2016**, *6*, 37949. [CrossRef] [PubMed]
9. Duncan, D.D.; Kirkpatrick, S.J.; Gladish, J.C. What is the proper statistical model for laser speckle flowmetry. Volume: Complex dynamics and fluctuations in biomedical photonics. In *Complex Dynamics and Fluctuations in Biomedical Photonics V*; SPIE: San Jose, CA, USA, 2008; Volume 6855.

10. Popov, I.; Weatherbee, A.; Vitkin, I.A. Statistical properties of dynamic speckles from flowing Brownian scatterers in the vicinity of the image plane in optical coherence tomography. *Biomed. Opt. Express* **2017**, *8*, 2004–2017. [CrossRef]
11. Zhu, Y.K.; Tian, G.Y.; Lu, R.S.; Zhang, H. A review of optical NDT technologies. *Sensors* **2011**, *11*, 7773–7798. [CrossRef] [PubMed]
12. Quino, G.; Chen, Y.; Ramakrishnan, K.R.; Martínez-Hergueta, F.; Zumpano, G.; Pellegrino, A.; Petrinic, N. Speckle patterns for DIC in challenging scenarios: Rapid application and impact endurance. *Meas. Sci. Technol.* **2020**, *32*, 015203. [CrossRef]
13. Jerabek, M.; Major, Z.; Lang, R.W. Strain Determination of Polymeric Materials Using Digital Image Correlation. *Polym. Test.* **2010**, *29*, 407–416. [CrossRef]
14. Passieux, J.C.; Bugarin, F.; David, C.; Périé, J.N.; Robert, L. Multiscale Displacement Field Measurement Using Digital Image Correlation: Application to the Identification of Elastic Properties. *Exp. Mech.* **2015**, *55*, 121–137. [CrossRef]
15. Arai, Y. Electronic speckle pattern interferometry based on spatial information using only two sheets of speckle patterns. *J. Mod. Opt.* **2014**, *61*, 297–306 [CrossRef]
16. Mazzoleni, P.; Matta, F.; Zappa, E.; Sutton, M.A.; Cigada, A. Gaussian pre-filtering for uncertainty minimization in digital image correlation using numerically-designed speckle patterns. *Opt. Lasers Eng.* **2015**, *66*, 19–33. [CrossRef]
17. Ding, X.; Wang, Z.; Hu, G.; Liu, J.; Zhang, K.; Li, H.; Ratni, B.; Burokur, S.N.; Wu, Q.; Tan, J. Metasurface holographic image projection based on mathematical properties of Fourier transform. *PhotoniX* **2020**, *1*, 16. [CrossRef]
18. Kim, J. Evaluation of resonance characteristics of thin film with improved opto acoustic method. In *Masters Abstracts International*; Ann Arbor: ProQuest Dissertations & Theses; Southeastern Louisiana University: Hammond, LA, USA, 2018; 81p.
19. Yoshida, S. Optical interferometric study on deformation and fracture based on physical mesomechanics. *Phys. Mesomech.* **1999**, *2*, 5–12.
20. Takahashi, S.; Yoshida, S.; Sasaki, T.; Hughes, T. Dynamic ESPI Evaluation of Deformation and Fracture Mechanism of 7075 Aluminum Alloy. *Materials* **2021**, *14*, 1530. [CrossRef] [PubMed]
21. Fransinski, L.J. Covariance mapping techniques. *Phys. B At. Mol. Opt. Phys.* **2016**, *49*, 152004. [CrossRef]
22. Yuan, Y.; Zhan, Q.; Xiong, C.; Huang, J. Digital image correlation based on a fast convolution strategy. *Optics Lasers Eng.* **2017**, *97*, 52–61. [CrossRef]
23. Rao, Y.R.; Prathapani, N.; Nagabhooshanam, E. Application of normalized cross correlation to image registration. *Int. J. Res. Eng. Technol.* **2014**, *3*, 12–16.
24. Hoq, M.E. An Investigation of Image Correlation for Real-Time of Wrapping Film Deformation. In *Masters Abstracts International*; Ann Arbor: ProQuest Dissertations & Theses; Southeastern Louisiana University: Hammond, LA, USA, 2020; 107p.

Article

Damage Progress Classification in AlSi10Mg SLM Specimens by Convolutional Neural Network and k-Fold Cross Validation

Claudia Barile, Caterina Casavola, Giovanni Pappalettera * and Vimalathithan Paramsamy Kannan

Dipartimento di Meccanica, Matematica e Management, Politecnico di Bari, Via E. Orabona 4, 70125 Bari, Italy; claudia.barile@poliba.it (C.B.); casavola@poliba.it (C.C.); pk.vimalathithan@poliba.it (V.P.K.)
* Correspondence: giovanni.pappalettera@poliba.it

Abstract: In this study, the damage evolution stages in testing AlSi10Mg specimens manufactured using Selective Laser Melting (SLM) process are identified using Acoustic Emission (AE) technique and Convolutional Neural Network (CNN). AE signals generated during the testing of AlSi10Mg specimens are recorded and analysed to identify their time-frequency features in three different damage evolution stages: elastic stage, plastic stage, and fracture stage. Continuous Wavelet Transform (CWT) spectrograms are used for the processing of the AE signals. The AE signals from each of these stages are then used for training a CNN based on SqueezeNet. Moreover, k-fold cross validation is implemented while training the modified SqueezeNet to improve the classification efficiency of the network. The trained network shows promising results in classifying the AE signals from different damage evolution stages.

Keywords: AlSi10Mg; SLM; NDE; acoustic emission; deep learning; CNN; k-fold cross validation

Citation: Barile, C.; Casavola, C.; Pappalettera, G.; Kannan, V.P. Damage Progress Classification in AlSi10Mg SLM Specimens by Convolutional Neural Network and k-Fold Cross Validation. *Materials* **2022**, *15*, 4428. https://doi.org/10.3390/ma15134428

Academic Editor: Thomas Niendorf

Received: 26 May 2022
Accepted: 21 June 2022
Published: 23 June 2022

Publisher's Note: MDPI stays neutral with regard to jurisdictional claims in published maps and institutional affiliations.

Copyright: © 2022 by the authors. Licensee MDPI, Basel, Switzerland. This article is an open access article distributed under the terms and conditions of the Creative Commons Attribution (CC BY) license (https://creativecommons.org/licenses/by/4.0/).

1. Introduction

The evolution of Additive Manufacturing (AM) technique is a narrative of its own. Its application began with complicated geometries and functional prototypes. In the last decade, it has evolved into one of the key manufacturing systems for many industrial and aeronautic components. The functionality of the AM components has made them enter into the maritime applications [1–4]. Consequently, an intensive scrutinizing is required for validating the safety of the AM components.

Selective Laser Melting (SLM) is a predominantly used AM technique, known for its good quality, short lead times, limited restriction and high resolution for complex shapes and structures [5]. A very fine-grained structure can be achieved through the SLM process, which results in the improved mechanical properties of the fabricated components. A wide range of materials including Ni, Al, Ti alloys can be manufactured using the SLM process. The components manufactured from the SLM process have relatively improved mechanical properties, corrosion resistance and fatigue life compared to the traditionally manufactured components. The fine-grained structure of SLM manufactured components is supposed to give better isotropic properties, but there has been a long-standing debate on this. SLM components generally have isotropic strength; however, they have a higher anisotropy in the elongation at break [6,7]. The isotropic behaviour of the SLM components is generally related to their building direction. A homogenous microstructure provides high mechanical strength to the components and can accommodate strain which can reduce cracking. However, due to the high cooling rates in the SLM process, which is in the range of 10^6–10^8 °C/s, achieving equiaxed grain is a challenge [5,8–11]. Thus, it is essential to study the behaviour of SLM components built in different direction.

In this study, the mechanical characteristics of the SLM components built from AlSi10Mg alloy at different orientations are studied. Acoustic Emission (AE) technique, a Non-Destructive Evaluation (NDE) technique, is used for studying the damage behaviour of the SLM components. The AE technique is based on the acquisition of elastic waves generated due to the rapid release of stored elastic energy [12]. Microscopic displacements within a solid material due to microcracking, crack nucleation or crack growth generate elastic waves. Each of these microscopic damages within a material generates acoustic waves that differ in their time-frequency characteristics. Analysing these acoustic waves to identify the damage source and predicting the failure is the basis of the AE technique [13–15]. This is one of the few available passive techniques for investigating the damage characteristics of a material throughout its loading history.

Time-frequency analysis of stress waves generated from Fibre Reinforced Polymer (FRP) composites is relatively common [16]. However, it is seldom used for characterizing the damage modes in metallic components. In the authors' previous studies, time-frequency analysis of acoustic waves has been successfully used in the identification of damage sources in metallic components [17].

In the recent years, Deep Learning has been used contemporarily with the AE technique for damage characterization [18]. Convolutional Neural Networks (CNN) have been used for identifying damage modes in SiC composites, CORTEN steel and civil structures [19–25], while Artificial Neural Networks (ANN) have been used for corrosion monitoring of steel. While both these networks can relate a large number of parameters and building a model for classification and prediction in real-time, CNN is preferred generally for image-based analysis. While there is an argument that ANN such as Long Short-Term Memory (LSTM) have advantages over the feed-forward neural networks such as CNN, many literatures have used CNN for image-based analysis [26,27]. LSTM is suitable for handling temporal or sequential data, while CNN are used predominantly for image-based analysis. Apart from these, there are several other image-based and wavelet-based neural networks, which have been used successfully in the past [28].

Acoustic signals generated from different damage modes have their unique time-frequency signatures. This can be observed and analysed in their time-frequency spectrum using wavelet spectrograms. The spectrograms are images containing the spectral details of the analysed waveforms. In that context, they can be used as inputs to train neural networks. Previously, this technique has been used successfully in identifying the damage modes in FRP composites, SiC composites, and civil structures [21,23,29]. The aim of this research is to use the image-analysing capability of CNN for classifying the acoustic emission signals from different damage modes. This is possible by training a CNN with time-frequency spectrograms of AE signals generated from various damage sources of a material/structure.

The objective of this research is to investigate the mechanical properties of SLM manufactured AlSi10Mg components in different orientations. In addition to that, the AE technique is used for identifying the damage sources and a Deep Learning neural network is modelled and trained for real-time identification of damage stages.

2. Materials and Methods

2.1. Materials

The AlSi10Mg alloy used in this study for manufacturing has a density of 2.68 g/cm^3 and a melting range of 570 °C to 590 °C. The chemical composition of the feed material is presented in Table 1.

Table 1. Chemical composition of the feed material AlSi10Mg alloy.

Element	Al	Si	Mg	Fe	N	O	Ti	Zn	Mn	Ni	Cu	Pb	Sn
Mass (%)	Bal *	11	0.45	<0.25	<0.2	<0.2	<0.15	<0.1	<0.1	<0.05	<0.05	<0.02	<0.02

* Balance percentage.

The SLM manufacturing system is RenAM 500 M (Renishaw S.p.A., Torino, Italy), in which an Nd:YAG laser (wavelength 1.064 µm) is used for melting the feed material. Dog-bone shaped tensile specimens are prepared on the heated bed inside the SLM chamber.

The recoater moved along the Y axis to coat the powder, then the laser moved along the X axis to melt the powder. The laser source of power 400 W created a single-track energy density of 20 J/mm² melted the coated powder. The laser beam of spot diameter 200 µm and the speed of the laser movement along the X axis is 100 mm/s. The powder is melted for form a uniform layer of 20 µm thickness, before the recoater coats the powder again. The process is repeated until the complete specimen is built.

Dog-bone shaped specimens as per ASTM E8M configurations are built along four different orientations, which are displayed in Figure 1. The different orientations are achieved by moving the powder bed, while keeping the laser axis the same throughout the process. At the end of the process, the specimens are kept inside the environmental chamber of the SLM system at 300 °C for 2 h and then they are air-cooled.

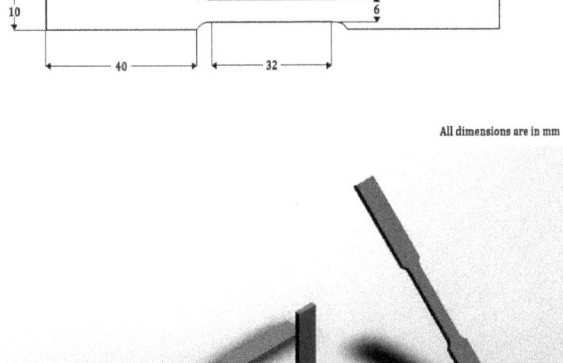

Figure 1. Dog-bone shaped specimens built in four different configurations in SLM building platform and their dimensions.

2.2. Test Methods

For testing the mechanical properties of the specimens, they are mounted on an INSTRON servo-hydraulic testing machine (Norfolk County, MA, USA), with a maximum loading capacity of 10 kN. The tensile tests are carried out at a speed of 1 mm/min as per ASTM E8M standard.

For recording the acoustic waves/stress waves generated due the damage evolution during the tensile test, a piezoelectric sensor is coupled to the surface of the specimen. A uniform thin layer of silicone grease is applied between the surfaces of the sensor and the specimen. The PICO sensor (Physical Acoustics Corporation, Princeton JCT, USA) is a wideband sensor with the maximum sensitivity in the acquisition range of 200 kHz to 750 kHz and it has a resonant frequency of 250 kHz. The signals recorded by the sensors are amplified by 40 dB using a preamplifier and filtered through 1 kHz/1 MHz low/high-band pass filters. The signals are recorded at a sampling rate of 1 MHz.

2.3. Time-Frequency Analysis of AE Signals

The acoustic signals recorded over the entire loading history of the tensile test of AlSi10Mg specimens are analysed in their time-frequency domain to understand their damage sources. The time-frequency characteristics of the acoustic signals are analysed using the signal processing technique, Continuous Wavelet Transform (CWT) [30,31]. First, CWT is used for identifying the damage modes; second, the spectrograms are used for training and testing the image-based classification neural network. Researchers commonly use waveforms in their time-series representations or spectrograms of CWT, Short-time Fourier Transform (STFT) or Mel Spectrograms [21–23]. The independence of selecting a user-defined wavelet for decomposing the signal makes CWT a formidable signal processing tool. It has been used successfully by several researchers for damage characterization of FRP composites using AE technique.

In CWT, the original signal $f(x)$ is decomposed into wavelet coefficients using a mother wavelet. The wavelet coefficients give information about the time-frequency localization of the spectral components of the original signal [31]. CWT can be explained by Equation (1).

$$CWT_f(a,b) = \int_{-\infty}^{\infty} f(x)\overline{\zeta_{a,b}(x)}dx, \ a > 0, \tag{1}$$

a in Equation (1) is the scaling factor with which the mother wavelet $\zeta_{a,b}(x)$ is dilated or compressed and b is the translation factor, which gives the wavelet components in different time domains. The mother wavelet used in this study is an analytical Morlet Wavelet, which can be defined by Equation (2) [32].

$$\zeta(x) = \exp(-x^2/2)\cos(5x), \tag{2}$$

Literatures are enriched with the basic principle and implementation of CWT technique for time-frequency analysis [30,31]. In this research, this process is carried out in MATLAB® (2020b).

2.4. Convolutional Neural Network

Generally, CNN consists of an input layer, a classification layer and a set of hidden layers. One of the most important hidden layers is a convolutional layer. Convolutional layer features a set of weighted filters, which extract the feature map from its input. The output of the convolutional layer is generally followed by a pooling layer and an activation function. The pooling layer extracts the most representative features of the convolutional output by padding and striding operations. Max pooling and average pooling are the commonly used pooling operations in a CNN. The most used activation functions are sigmoid, tanh and ReLu activations. Stochastic Gradient Descent (SGD) is typically used for training the models. ReLu activation function is preferred for SGD training algorithms because of its non-saturation of gradients and the efficient convergence in SGD [33].

Typically, the number of hidden layers is based on the depth of the features to be extracted from the input image. However, if more layers are used, overfitting may occur, and the classification accuracy is ultimately affected. To avoid the overfitting of the results, a dropout layer can be used.

Branching and exploring multiple paths of the hidden layers are explored by many researchers to obtain the highest classification accuracy. This also can extract different levels of abstraction and enables the network to learn the information in the early stages, which flows the information more easily into the classification layer.

More deep layers and branching makes the computation time longer and requires large computation power. To overcome this problem, Iandola et al. introduced a fire module [34]. This contains a squeeze convolutional layer, which has a 1 × 1 filter, whose output is fed to an expanding convolutional layer, which is a mix of two convolutional layers having 1 × 1 filter and 3 × 3 filter, respectively. The idea is to use 1 × 1 filters for most of the convolutional layer and create a large activation data pool by delaying the downsampling of data. Most commonly the downsampling is done by setting the stride value in the CNN architecture to be greater than 1. In SqueezeNet, however, the stride value is kept to 1. The pooling layer always follows the convolutional and activation layer. More details about this SqueezeNet architecture can be found in the source paper.

In this research, a similar CNN based on the SqueezeNet is used for classifying the waveforms with high accuracy. To avoid the overfitting of data, some of the deeper convolutional layers are removed from the original network, but the sizes of filters are increased in the deeper convolutional layers. In the original network, there were 68 layers and 75 connections. It had 8 fire modules for extracting the deeper features. In the modified network, there are 39 layers with 42 connections and 4 fire modules. However, large filter sizes are used in the deeper convolutional layers, similar to the original SqueezeNet. Subsequently, a moderately large activation pool is obtained. The dropout layer is used for avoiding the overfitting of data.

In this study, to improve the training efficiency of the SqueezeNet, k-fold cross validation is implemented. k-fold partitions the input data to the network into k number of subsets. For each iteration of training the network, $(k-1)$ number of subsets are used as a training data and the remaining subset is used as a test data [35,36]. This reduces the bias as most of data are used for training the network for k iterations. Besides, the network weights of the convolutional layers are updated constantly for each iteration, thereby improving the efficiency of training. Typical k-fold configuration is presented in Figure 2. In this study, the input data is split into $k = 5$ subsets and the network is trained for 5 iterations.

Figure 2. Schematic of k-fold cross validation.

The architectural details of the CNN built for this study are presented in Figure 3. The configuration of a typical fire module consisting of squeezing and expansion of the convolutional output is explained in Table 2.

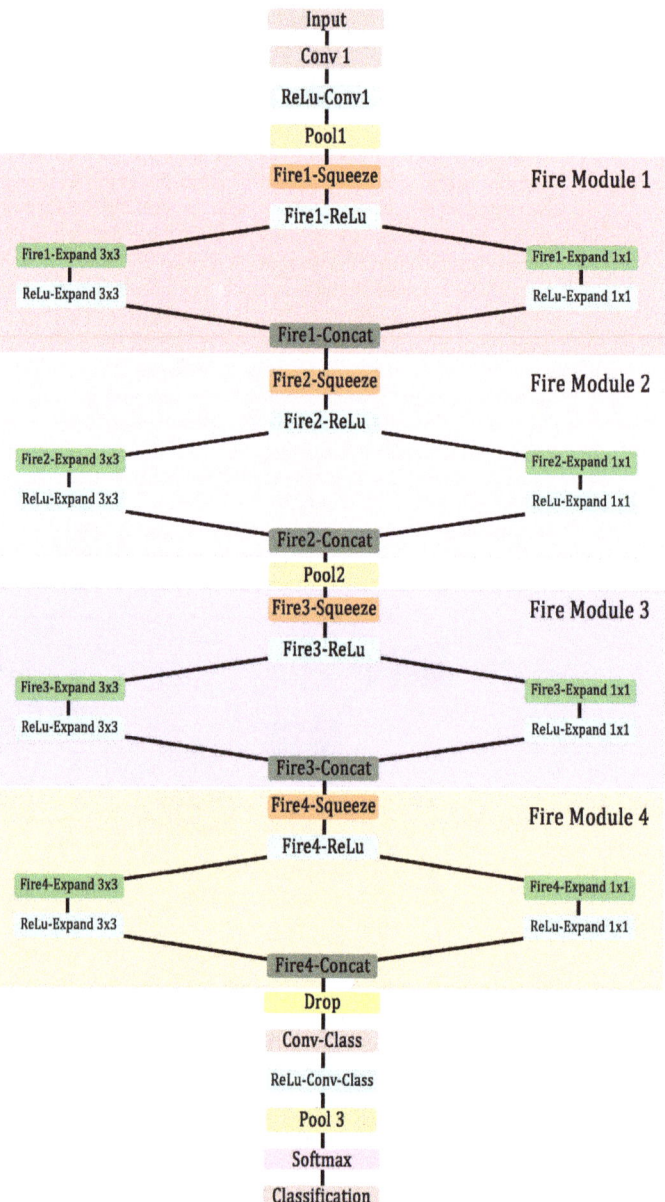

Figure 3. Architecture of the modified SqueezeNet (data flowing in one direction—from input layer to classification layer).

Table 2. Typical fire module consisting of squeeze and expand convolutional layers.

Fire Module n	Layer Name	Layer Description
Squeeze	Fire$n$-Squeeze	Number of Filters: X Filter Size: 1 × 1 Stride: 2 × 2 ReLU activation
Expand	Fire$n$-Expand 1 × 1	Number of Filters: X Filter Size: 1 × 1 Stride: 2 × 2 ReLU activation
	Fire$n$-Expand 3 × 3	Number of Filters: X Filter Size: 1 × 1 Stride: 2 × 2 ReLU activation
Concatenation	Fire$n$-Concat	

In Table 2, the n represents the number of the fire module layer, which can be seen in Figure 3. X is the number of filters in each layer. More details about the number of filters in each layer and other features such as filter size and stride are presented in Table 3.

Table 3. Layer details and descriptions of the modified SqueezeNet.

Fire Module n	Layer Name	Layer Description
Input Layer	Input	32 × 32 × 3 Spectrograms
Convolutional Layer	Conv1	Number of Filters: 32 Filter Size: 3 × 3 Stride: 2 × 2 ReLU activation
Max Pooling Layer	Pool1	Pool Size 3 × 3 Stride: 2 × 2
Fire Module	Fire Module 1	Number of Filters in Squeeze: 16 Number of Filters in Expand: 32
Fire Module	Fire Module 2	Number of Filters in Squeeze: 32 Number of Filters in Expand: 64
Max Pooling Layer	Pool2	Filter Size: 3 × 3 Padding: 0,0,0,0 Stride: 2 × 2
Fire Module	Fire Module 3	Number of Filters in Squeeze: 64 Number of Filters in Expand: 128
Fire Module	Fire Module 4	Number of Filters in Squeeze: 128 Number of Filters in Expand: 256
Dropout Layer	Drop	Probability: 0.5
Convolutional Layer	Conv-Class	Number of Filters: 6 Filter Size: 1 × 1 Stride: 2 × 2 ReLU activation
Global Average Pooling Layer	Pool3	-
Softmax Layer	-	-
Classification Layer	-	-

The data used for training the CNN and for evaluating its efficiency are presented in the Results and Discussions section.

3. Results and Discussions

3.1. Tensile Test Results

Four different configurations of SLM specimens based on different build orientations (named as Tx, Ty, Tz and T45) are tested. Six specimens from each group are tested. The mechanical properties, ultimate tensile strength, yield strength, Young's modulus, and elongation at break of the specimens are reported in Table 4. These properties are calculated according to the ASTM E8 standard.

Table 4. Tensile test results of AlSi10Mg specimens built in different orientations.

Specimen Name	Ultimate Tensile Strength	Yield Strength	Young's Modulus	Elongation at Break
	MPa	MPa	GPa	%
Tx	217.2 ± 2.4	137.0 ± 1.5	70.7 ± 3.8	14.2 ± 0.5
Ty	213.6 ± 4.2	132.4 ± 2.4	65.2 ± 1.9	11.2 ± 4.9
Tz	214.4 ± 2.5	126.7 ± 2.8	65.8 ± 1.1	8.9 ± 1.1
T45	218.7 ± 2.4	132.0 ± 2.5	67.4 ± 3.2	7.4 ± 18

For all the four groups of specimens, the ultimate tensile strength, yield strength and Young's modulus are quite similar. However, the elongation at break of specimen groups Tz and T45 specimens are quite low in comparison with Tx and Ty. It has been reported by several researchers that the changing in build orientation can result in the microstructural heterogeneity of the SLM components [10]. Dong et al. studied the thermal transfer mechanisms of AlSi10Mg specimens and derived the microstructural variations due to the build orientations of the components [8,9,11]. This probably could be the reason for the large variation in elongation at break between the specimens.

More details about the mechanical results and their dependence on the build orientations can be found in the authors' previous research works [17,37,38]. In this work, the different damage evolution stages in these specimens are analysed using the AE technique and deep neural network model.

Representative load-displacement curves of the four different specimen groups are presented in Figure 4. All the four specimens apparently show a similar load response during the tensile tests. Until the yield point, these specimens show a linear elastic behaviour, which is followed by a yield phase. As observed in Table 4, the duration of the yielding phase varies between the specimens. Nonetheless, the duration between the commencement of fracture and the final failure is quite similar. It can be assumed that these specimens have three damage evolution stages: elastic stage, plastic stage, and fracture stage.

First, the yield point is selected from the stress–strain curve of the tensile test data as per the instructions in ASTM E8 standard. The region until the yield point is considered as the elastic stage, where there is a linear elastic stress response by the specimen to the applied load. Second, the plastic stage is selected from the yield point until the region where the slope of the load response starts to decrease. Finally, the region beyond the plastic stage until the final fracture is considered as the fracture region. A schematic representation of the different damage evolution stages is presented in Figure 5. AE signals recorded from each of these stages are collected and are analysed in their time-frequency domain.

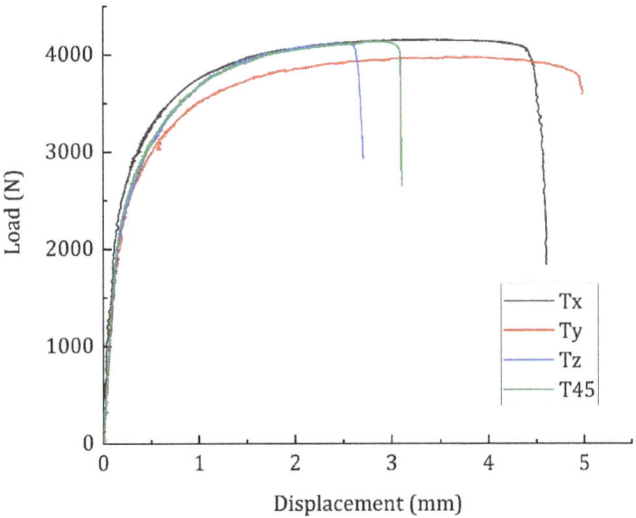

Figure 4. Load-displacement curves of the AlSi10Mg specimens built in four different orientations.

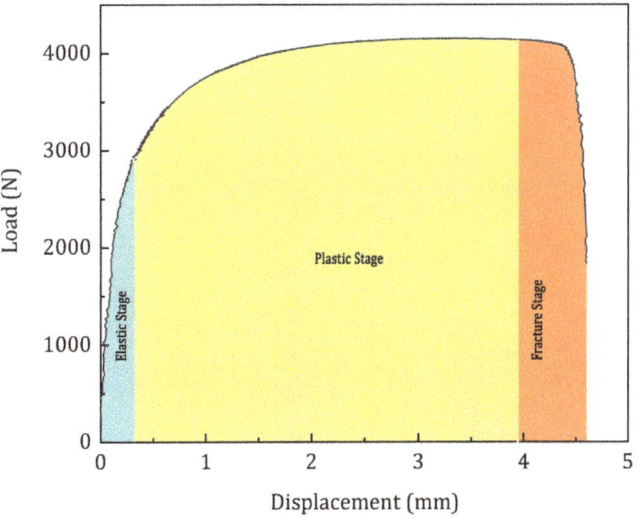

Figure 5. Damage stages considered for the AE analysis.

3.2. Time-Frequency Characteristics of AE Signals from Different Damage Modes

AE signals from the elastic stage of all four groups of specimens Tx, Ty, Tz and T45 are collected and analysed in their time-frequency domain using CWT. Interestingly, all the AE signals from the elastic stage can be grouped into two categories based on their time-frequency characteristics. The CWT spectrograms of these two categories of signals are presented in Figure 6.

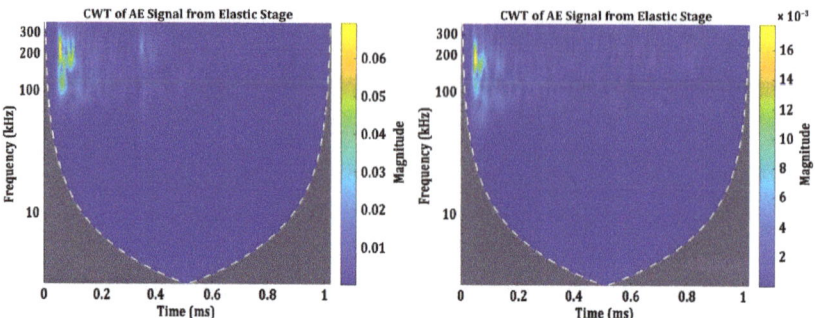

Figure 6. CWT Spectrograms of the AE signals from the elastic stage.

It can be observed that in both the categories, most part of the spectral components are present in the same frequency (between 200 kHz and 250 kHz) [39] and are localized in a similar time domain. The only difference between these two categories is their magnitude. The maximum magnitude of AE signals in the first category is between 0.06 and 0.08, while the second category is generally between 0.01 to 0.02. Nonetheless, these are the typical characteristics of the AE signals from the elastic stage. There are very few literatures available to compare the frequency components and the magnitudes of AE signals generated during the linear elastic response of metallic components [40]. However, it has been reported by some researchers that during the elastic stage, the AE events are generated due to the dislocation motions, and they have frequency components between 200 kHz and 250 kHz. The presence of similar AE signals in all the four specimen groups show that the signals presented in Figure 6 are the general characteristics of the AE signals generated during the elastic stage.

During the plastic stage, however, AE signals are generated due to various damage evolutions in the specimens. Grain boundary movements, local yielding around the inclusions, local plastic deformation around the pores and the plastic deformation are some of the sources of AE signals during the plastic deformation stage [39–43]. While analysing the AE signals in the plastic stage of all four groups of specimens, three different categories of AE signals are observed. These signals can be categorized based on their time-frequency characteristics. The CWT spectrograms of the signals from the plastic stage are presented in Figure 7.

The three categories of AE signals in the plastic stage have obvious differences in their frequency characteristics. The first category of the signals in the plastic stage has two frequency components, one localized around 100 kHz and the other around 200 kHz. The second category of AE signals has its spectral energy centred in the frequency band above 300 kHz and the signals are localized in a very narrow time domain compared to the other categories. The third category has its spectral energy centred between 200 kHz and 250 kHz. However, these signals have lots of reverberations, which can be seen by the similar spectral components with lesser magnitude appearing up to the signal length of 0.4 ms. In the literature, some of these signal characteristics are observed in the plastic stage of loading. Abkari and Ahmed have observed AE signals with frequencies above 300 kHz in the plastic region [42]. Nonetheless, it can be said that the AE signals generated during the plastic stage have three specific characteristics.

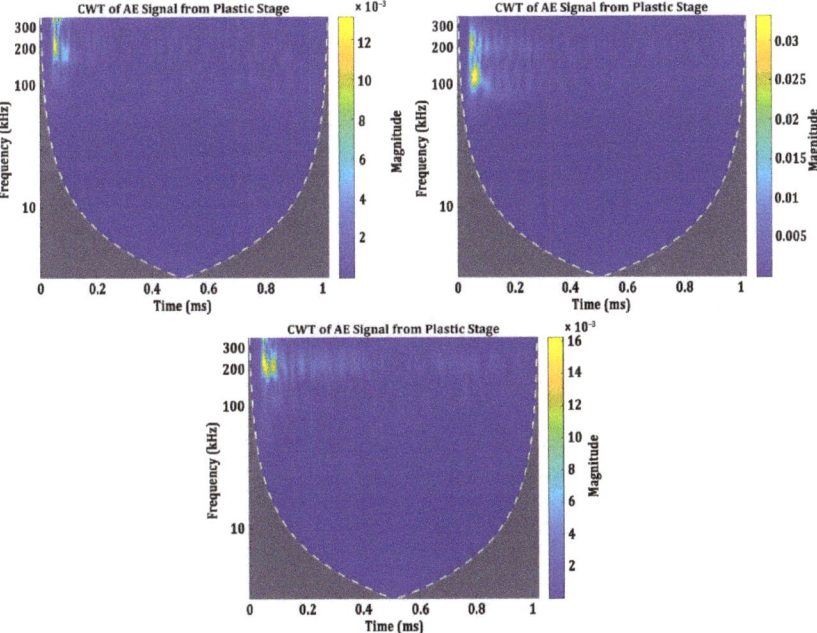

Figure 7. CWT Spectrograms of the AE signals from the plastic stage.

AE signals from the fracture stage show two distinct time-frequency characteristics. The CWT spectrograms of these signals are presented in Figure 8. It can be observed that the CWT spectrograms of the signals in Figure 8 have more reverberations than the signals shown in Figures 6 and 7. This possibly could be due to the overlapping of several signals generated in short intervals in the fracture stage. The final fracture occurs in a very short duration under loading, and this possibly could have generated signals with lots of reverberations. The first category of these signals has its spectral energy distributed in two frequency bands: one at 200 kHz and another above 300 kHz. The second category of the signals have lots of reverberations, but the spectral component is localized at 200 kHz. The reverberations extend up to the length of 0.4 ms of the signal.

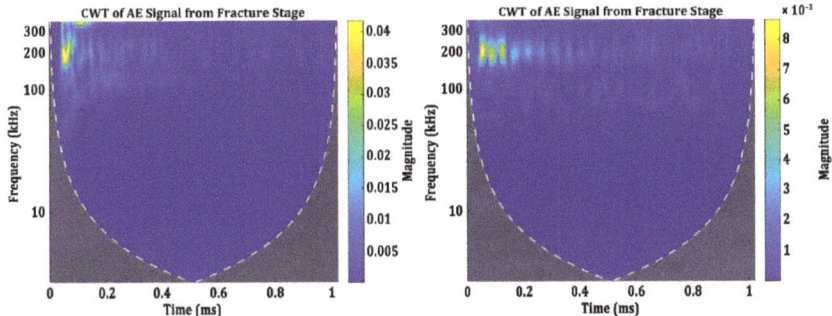

Figure 8. CWT Spectrograms of the AE signals from the fracture stage.

3.3. Convolutioanl Neural Network Training and Test Results

In the previous section, AE signals from all the specimens are recorded and analysed in their time-frequency domain. Based on their time-frequency characteristics, two categories of signals in the elastic stage, three categories in the plastic stage and two categories of signals in the fracture stage are observed. AE signals having the similar time-frequency characteristics as those displayed in Figures 6–8 are used for training the CNN. The objective is to train the CNN to classify these groups of signals automatically and thereby classifying the damage evolution stages of the AlSi10Mg specimens under loading.

The CNN can be trained more efficiently if the number of training data is high. The number of AE signals recorded during the tensile tests of four groups of AlSi10Mg specimens is less than 5000. When these signals are classified into three damage evolution stages, the plastic stage contains around 3000 signals, while the elastic stage and fracture stage has 1000 signals each. This is not sufficient to train the CNN. Therefore, these signals are augmented by adding random noises. In all, 15,000 signals for each damage evolution stage, a total of 45,000 signals, are generated. A random selection of 30,000 signals is used for training and the remaining 15,000 signals are used for validating its classification efficiency.

As indicated in Section 2.3, the CWT spectrograms of these signals are used for training and testing the CNN. The CNN is trained for a maximum of 10 epochs with a minibatch size of 100.

During the initial analysis, the default SqueezeNet from the MATLAB Deep Network Designer toolbox was used. The training efficiency reached 100% at the end of 10 epochs. The total time for training the original SqueezeNet is 54 min. Although the training efficiency reached 100%, the classification efficiency was merely 48.7% (Figure 9). This is due to the overfitting of data. The original SqueezeNet is more efficient in extracting deeper features in complex images of objects or beings. The input data used in this study are spectrograms. Extracting deeper features in the spectrograms often results in overfitting of data. Therefore, a modified SqueezeNet, as explained in Section 2.4, is trained in this study.

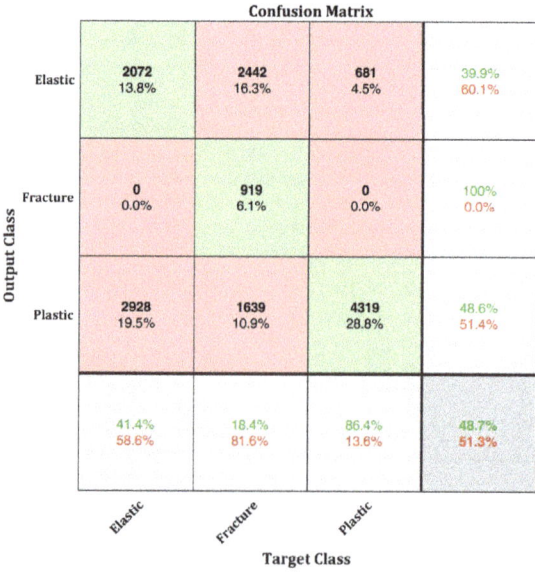

Figure 9. Confusion matrix of the original SqueezeNet without k-fold cross validation.

After implementing k-fold cross validation into the training module, the classification efficiency of the modified SqueezeNet increased to 100%. The data is partitioned into k = 5 and trained for 5 iterations. Since the training is continued for 5 iterations to improve the classification efficiency, the total time for training this network is approximately 5 times the time taken for training the original SqueezeNet. Nonetheless, considering the increase in classification efficiency from 48.7% to 100%, the total time consumption of 4 h and 12 min is within the acceptable limit. Thus, the modified SqueezeNet with k-fold implementation is considered a relative success. The confusion matrix of the modified SqueezeNet with k-fold implementation is presented in Figure 10.

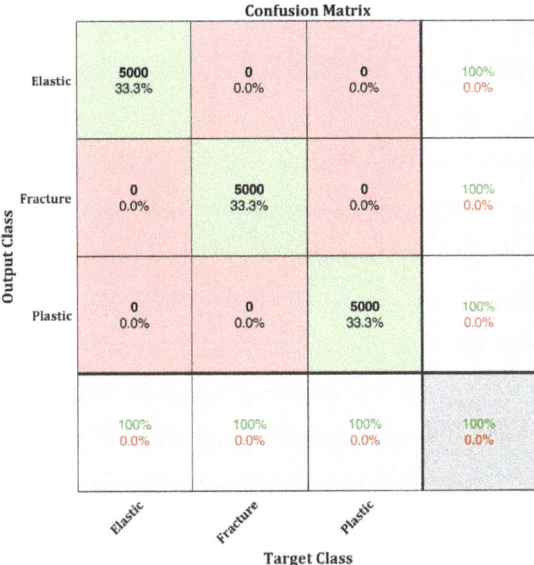

Figure 10. Confusion matrix of the modified SqueezeNet after implementing k-fold cross validation.

The results show that the modified SqueezeNet can explicitly classify the AE signals from the three different damage evolution stages when the k-fold cross validation is implemented. Using this neural network model, the AE signals generated from different damage evolution stages can be identified.

4. Conclusions

A modified SqueezeNet neural network with k-fold cross validation is proposed in this research work for identification and classification of AE signals generated from different damage evolution stages of AlSi10Mg test specimens. Four different configurations of AlSi10Mg specimens prepared using SLM process are tested. AE signals generated during the test are recorded and analysed. First, the damage evolution of the specimens is classified into three stages: elastic stage, plastic stage, and fracture stage. AE signals from each stage are extracted and analysed in their time-frequency domain using CWT. AE signals from each of the damage stage showed differences in their time-frequency characteristics, which are used for training the modified SqueezeNet neural network built for this study. The classification efficiency of 48.7% is obtained for the original SqueezeNet without the k-fold implementation and it increased to 100% when k-fold cross validation is implemented for training. The proposed network can efficiently classify the AE signal generated from different damage evolution stages of AlSi10Mg specimens.

Author Contributions: Conceptualization, V.P.K.; methodology, V.P.K., C.B. and G.P.; software, V.P.K.; validation, V.P.K., C.B. and G.P.; formal analysis, V.P.K., C.B. and G.P.; investigation, V.P.K., C.B. and G.P.; resources, C.B. and C.C.; data curation, V.P.K. and C.B.; writing—original draft preparation, V.P.K.; writing—review and editing, V.P.K., C.B. and G.P.; supervision, C.C.; project administration, C.B. All authors have read and agreed to the published version of the manuscript.

Funding: This research received no external funding.

Institutional Review Board Statement: Not applicable.

Informed Consent Statement: Not applicable.

Data Availability Statement: Not applicable.

Conflicts of Interest: The authors declare no conflict of interest.

References

1. Frazier, W.E. Metal Additive Manufacturing: A Review. *J. Mater. Eng. Perform.* **2014**, *23*, 1917–1928. [CrossRef]
2. Gibson, I.; Rosen, D.W.; Stucker, B.; Khorasani, M.; Rosen, D.; Stucker, B.; Khorasani, M. *Additive Manufacturing Technologies*; Springer: Berlin/Heidelberg, Germany, 2021; Volume 17.
3. Gardan, J. Additive Manufacturing Technologies: State of the Art and Trends. *Addit. Manuf. Handb.* **2017**, *54*, 149–168. [CrossRef]
4. Campbell, F.C., Jr. *Manufacturing Technology for Aerospace Structural Materials*; Elsevier: Amsterdam, The Netherlands, 2011; ISBN 0080462359.
5. Yusuf, S.M.; Hoegden, M.; Gao, N. Effect of Sample Orientation on the Microstructure and Microhardness of Additively Manufactured AlSi10Mg Processed by High-Pressure Torsion. *Int. J. Adv. Manuf. Technol.* **2020**, *106*, 4321–4337. [CrossRef]
6. Maamoun, A.H.; Elbestawi, M.; Dosbaeva, G.K.; Veldhuis, S.C. Thermal Post-Processing of AlSi10Mg Parts Produced by Selective Laser Melting Using Recycled Powder. *Addit. Manuf.* **2018**, *21*, 234–247. [CrossRef]
7. Rafieazad, M.; Chatterjee, A.; Nasiri, A.M. Effects of Recycled Powder on Solidification Defects, Microstructure, and Corrosion Properties of DMLS Fabricated AlSi10Mg. *JOM* **2019**, *71*, 3241–3252. [CrossRef]
8. Chen, H.; Gu, D.; Dai, D.; Ma, C.; Xia, M. Microstructure and Composition Homogeneity, Tensile Property, and Underlying Thermal Physical Mechanism of Selective Laser Melting Tool Steel Parts. *Mater. Sci. Eng. A* **2017**, *682*, 279–289. [CrossRef]
9. Dong, Z.; Liu, Y.; Li, W.; Liang, J. Orientation Dependency for Microstructure, Geometric Accuracy and Mechanical Properties of Selective Laser Melting AlSi10Mg Lattices. *J. Alloys Compd.* **2019**, *791*, 490–500. [CrossRef]
10. Yao, Y.; Wang, K.; Wang, X.; Li, L.; Cai, W.; Kelly, S.; Esparragoza, N.; Rosser, M.; Yan, F. Microstructural Heterogeneity and Mechanical Anisotropy of 18Ni-330 Maraging Steel Fabricated by Selective Laser Melting: The Effect of Build Orientation and Height. *J. Mater. Res.* **2020**, *35*, 2065–2076. [CrossRef]
11. Dong, Z.; Liu, Y.; Zhang, Q.; Ge, J.; Ji, S.; Li, W.; Liang, J. Microstructural Heterogeneity of AlSi10Mg Alloy Lattice Structures Fabricated by Selective Laser Melting: Phenomena and Mechanism. *J. Alloys Compd.* **2020**, *833*, 155071. [CrossRef]
12. Barile, C.; Casavola, C.; Pappalettera, G.; Kannan, V.P. Application of Different Acoustic Emission Descriptors in Damage Assessment of Fiber Reinforced Plastics: A Comprehensive Review. *Eng. Fract. Mech.* **2020**, *235*, 107083. [CrossRef]
13. Tetelman, A.S.; Chow, R. Acoustic Emission Testing and Microcracking Processes. *Acoust. Emiss. ASTM STP* **1972**, *505*, 30–40.
14. Harris, D.O.; Bell, R.L. The Measurement and Significance of Energy in Acoustic-Emission Testing. *Exp. Mech.* **1977**, *17*, 347–353. [CrossRef]
15. Grosse, C.U.; Ohtsu, M.; Aggelis, D.G.; Shiotani, T. *Acoustic Emission Testing: Basics for Research-Applications in Engineering*; Springer Nature: Berlin/Heidelberg, Germany, 2021; ISBN 3030679365.
16. Sause, M.G.R.; Müller, T.; Horoschenkoff, A.; Horn, S. Quantification of Failure Mechanisms in Mode-I Loading of Fiber Reinforced Plastics Utilizing Acoustic Emission Analysis. *Compos Sci. Technol.* **2012**, *72*, 167–174. [CrossRef]
17. Barile, C.; Casavola, C.; Pappalettera, G.; Vimalathithan, P.K. Acoustic Emission Descriptors for the Mechanical Behavior of Selective Laser Melted Samples: An Innovative Approach. *Mech. Mater.* **2020**, *148*, 103448. [CrossRef]
18. Venkatesan, R.; Li, B. *Convolutional Neural Networks in Visual Computing: A Concise Guide*; CRC Press: Boca Raton, FL, USA, 2017; ISBN 1315154285.
19. Sikdar, S.; Liu, D.; Kundu, A. Acoustic Emission Data Based Deep Learning Approach for Classification and Detection of Damage-Sources in a Composite Panel. *Compos. Part B Eng.* **2022**, *228*, 109450. [CrossRef]
20. McCrory, J.P.; Al-Jumaili, S.K.; Crivelli, D.; Pearson, M.R.; Eaton, M.J.; Featherston, C.A.; Guagliano, M.; Holford, K.M.; Pullin, R. Damage Classification in Carbon Fibre Composites Using Acoustic Emission: A Comparison of Three Techniques. *Compos. Part B Eng.* **2015**, *68*, 424–430. [CrossRef]
21. Nasiri, A.; Bao, J.; Mccleeary, D.; Louis, S.-Y.M.; Huang, X.; Hu, J. Online Damage Monitoring of SiC F-SiC m Composite Materials Using Acoustic Emission and Deep Learning. *IEEE Access* **2019**, *7*, 140534–140541. [CrossRef]
22. Lin, Y.; Nie, Z.; Ma, H. Structural Damage Detection with Automatic Feature-extraction through Deep Learning. *Comput.-Aided Civ. Infrastruct. Eng.* **2017**, *32*, 1025–1046. [CrossRef]
23. Cha, Y.; Choi, W.; Büyüköztürk, O. Deep Learning-based Crack Damage Detection Using Convolutional Neural Networks. *Comput.-Aided Civ. Infrastruct. Eng.* **2017**, *32*, 361–378. [CrossRef]
24. Sadowski, L. Non-Destructive Investigation of Corrosion Current Density in Steel Reinforced Concrete by Artificial Neural Networks. *Arch. Civ. Mech. Eng.* **2013**, *13*, 104–111. [CrossRef]
25. Barile, C.; Casavola, C.; Pappalettera, G.; Kannan, V.P. Designing a Deep Neural Network for an Acousto-Ultrasonic Investigation on the Corrosion Behaviour of CORTEN Steel. *Procedia Struct. Integr.* **2022**, *37*, 307–313. [CrossRef]
26. Elsheikh, A.H.; Abd Elaziz, M.; Vendan, A. Modeling Ultrasonic Welding of Polymers Using an Optimized Artificial Intelligence Model Using a Gradient-Based Optimizer. *Weld. World* **2022**, *66*, 27–44. [CrossRef]

27. Elsheikh, A.H.; Katekar, V.P.; Muskens, O.L.; Deshmukh, S.S.; Abd Elaziz, M.; Dabour, S.M. Utilization of LSTM Neural Network for Water Produauction Forecasting of a Stepped Solar Still with a Corrugated Absorber Plate. *Process Saf. Environ. Prot.* **2021**, *148*, 273–282. [CrossRef]
28. Elsheikh, A.H.; Sharshir, S.W.; Abd Elaziz, M.; Kabeel, A.E.; Guilan, W.; Haiou, Z. Modeling of Solar Energy Systems Using Artificial Neural Network: A Comprehensive Review. *Sol. Energy* **2019**, *180*, 622–639. [CrossRef]
29. Barile, C.; Casavola, C.; Pappalettera, G.; Kannan, V.P. Damage Monitoring of Carbon Fibre Reinforced Polymer Composites Using Acoustic Emission Technique and Deep Learning. *Compos. Struct.* **2022**, *292*, 115629. [CrossRef]
30. Lilly, J.M.; Olhede, S.C. Higher-Order Properties of Analytic Wavelets. *IEEE Trans. Signal Processing* **2008**, *57*, 146–160. [CrossRef]
31. Lilly, J.M.; Olhede, S.C. Generalized Morse Wavelets as a Superfamily of Analytic Wavelets. *IEEE Trans. Signal Processing* **2012**, *60*, 6036–6041. [CrossRef]
32. Misiti, M.; Misiti, Y.; Oppenheim, G.; Poggi, J.-M. *Wavelet Toolbox*; MathWorks Inc.: Natick, MA, USA, 1996; Volume 15, p. 21.
33. Krizhevsky, A.; Sutskever, I.; Hinton, G.E. Imagenet Classification with Deep Convolutional Neural Networks. *Adv Neural Inf Process Syst.* **2012**, *25*, 84–90. [CrossRef]
34. Iandola, F.N.; Han, S.; Moskewicz, M.W.; Ashraf, K.; Dally, W.J.; Keutzer, K. SqueezeNet: AlexNet-Level Accuracy with 50x Fewer Parameters And <0.5 MB Model Size. *arXiv* **2016**, arXiv:1602.07360.
35. Tilekar, P.; Singh, P.; Aherwadi, N.; Pande, S.; Khamparia, A. Breast Cancer Detection Using Image Processing and CNN Algorithm with K-Fold Cross-Validation. In *Proceedings of Data Analytics and Management*; Springer: Berlin/Heidelberg, Germany, 2022; pp. 481–490.
36. Xiong, Z.; Cui, Y.; Liu, Z.; Zhao, Y.; Hu, M.; Hu, J. Evaluating Explorative Prediction Power of Machine Learning Algorithms for Materials Discovery Using K-Fold Forward Cross-Validation. *Comput. Mater. Sci.* **2020**, *171*, 109203. [CrossRef]
37. Barile, C.; Casavola, C.; Moramarco, V.; Vimalathithan, P.K. A Comprehensive Study of Mechanical and Acoustic Properties of Selective Laser Melting Material. *Arch. Civ. Mech. Eng.* **2020**, *20*, 3. [CrossRef]
38. Barile, C.; Casavola, C.; Pappalettera, G.; Kannan, V.P. Novel Method of Utilizing Acoustic Emission Parameters for Damage Characterization in Innovative Materials. *Procedia Struct. Integr.* **2019**, *24*, 636–650. [CrossRef]
39. Raj, B.; Jha, B.B.; Rodriguez, P. Frequency Spectrum Analysis of Acoustic Emission Signal Obtained during Tensile Deformation and Fracture of an AISI 316 Type Stainless Steel. *Acta Metall.* **1989**, *37*, 2211–2215. [CrossRef]
40. Rouby, D.; Fleischmann, P. Spectral Analysis of Acoustic Emission from Aluminium Single Crystals Undergoing Plastic Deformation. *Phys. Status Solidi* **1978**, *48*, 439–445. [CrossRef]
41. Kiesewetter, N.; Schiller, P. The Acoustic Emission from Moving Dislocations in Aluminium. *Phys. Status Solidi* **1976**, *38*, 569–576. [CrossRef]
42. Akbari, M.; Ahmadi, M. The Application of Acoustic Emission Technique to Plastic Deformation of Low Carbon Steel. *Phys. Procedia* **2010**, *3*, 795–801. [CrossRef]
43. Vinogradov, A.; Patlan, V.; Hashimoto, S. Spectral Analysis of Acoustic Emission during Cyclic Deformation of Copper Single Crystals. *Philos. Mag. A* **2001**, *81*, 1427–1446. [CrossRef]

Review

Organic Anode Materials for Lithium-Ion Batteries: Recent Progress and Challenges

Alexander A. Pavlovskii, Konstantin Pushnitsa, Alexandra Kosenko *, Pavel Novikov and Anatoliy A. Popovich

Institute of Machinery, Materials and Transport, Peter the Great Saint Petersburg Polytechnic University, Politechnicheskaya ul. 29, 195251 Saint Petersburg, Russia
* Correspondence: alxndra.kosenko@gmail.com; Tel.: +7-981-826-3338

Abstract: In the search for novel anode materials for lithium-ion batteries (LIBs), organic electrode materials have recently attracted substantial attention and seem to be the next preferred candidates for use as high-performance anode materials in rechargeable LIBs due to their low cost, high theoretical capacity, structural diversity, environmental friendliness, and facile synthesis. Up to now, the electrochemical properties of numerous organic compounds with different functional groups (carbonyl, azo, sulfur, imine, etc.) have been thoroughly explored as anode materials for LIBs, dividing organic anode materials into four main classes: organic carbonyl compounds, covalent organic frameworks (COFs), metal-organic frameworks (MOFs), and organic compounds with nitrogen-containing groups. In this review, an overview of the recent progress in organic anodes is provided. The electrochemical performances of different organic anode materials are compared, revealing the advantages and disadvantages of each class of organic materials in both research and commercial applications. Afterward, the practical applications of some organic anode materials in full cells of LIBs are provided. Finally, some techniques to address significant issues, such as poor electronic conductivity, low discharge voltage, and undesired dissolution of active organic anode material into typical organic electrolytes, are discussed. This paper will guide the study of more efficient organic compounds that can be employed as high-performance anode materials in LIBs.

Keywords: lithium-ion batteries; redox-active organic anode; organic anode materials

Citation: Pavlovskii, A.A.; Pushnitsa, K.; Kosenko, A.; Novikov, P.; Popovich, A.A. Organic Anode Materials for Lithium-Ion Batteries: Recent Progress and Challenges. *Materials* **2023**, *16*, 177. https://doi.org/10.3390/ma16010177

Academic Editors: Giovanni Pappalettera and Sanichiro Yoshida

Received: 18 November 2022
Revised: 8 December 2022
Accepted: 20 December 2022
Published: 25 December 2022

Copyright: © 2022 by the authors. Licensee MDPI, Basel, Switzerland. This article is an open access article distributed under the terms and conditions of the Creative Commons Attribution (CC BY) license (https:// creativecommons.org/licenses/by/ 4.0/).

1. Introduction

Rechargeable lithium-ion batteries (LIBs), one of the most successful commercialized secondary batteries, play a pivotal role in every aspect of our lives thanks to their suitable energy density, long service life, excellent rechargeability, and prolonged storage capacity. Nowadays, LIBs are an integral part of different consumer electronic devices, such as smartphones, laptops, watches, power banks, digital cameras, tablets, etc., and are expected to fuel plug-in hybrid electric vehicles and to be applied in smart grids in the near future. With worldwide industrial development and techno-economic growth, the energy density requirements for the above-mentioned energy-storage devices have continuously increased in the past few decades. In order to satisfy the ever-growing demand for consumer electronics and electric vehicles, achieving higher energy/power densities and longer cycle lives of LIBs has risen as a major issue for the development next generations of high-energy rechargeable Li-ion batteries.

In LIBs, the choice of electrode active materials is the dominant factor influencing the electrochemical performance of the batteries to a large extent. However, presently available commercial LIBs assembled with inorganic layered transition metal oxides, such as $LiCoO_2$ (LCO), $LiMn_2O_4$ (LMO), $LiFePO_4$ (LFP), $LiNi_xCo_yMn_{1-x-y}$ (NCM), etc. [1], as a cathode material and graphite or Si/C as the most commonly used anode material reach their limit in terms of sustainability, capacity, and output voltage. Graphite usually applied as a commercial anode in the current state-of-the-art LIB has poor rate capability and a low theoretical specific capacity of 372 mAh/g which significantly hampers the further development of LIBs [2–4]. Great research efforts have been put forth to find suitable alternatives to conventional graphite anodes with enhanced energy and power density. Thus, a variety of alloy-based materials (e.g., Sn, P, Ge, Si, Bi, Sb) and transition metal oxides (M_xO_y, M = Ni, Mn, Fe, Cr, Mo, Co, Nb, etc.) have been extensively investigated as high specific capacity LIBs anodes [5–11]. However, these anode materials for LIBs have poor electrochemical stability and fast capacity fading which impede their commercial application for Li-ion batteries.

Organic compounds with electroactive functional groups or moieties, either a polymer or small organic molecule, can undergo an electron transfer through a reversible electrochemical redox reaction. Therefore, they are suitable for the conversion and storage of chemical and electrical energy.

Compared with conventional inorganic anode materials, redox-active organic anode materials are endowed with obvious advantages, such as low cost, structural diversity, environmental friendliness, outstanding flexibility in their molecular design, low weight, and higher theoretical capacity, which can meet the rising demand for the energy density of next-generation battery systems [12–15]. In contrast to inorganic compounds, organic materials are mainly composed of such chemical elements that are widely distributed in nature (e.g., C, H, O, N, S). Thus, organic anode materials can be easily synthesized from renewable resources in facile steps [16] that have a minimal environmental footprint [17]. Moreover, both the tunable molecular structure and high structural diversity of organic compounds, unlike those of their metal-ion inorganic counterparts, allow for optimizing the electrochemical performance of organic electrode materials, such as redox activity and operating voltage in various metal-ion batteries [18–21].

For instance, the electronegativity of substituents in organic compounds could tune the redox potential. It was corroborated that the introduction of an electron-withdrawing group into the redox active organic molecule enhances the redox potential because the electron cloud around the redox center of the organic material is attracted toward higher electronegative groups, such as heteroatoms [22]. On the contrary, attaching an electron-donating group to the redox-active organic molecule reduces the redox potential [23]. It should be emphasized that some organic compounds enable the batteries to be exploited in extreme conditions, such as a high pH value, an extended temperature range from −70 to 150 °C, and the presence of oxygen [24]. With these advantages of organic electrode materials, LIBs based on organic electrodes have attracted considerable research interest and have made magnificent progress [20,24–29].

Due to the versatile properties of organic electrodes, various types of organic materials have been used in all kinds of rechargeable batteries, comprising all-solid-state batteries, nonaqueous sodium-ion, lithium-ion, potassium-ion, dual-ion batteries, multivalent-metal batteries, redox flow batteries, and aqueous batteries [24]. Furthermore, apart from small organic molecules, different polymers, polymer frameworks, and 2D organic materials can also be adopted as organic electrodes in rechargeable batteries [30–33].

For a long time, organic anode materials have received much less attention compared to their inorganic counterparts mostly because of their relatively low electronic conductivity and the tremendous success of inorganic anode materials in either application or research. Although there are already numerous reviews of organic electrode materials used in different kinds of rechargeable batteries [14,16,17,24,25,34–37], a review article that briefly, simply, and systematically introduces the fundamental knowledge of organic anode materials for LIBs is still absent.

Inspired by the advantages of organic electrode materials, we decided to write this review paper, as a brief introduction to the representative organic anode materials in LIBs, as well as the fundamental principles, state-of-the-art developments in organic anodes, and outlooks of these perspective materials for LIBs, aiming to cause more interest and innovation in the battery industry.

2. Fundamentals of Organic Anode Materials

2.1. Basic Components of an Electrochemical Cell

Rechargeable (also called "secondary") batteries consist of two electrodes, a cathode (or positive electrode) with high redox potential and an anode (or negative electrode) with lower redox potential, in contact with an electrolyte (either solid or liquid). The electrolyte system with low viscosity, high stability, and ionic conductivity at a large potential window is preferred. Furthermore, electrolytes must be inert towards both the cathode and active anode materials. The cathode and anode of the cell are connected through an external circuit. To avoid short circuits, the cathode is mechanically separated from the anode by an electronically insulating porous membrane (usually named separator). These are the widespread cell components for a battery. Figure 1 depicts a schematic of a typical battery design.

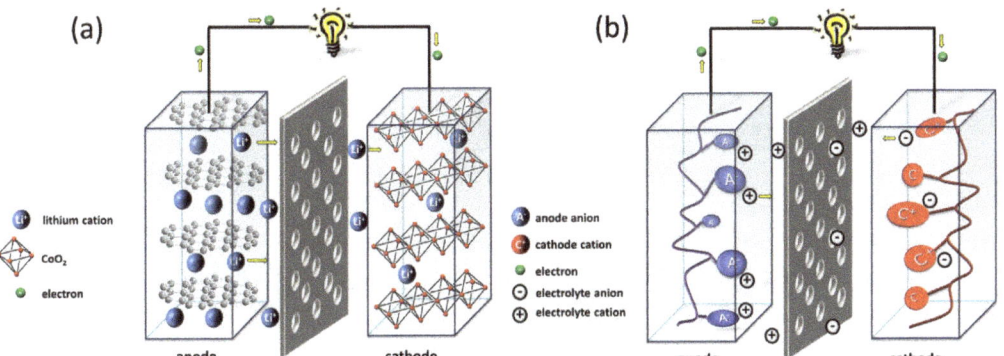

Figure 1. Schematic showing of the processes occurring during discharging of (**a**) traditional inorganic and (**b**) organic electrode-based Li-ion batteries. Instead of intercalation-based reactions in inorganic compounds, organic compounds undergo surface-based redox reactions. Reprinted with permission from Ref. [38], Copyright 2016, American Chemical Society.

2.2. Mechanism of Charge Storage in Lithium-Ion Batteries and Working Principles of Organic Anode Materials

During the charging process of a battery, electrons are forced (by an applied potential or current) from the positive electrode to the negative electrode, oxidizing the cathode and reducing the anode. During the discharge process of a battery, electrons spontaneously flow in the opposite direction (from the negative electrode to the positive electrode), reversing the previous redox reactions. Thus, both positive and negative electrodes, with various redox potentials, must be capable of reversible electrochemical redox reactions. The difference in chemical potential between these reactions constitutes the operating voltage of the cell, also known as the open circuit voltage.

In contrast to inorganic electrode materials, with redox reactions that are based on the valence change of the elemental substance or transition metal, the redox reaction of organic electrode materials is based on the change in the charge state of the redox active organic species or an electroactive organic group. When an active electrode material changes its state of charge, oppositely charged ions in the electrolyte diffuse into the active material to charge compensate. Depending upon the potential of this redox process, organic materials can either be suited for use as a cathode or anode active material. The polarized state of organic active material is used for interaction with the mobile ion.

In other words, a reversible metal ion intercalation-based electrochemical redox reaction is employed to store the charge in conventional inorganic-based electrode materials. In organic electrode materials, a change in the charge state of the redox active moieties is utilized to store the charge. During the charge and discharge electrochemical process, the inorganic-based electrode structure is stabilized by the change in the oxidation states of the transition metals from inorganic cathode materials for LIBs, while in organic electrode material, the polarized state of organic active material is commonly balanced by its interaction with metal ions from the electrolyte.

2.3. Types of Organic Anode Materials and Their Advantages

According to different redox reactions of organic electrode materials, they can be generally classified as N-type, P-type, or bipolar (Figure 2). N-type organics obtain a negative charge (N^-) from the original neutral state (N) during the electrochemical reduction reaction and revert to their initial state by oxidation. P-type organic compounds (P) are oxidized during the redox process yielding positively charged cations. B-type (or bipolar) organics are another type of organic compound, for which the initial neutral state (B) can be either oxidized to a positively charged state (B^+) or reduced to a negatively charged state (B^-), which depends on the applied voltage. B-type organic compounds can be also considered N- or P-type organics since, in general, only half of the reaction takes place on one electrode.

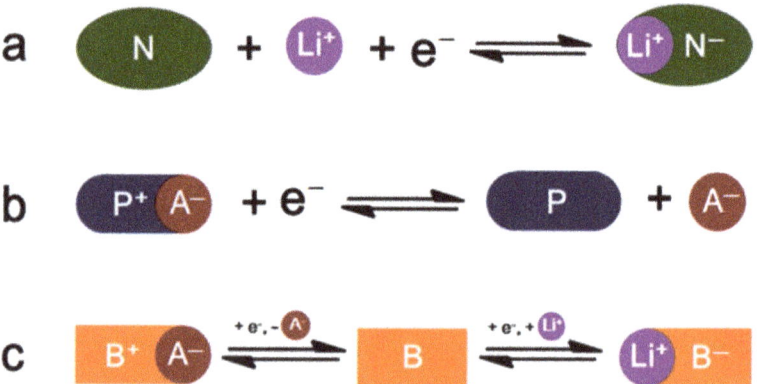

Figure 2. The reversible redox reaction of three types of redox-active organic compounds: (**a**) N-type; (**b**) P-type; (**c**) bipolar. The anion of the electrolyte is denoted as A^-. Used with permission from the Royal Society of Chemistry, from Ref. [39]; permission conveyed through Copyright Clearance Center, Inc.

In the electrochemical oxidation reaction of P-type or reduction reaction of N-type, anion (A^-) or cation (Li^+) are, respectively, required to neutralize the positive charge of P^+ or the negative charge of N^-. In the reverse redox reaction, lithium cations or A^- anions will move back from the electrode to the electrolyte.

The noticeable difference between N-type organic materials and inorganic intercalation compounds is that lithium ions can be substituted by other alkali metals, such as K^+ and Na^+ or even H^+ cations, which will not significantly influence the electrochemical behavior of the material. Inorganic intercalation compounds, on the contrary, are very sensitive to the ionic radius of the cation.

For P-type organic materials, many kinds of anions can be adopted as A^-, e.g., NO_3^- and Cl^- in aqueous electrolytes, or $TFSI^-$, PF_6^-, BF_4^-, and ClO_4^- in non-aqueous electrolyte.

B-type organic compounds can act simultaneously as an anode and cathode active material. For instance, Yang and coworkers [40] reported such a bipolar organic compound as polyparaphenylene, which has a large potential gap of 3.0 V and electrochemical reversibility for the electrochemical p- and N-type redox reactions. They indicated that this bipolar redox-active polymer is the most attractive candidate for application in all-organic batteries with the same cathode and anode active material. The findings demonstrated that polyparaphenylene could be either p-doped with a specific capacity of 80 mAh/g at high potentials of >3.9 V or n-doped with a superior specific capacity of 350 mAh/g at a quite low potential region of <1.5 V [40].

2.4. Advantages of Organic Anode Materials

It should be noted that in traditional inorganic intercalation-based electrode materials, lithium-ion insertion leads to the transformation of the cathode lattice structure. This result in heat generation and slow reaction kinetics during the charge/discharge process. At the same time, fast insertion and de-insertion of lithium-ion may result in the structural degradation of the cathode material. These challenges substantially affect both the rate and cyclic performance of the typical LIBs, impeding their role in long-life and high-power applications. In contrast, organic electrode materials undergo fast and simple redox reactions without structural degradation of the electrode active material. Thus, they exhibit excellent cyclic and rate performance. Moreover, as previously stated, inorganic intercalation-based materials are very sensitive to the ionic radius of cation, whereas, in organic electrode materials, lithium cation can be easily substituted with various alkali metal cations with a bigger ion radius, such as Na^+, K^+, etc., without any significant fade of the capacity of batteries. In addition, both the larger interlayer spacings and greater flexible structures of organic electrode materials (Figure 3) afford faster ion diffusion and reduced structural and volumetric changes upon charge/discharge processes, enhancing rate capabilities [25].

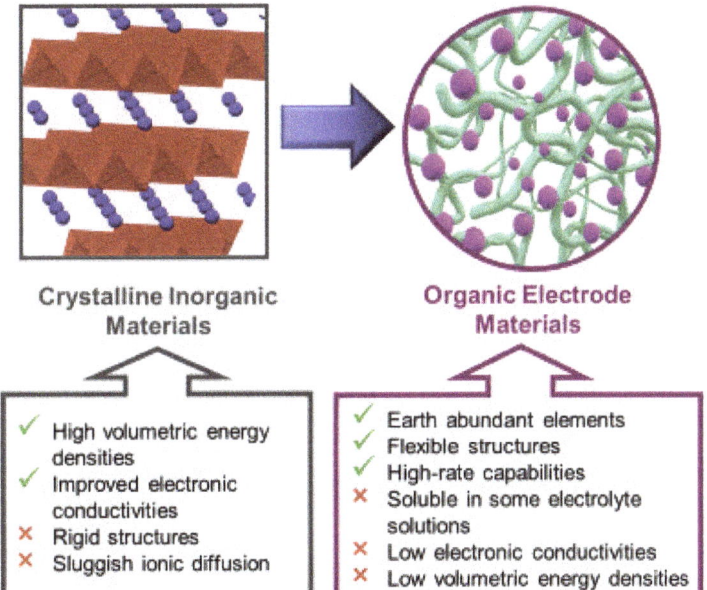

Figure 3. Schematic illustration of the unique properties of organic electrode materials which make them particularly attractive for use as high-performance electrode materials in rechargeable LIBs [25]. Ref. [25] is an open-access article distributed under the terms of the Creative Commons CC BY license, which permits unrestricted use, distribution, and reproduction in any medium provided the original work is properly cited.

3. Organic Anode Materials Based on Organic Carbonyl Compounds

To date, different kinds of redox-active moieties, such as phenazine, carbonyl, triazine, imide, and aromatic ring structures, have been acknowledged as the main functional groups in redox-active organic anodes [41–48]. Additionally, conducting polymers, thioethers, organodisulfides, conjugated carbonyl compounds, radical polymers, and other organic compounds have been reported as anode active materials [49–59].

Among all reported redox-active organic compounds, carbonyl compounds, such as imides, quinones, carboxylates, ketones, anhydrides, and their derivatives have become a class of attractive organic materials for use as potentially high-capacity anode or cathode materials for LIBs during the past decades [58,60–65]. This is because of their highly active redox carbonyl centers providing wide structural diversity, higher reversible capacity, and fast kinetics. For instance, the simplest quinone (1,4-benzoquinone) can provide a reversible capacity of nearly 500 mAh/g [66]. Carbonyl functional groups (C=O) can be found in numerous organic compounds with different forms, while most of the investigated carbonyl-based organic electrodes are primarily based on either the derivatives of aromatic carboxylic acid or quinone [67–69].

According to the literature, conjugated carboxylates, aromatic imides, quinones, and aromatic anhydrides are the four categories of conventional organic carbonyl electrode materials which have been successfully applied in LIBs [47]. Nearly all carbonyl compounds are N-type electrode materials and are reduced during the charge/discharge process, forming a negatively charged anion. Their redox mechanism can be presented as a nucleophilic addition reaction of electrochemically active C=O functionalities, in which the enolization reaction is the main step [39].

Although most quinone derivatives have been primarily utilized as cathode materials for LIBs (because these quinone systems have voltages of $\gtrsim$ 2.0 V versus Li/Li$^+$, typical for LIB cathodes), some conjugated carbonyl materials with low redox potentials could be used as negative electrodes for LIBs, which stems from the significant charge repulsion interaction between CO− moieties [70]. Carboxylates with electron-donating groups (such as -OLi) also reveal a redox reaction at a very low potential and, thus, they can be also utilized as anode materials [71].

Lithium *trans-trans*-muconate ($Li_2C_6H_4O_4$) and lithium terephthalate ($Li_2C_8H_4O_4$) were the first two examples of carbonyl anode materials for LIBs proposed by Tarascon's research group in 2009 [71]. Remarkably, both $Li_2C_6H_4O_4$ and $Li_2C_8H_4O_4$ as a negative-electrode material exhibit very flat discharge plateaus at 1.4 and 0.8 V versus Li/Li$^+$, respectively, which allows the use of a cheaper aluminum current collector for the anodes of LIBs. A noteworthy advantage of the organic anode material based on these two conjugated dicarboxylates is their higher thermal stability over typical carbon negative electrodes in typical carbonate electrolytes (1 M LiPF$_6$ in ethylene carbonate (EC)/dimethyl carbonate (DMC)), which should lead to safer Li-ion batteries [71].

$Li_2C_6H_4O_4$ revealed a reversible capacity of 170 mAh/g with 73.5% ($\approx$125 mAh/g) capacity retention after 80 cycles at 0.1C rate charge/discharge, while $Li_2C_8H_4O_4$ delivered a greater reversible capacity of 300 mAh/g with 78% ($\approx$234 mAh/g) capacity retention after 50 cycles at a current density of 15 mA/g [71].

It should be emphasized that only the trans versions of conjugated carboxylates $LiO_2C(CH=CH)_nCO_2Li$ with n = 2, 3, and 4 (Figure 4) showed reversible electrochemical reactivity towards lithium [72].

Ethoxycarbonyl-based lithium salt ($Li_2C_{18}H_{12}O_8$) delivered a capacity of 125 mAh/g in the initial cycle and about 110 mAh/g in the 50th cycle over two main voltage plateaus at ~1.96 and 1.67 V [68].

Wang and co-workers [73] prepared alkaline earth metal terephthalates $MC_8H_4O_4$ (M = Ca, Sr, Ba) via eco-friendly and facile displacement reactions and adopted these organic compounds as anodes for LIBs. The scheme of the lithium storage mechanism of carbonyl-based terephthalates on the level of molecular structure is presented in Figure 5.

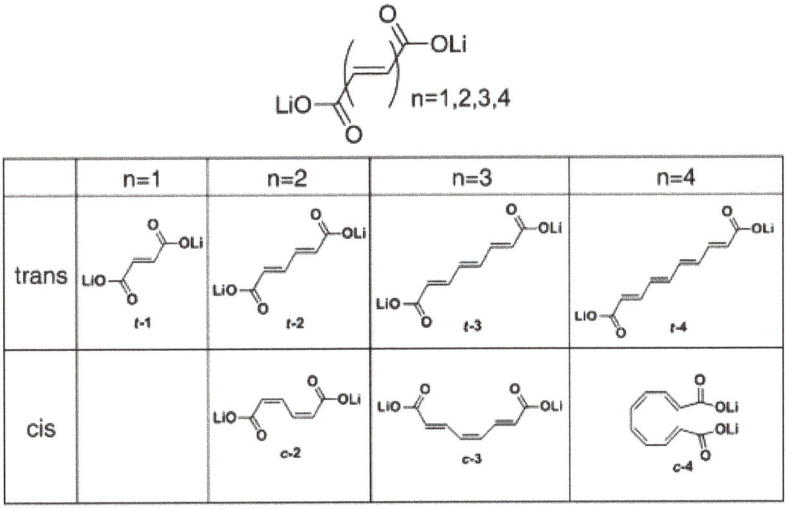

Figure 4. Chemical structures of the conjugated carboxylates with conjugation pathways of 2, 3, and 4 units. Reprinted from Ref. [72], Copyright 2010, with permission from Elsevier.

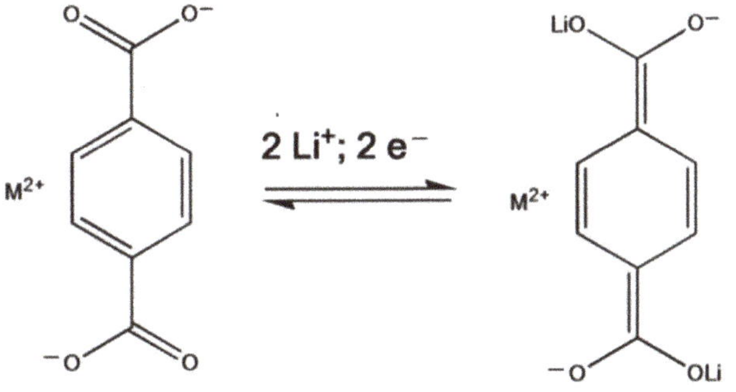

Figure 5. Schematic depicting the lithium storage/release mechanism of carbonyl-based terephthalates at the molecular structure level. Reprinted from Ref. [73], Copyright 2016, with permission from Elsevier.

In the discharge process, the calcium terephthalate displays a flat plateau at nearly 0.8 V vs. Li/Li$^+$. Overall, metal terephthalates produced by Wang and colleagues exhibited a reversible capacity of nearly 130 mAh/g at 0.2 A/g. They also found that the main factors determining the various electrochemical properties (operational voltage, rate performance, cyclic lifetime, and specific capacity) of the obtained organic anodes are the electrostatic interactions between the terephthalate anions and metal cations leading to the formation of metal-organic frameworks (MOFs) with high crystalline structure [73].

Organic metal salts, such as lithium tannic acid (LiTA), have been also applied as advanced energy storage anode materials for LIBs. The LiTA anode shows a reversible capacity of 133.5 mAh/g at a current density of 100 mA/g and maintains 100.5 mAh/g after 100 charge–discharge cycles. Besides, LiTA demonstrates excellent rate properties, surprisingly long cycle performance, and high coulombic efficiency resulting from reversible and stable redox reactions between C-O, and C=O chemical bonds [74].

Xiao and co-workers [75] synthesized copper maleate hydrate (CMH) and utilized this organic material as an organic anode material for rechargeable LIBs for the first time. This small molecular organic compound demonstrated superior electrochemical performance.

It exhibited an initial reversible capacity of 404.6 mAh/g at a current density of 0.2 A/g and could maintain a reversible capacity of nearly 383.2 mAh/g after 250 cycles [75].

Table 1 summarizes the above-stated information about typical organic anode materials for LIBs based on carbonyl compounds.

Table 1. Structural formulas of typical organic anode materials based on carbonyl compounds and their key electrochemical parameters.

Organic Anode Material	Current Density/Specific Capacity (C or mAg^{-1}/mAhg^{-1})	Structural Formula of Organic Active Anode Material	Refs.
Lithium perylene-3,4,9,10-tetracarboxylate (Li-PTCA)	24 mAg^{-1}/195 (with 61.5% capacity retention after 50 cycles) 240 mAg^{-1}/200 (with 60% capacity retention after 50 cycles)		[29]
Lithium 2,6-bis(ethoxycarbonyl)-3,7-dioxo-3,7-dihydro-s-indacene-1,5-bis(olate)	0.05C/125		[68]
Lithium trans-trans-muconate	0.1C/170 (with 73.5% capacity retention after 80 cycles)		[71]
Lithium terephthalate	0.1C/300 (with 78% capacity retention after 50 cycles)		[71]
Alkaline earth metal terephthalates	200 mAg^{-1}/130	M = Ca, Sr, Ba	[73]
Lithium tannic acid (LiTA)	100 mAg^{-1}/133.5 (with 75.3% capacity retention after 100 cycles)	M = Li	[74]
Copper maleate hydrate (CMH)	200 mAg^{-1}/404.6 (with 94.7% capacity retention after 250 cycles)		[75]

Table 1. *Cont.*

Organic Anode Material	Current Density/Specific Capacity (C or mAg^{-1}/mAhg^{-1})	Structural Formula of Organic Active Anode Material	Refs.
4-Nitrobenzoic acid lithium salt (NBALS)	0.5C (1C = 155 mAg^{-1}) /153 (with 85.6% capacity retention after 100 cycles)		[76]
Dilithium 2,5-dibromoterephthalate (Li$_2$-DBT)	0.1C/122, after 50 cycles		[77]
Dilithium 2,5-dimethoxyterephthalate (Li$_2$-DMoT)	0.1C/95, after 50 cycles		[77]
Dilithium 2,5-diaminoterephthalate (Li$_2$-DAT)	0.1C/98, after 50 cycles		[77]

It should be noted that all carboxylate-based derivatives featuring straight chains could only achieve one-Li uptake, thus resulting in low capacity (less than 170 mA/g). In sharp contrast, Li$_2$C$_8$H$_4$O$_4$ consisting of a terephthalate system that does not have a straight carbon chain could realize a full two-Li uptake, giving a stable specific capacity of nearly 300 mAh/g [71]. Therefore, further efforts were primarily made in enhancing the electrochemical performance of the terephthalate systems, including carbon coating, nanosizing, constructing metal-organic frameworks, modification of the aromatic core, etc. [78–81].

4. Organic Anode Materials Based on Covalent Organic Frameworks (COFs)

Covalent organic frameworks (COFs) materials are a subclass of crystalline microporous organic polymers featuring atomically precise self-assembled two- (2D) or three-dimensional (3D) extended structures with ordered nanopores, strong covalent bonds between lightweight elements (C, N, O, B, etc.) of rigid building blocks, and periodic skeletons [82,83]. Apart from cathode materials, COFs have been heralded as perspective organic anode materials [35].

Compared with typical polymers and small organic molecules, COFs are more preferable candidates for use as electrode materials in LIBs owing to their structural diversity, porous structure, framework tunability, low electrode volume change, large specific surface area, and functional versatility [20,35,84–86].

The capacitance capacity of COFs is caused by their porous structure, whereas their specific capacity comes from the insertion of lithium ions in the active redox site. A large molecular weight of COFs coupled with the robustness of their framework inhibits the dissolution of this material in electrolytes, thus enabling a stable cycle performance. At the same time, the porous framework structure reduces the path of ion diffusion, leading to enhanced kinetics of lithium ions [35].

Among the diversity of organic framework materials, COFs based on nitrogen-containing active sites have gradually garnered the merited attention of researchers in recent years [87].

Yang and coworkers [88] reported the first example of an organic 2D COF material for an anode of LIB. Their work paved the way for the utilization of COFs in the next-generation high-capacity LIB. Researchers designed 2D COFs based on covalently linked porphyrin with 4-thiophenephenyl groups (TThPP) and used the TThPP films grown on the copper foil without adding any binders as the anode of LIBs. Due to the alignment of 2D polyporphyrin nanosheets, this material demonstrated high electronic conductivity, long cycle lives, insolubility, high specific capacities (up to 666 mAh/g), and outstanding rate performances. The high capacity and excellent rate performance of TThPP have been explained by the presence of many sites for efficient adsorption of lithium atoms and the existence of both open nanopores holding electrolyte and short-ended paths for the rapid lithium-ion diffusion in the TThPP structure [88].

Bai and coworkers [89] presented two fully conjugated porous COFs as the organic anode of LIB, which demonstrated stable long life (500 cycles) and excellent specific capacity of ca. 700 mAh/g. Such superior charge–discharge performance corroborated the high potential of COFs to be utilized for eco-friendly energy storage. More importantly, utilizing conjugated COFs not only avoids complicated synthesis of typically conjugated polymers and doped or multilayered hybrids but also embodies their outstanding properties, such as flexibility, high electrochemical activity, structural diversity, etc., into energy conversion applications.

Zhang's research group [90] opened a window for the promising applications of high-performance covalent organic framework (COF) structured polymers in flexible electronics and optoelectronic devices in the future anode market. They produced a two-dimensional (2D) nitrogen-rich graphene-like holey conjugated polymer (NG-HCP) through nanoengineering (Figure 6) and evaluated the as-prepared NG-HCP nanosheets as the anode material of Li-ion batteries. The findings demonstrated that the obtained NG-HCP nanosheets displayed an extremely high initial charge capacity of nearly 1320 mAh/g at a current density of 20 mA/g in the voltage range of 3–0 V vs. Li$^+$/Li, without noticeable capacity loss during the first 20 cycles, and ultralong-term cycling life of 600 cycles [90]. Remarkably, after 230 cycles, the specific capacity of the NG-HCP after nano-engineering remained at a high value of nearly 1015 mAh/g under a current density of 100 mA/g. For comparison, the as-fabricated NG-HCP for the LIB anode without nano-engineering exhibits a specific capacity of less than 300 mAh/g. On the one hand, this outstanding rate performance of NG-HCP after nanoengineering is achieved due to both the large specific surface area and increased reaction contact area, which promote the charge-transfer reaction. On the other hand, the doping of hetero-atoms (such as nitrogen atoms) into the chain of the conjugated COF structure provides the increased electrochemical reactivity of the NG-HCP material, resulting in the superior performance of the electrode material. To sum up, the good dispersity of NG-HCP nanosheets with high electrical conductivity and the synergistic effects of N-doping and high porosity make NG-HCP nanosheets suitable as excellent anodes for LIBs.

Lei and coworkers [91] synthesized a composite electrode material (COF@CNTs) through the controllable growth of few-layered imine-based 2D COFs on the exterior surface of carbon nanotubes (CNT) for application as the anode of LIBs. Imine-based C=N functional group coupled to COF is shown in Figure 7a, while the graphical illustration of obtaining COF@CNTs composite featuring few COF layers covered on the exterior surface of CNTs is illuminated in Figure 7b.

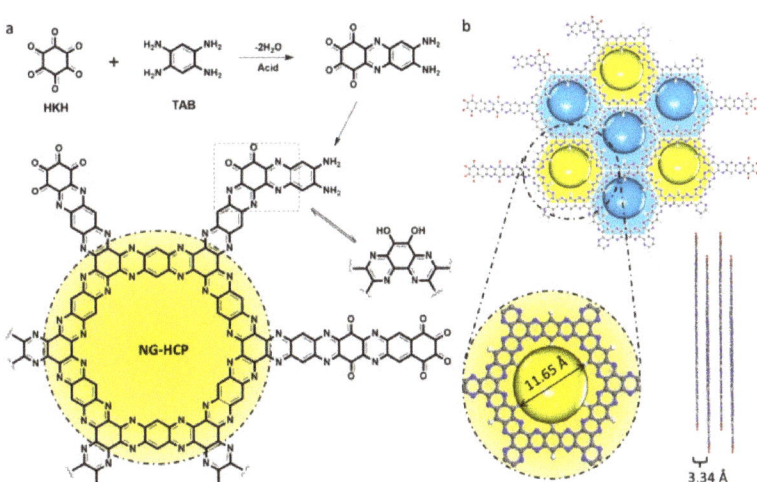

Figure 6. (**a**) Synthetic route of obtaining HG-HCP by a dehydration–condensation reaction between hexaketocyclohexane (HKH) octahydrate and 1,2,4,5-tetraaminobenzene (TAB) tetrahydrochloride in NMP solution with sulphuric acid as a catalyst and a schematic illustration of the alternative pristine NG-HCP structure, where many excessive amino and carbonyl functional groups exist at the edge; (**b**) molecular configuration of a single layer of the NG-HCP with a pore size of 11.65 Å and a packing distance of 3.34 Å. Reprinted from Ref. [90], Copyright 2017, with permission from Elsevier.

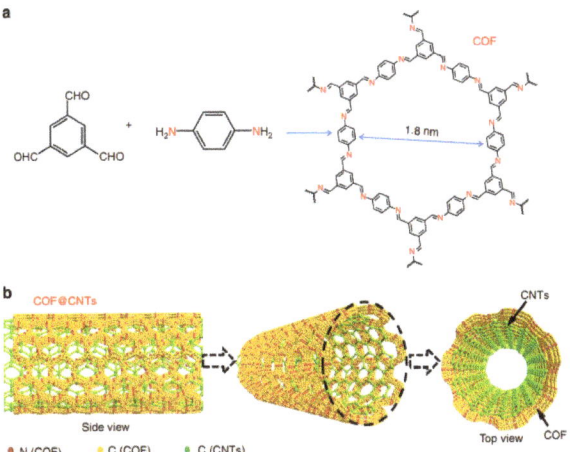

Figure 7. Preparation of the COF and COF@CNTs composite. (**a**) Preparation of COF with a 2D structure and an average pore size of 1.8 nm; (**b**) graphical representation of the COF@CNTs anode material with few COF layers covered on the exterior surface of carbon nanotubes [91]. Ref. [91] is an open-access article distributed under the terms of the Creative Commons CC BY license, which permits unrestricted use, distribution, and reproduction in any medium provided the original work is properly cited.

Remarkably, upon activation, COF in the composite (COF@CNTs) proposed by Lei et al. [91] as an organic anode material for LIBs delivers an extremely large reversible capacity of 1536 mAh/g (corresponding to the 14-lithium-storage mechanism for a COF monomer), which even exceeds the theoretical capacity (700–1000 mAh/g) of conventional Sn-based anodes and such inorganic anode materials as transition metal oxides (Co_3O_4,

Fe$_2$O$_3$, CuO, NiO, etc.). More importantly, after 500 cycles, the reversible capacity of COF@CNTs material is sustained at a high value of 1536 mAh/g under a current rate of 100 mA/g.

The underlying reason for the ultrahigh capacity of COF@CNTs composite negative electrode might be explained by the efficient utilization of active centers of the COF@CNTs anode for lithium-ion storage, which includes one lithium ion per C=N functional group and six lithium ions per benzene rings. Unlike the conventional C=O or C=N chemical bonds which have limited lithium-ion storage capability, the six lithium ions per benzene ring provide a much higher specific capacity during the lower voltage window [91]. This 14-lithium storage mechanism based on the five-step lithium-ion insertion/extraction process for a COF monomer in the COF@CNTs anode is presented in Figure 8.

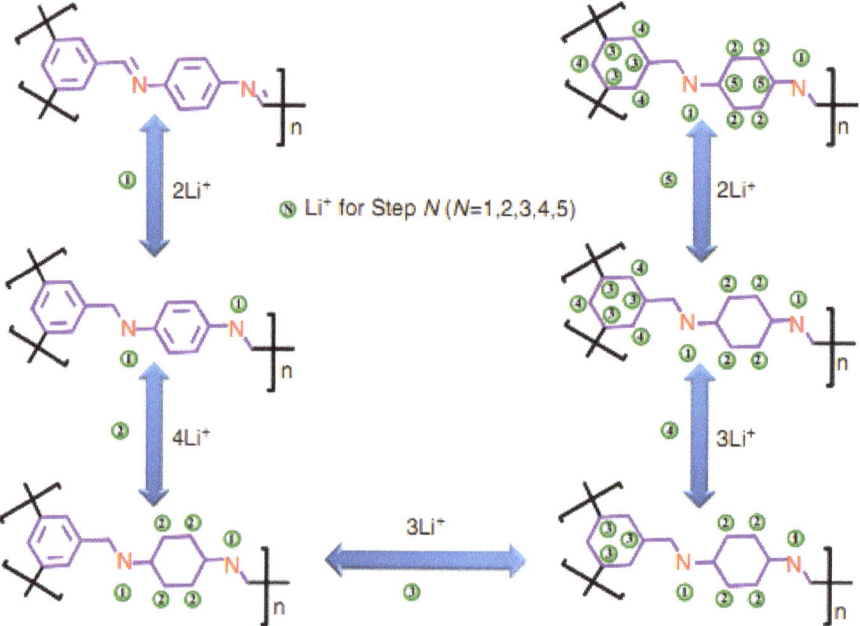

Figure 8. Schematical representation demonstrating the stepwise lithium-storage mechanism for a COF monomer in the COF@CNTs composite anode material [91]. Ref. [91] is an open-access article distributed under the terms of the Creative Commons CC BY license, which permits unrestricted use, distribution, and reproduction in any medium provided the original work is properly cited.

Lei and coworkers [92] produced two types of triazine-based covalent organic nanosheets (CONs) with various pore sizes: Schiff base networks-1 (SNW-1) with a pore size of 0.5 nm and covalent imine network-1 (CIN-1) with a pore size of 2 nm. In order to obtain increased Li-ion storage capacity, the as-exfoliated CONs were further composited with carbon nanotubes (CNTs) (denoted as SNW-1/CNT and CIN-1/CNT composites) to expose more active sites for lithium storage and enhance the ions/electrons diffusion. It is worth noting that the obtained SNW-1/CNT and CIN-1/CNT composites are porous polymers connected by covalent chemical bonds, which can be also considered amorphous COF materials having disordered pore structures. After modification of these composites by mechanical stripping, exfoliated E-SNW-1/CNT and E-CIN-1/CNT exhibit outperforming lithium storage charge–discharge performances thanks to the facilitated surface-control kinetic coupled with the thin-layered 2D structures of these exfoliated CON materials. When employed as an organic anode material for LIBs, E-SNW-1 and E-CIN-1 materials demonstrate extremely high reversible capacities after conducting 250 cycles: 920 mAh/g and

1005 mAh/g, respectively (at a current density of 100 mA/g). Moreover, the mechanism of Li-storage has been found to have an intriguing 11 (for E-CIN-1/CNT electrode) and 16 (for E-SNW-1/CNT electrode) electrons involved in the electrochemical redox reactions, associated with not only typical organic functional groups (-NH- or C=N groups) but also unusual functional groups, such as triazine, piperazine, and benzene rings.

In 2021, Wu and coworkers [93] reported a high-rate anode material for LIBs based on redox-active piperazine-terephthalaldehyde COF (denoted as PA-TA) featuring ultra-large interlayer distance. Their research indicates that increasing the interlayer distance of COFs by a reasonable molecular design strategy would be of particular importance to producing high-performance anode materials for LIBs. The proposed PA-TA COF anode material by Wu and coworkers [93] exhibited remarkable cycling and rate performance delivering a high lithium storage capacity of 543 mAh/g even after 400 charging-discharging cycles at a current density of 1 A/g. Most importantly, the PA-TA COF anode could display a reversible capacity of 207 mAh/g even at a high current rate of 5 A/g.

It is worth noting that there are only several COF-related compounds that have been proposed as potential electrode materials for Li-ion batteries due to their redox-active centers with lithium ions, such as C=O and C=N functional groups, and micro/mesoporous structure [88,89,94–96]. The COF materials bearing carbonyl moieties are commonly investigated as cathodic materials for LIBs based on the lithium redox reaction with C=O species [94–96]. The existence of aromatic heterocyclic functional groups and/or imine moieties promotes the application of COF materials as the anode materials for LIBs [88,89].

The key electrochemical properties of COF-based organic anode materials for LIBs listed above are compared and summarized in Table 2.

Table 2. Summary of the main electrochemical performance of various COF-based anode materials.

Organic Anode Material	Discharge Capacity or Capacity Retention (mAhg^{-1} or %)/Current Rate (mAg^{-1})/after Cycles	Discharge Capacities (mAh/g) with High Current Rate/Current Rate (mA/g)	Refs.
2D COF based-polyporphyrin (TThPP) film	381 mAhg^{-1} or 61.1%/1000/200	195/4000	[88]
N2-COF	600 mAhg^{-1} or 82%/1000/500	Unspecified	[89]
N3-COF	593 mAhg^{-1} or 81%/1000/500	Unspecified	[89]
Free-standing nitrogen-rich graphene-like holey conjugated polymers (NG-HCP nanosheets)	1015 mAhg^{-1}/100/230	237/2500	[90]
Bulk covalent organic framework (COF)	125 mAhg^{-1}/100/300	Unspecified	[91]
Two-dimensional covalent organic framework trapped by carbon nanotubes (COF@CNTs composite)	1536 mAhg^{-1}/100/500	217/5C (1C = 1Ag^{-1})	[91]
E-CIN-1/CNT	1005 mAhg^{-1} or 79.2%/100/250	97/5000	[92]
E-SNW-1/CNT	920 mAhg^{-1} or 62.6%/100/250	212/5000	[92]
PA-TA	543 mAhg^{-1} or %/1000/400	207/5000	[93]

5. Metal-Organic Framework-Based Anode Materials for LIBs and Their Practical Applications in Lithium-Ion Full Cells with Different Cathode Materials

Among the novel electrode materials, metal-organic framework (MOF) nanostructures have aroused intensive research interest as anodes for LIBs due to their extremely high surface area, controllable structures, large porosity, and presence of metal ions with redox activities which might be profitable to large lithium-ion accommodation space as well as an effective pathway for lithium ion and electron transport. Typically, MOFs consist of two major components: organic linkers (ligands) containing functional groups to build an open framework, and the metal ions (metal clusters) as redox-active centers which coordinate with organic ligands forming one- (1D), two- (2D), or three (3D)-dimensional structures. In particular, transition metals with high redox activity are used as metal ions of MOFs.

To date, numerous MOF-based compounds have been reported for various components of LIBs including cathodes, separators, electrolytes, and anodes. Many efforts have been made to construct MOF-based anode materials for LIBs in order to achieve higher capacity when working in LIB full cells with commercial cathode materials. Thus, in this section, different lithium battery systems composed of MOF-based anode materials are discussed.

In 2006, Chen and coworkers [97] first proposed a Zn-based MOF, $Zn_4O(1,3,5$-benzenetribenzoate$)_2$, named MOF-177, as an anode material for LIBs. They confirmed that different morphologies of this MOF can be regulated by a solvothermal route. However, the electrochemical performance of as-synthesized MOF-177 is found not satisfactory for practical application in reversible LIBs since the cycling capacity of MOF-177 was limited [97]. Besides that, in terms of anodes in LIBs, other valence metals have been used to design a variety of MOFs, such as Co-based MOF-71@300N [98], Cu-based MOF-199 [99], carbon-coated ZnO quantum dots-based MOF—MOF-5 [100], organic-coated ZIF-8 nanocomposites [101], Fe-based MIL-88 [102], Co-based zeolitic imidazolate frameworks—ZIF-67 [103], etc. However, all of these listed MOF-based anode materials still suffer from either poor stability or limited capacity in the first several charge–discharge cycles.

Song and coworkers [104] fabricated for the first time one-dimension (1D) cobalt-based MOF nanowires (named CoCOP) with a facile hydrothermal reaction and explored the as-prepared CoCOP nanowires as the anode material for coin-type full LIBs with commercial LFP cathode powders. They found that typical coin-type full Li-ion batteries with the CoCOP nanowires anode material and commercial LFP cathode material demonstrated a high capacity of 138 mAh/g at a current rate of 0.1C. Moreover, these LFP ∥ CoCOP full Li-ion batteries exhibited rapid charge–discharge capability and outstanding cyclability with a good capacity retention of 83% after conducting 300 cycles at a 1C charge–discharge rate, between 0.4 and 3.8 V [104].

In follow-up work, Song and coworkers [105] investigated a series of metal inorganic-organic hybrid composites (M-IOHCs, where M is Mn, Co, or Ni) as anode materials for Li-ion storage. Among all tested M-IOHC anode materials, Ni-IOHCs exhibited higher lithium storage capacity and showed an excellent power density of 297 W/kg at an energy density of 32 Wh/kg when assembled with a commercial $LiFePO_4$ cathode powder in a full coin-type cell system.

To achieve lower toxicity and lower cost, some researchers adopted Fe-based MOFs as anode materials for LIB full cells. For instance, Shen and coworkers [106] applied $[Fe_3O(BDC)_3(H_2O)_2(NO_3)]_n$ (named as Fe-MIL-88B) as a Fe-based MOF anode material for CR2032 full cells with commercial LFP powders as a cathode material. The full cells demonstrated an initial discharge capacity of 108 mAh/g at a current rate of 0.2C in a voltage cutoff of 0.4–3.8 V, and a large capacity of 89.8 mAh/g was retained after 100 repetitive cycles at a 0.25C charge–discharge rate. Overall, the findings obtained by Shen and coworkers [106] revealed that this Fe-MOF material is a prospective candidate for stable Li-ion insertion/de-insertion electrochemical processes.

Similarly, Sharma and coworkers [107] used another Fe-MOF configured with two ligands consisting of naphthalene dicarboxylic acid (H_2NDC) and terephthalic acid (H_2BDC) as an anode material for a full (coin) cell device assembled with LCO cathode material coated on aluminum foil. Impressively, this LCO ∥ Fe-MOF LIB full cell displayed a significant energy density of 360 Wh/kg (which is equivalent to a reversible capacity of 120 mAh/g) at a current density of 500 mA/g, with high cycling stability up to 1000 cycles.

It should be noted that the above-mentioned MOF anode material achieved interaction with lithium-ion through the valence change of its constituent transition metal. However, some conductive MOFs can interact with lithium-ion without the participation of their constituent transition metals in the lithiation/de-lithiation cycling processes.

For instance, Guo and coworkers [108] explored for the first time a bottom-up solvothermal technique to produce a one-dimensional (1D) Cu-based conductive MOF with 2,3,6,7,10,11-hexahydroxytriphenylene as the organic ligand (denoted as Cu-CAT) and found it as a competitive high-rate anode material for robust LIBs. A Cu-CAT nanowires-based full cell constructed with NCM811 cathode powder demonstrated outstanding cycling behavior, as well as a high energy density of ca. 275 Wh/kg. Remarkably, many nanowires of the Cu-CAT material of several micrometers in length are randomly connected together to build a three-dimension (3D) cross-linked porous network, which promotes the diffusion of electrolyte ions and convenient penetration of Li-ions.

In 2020, Weng and coworkers [109] reported for the first time such a Zn-based conductive MOF, $[Zn_2(py\text{-}TTF\text{-}py)_2(BDC)_2]\cdot 2DMF\cdot H_2O$, with terephthalic acid (H_2BDC) and 2,6-bis(4′-pyridyl)tetrathiafulvalene (py-TTF-py) as organic ligands and tested this material as an anode material for LIBs. The full cell of this Zn-based MOF with NCM622 cathode powder displayed a large reversible capacity of 131.9 mAh/g at a current rate of 100 mA/g with high Coulombic efficiency of 99.45% after 70 repetitive discharging/charging cycles. Moreover, this NCM622||$[Zn_2(py\text{-}TTF\text{-}py)_2(BDC)_2]\cdot 2DMF\cdot H_2O$ LIBs full cell demonstrated tolerance to high-current operation. Remarkably, the valence of Zn is not changed during the discharge/charge process, i.e., Zn^{2+} ions are not involved in the reversible electrochemical reaction occurring in LIBs, while Li-ions bond with sulfur atoms of the tetrathiafulvalene (denoted as TTF) moiety which provide the inserting sites of Li-ions during the discharging process.

It is worth noting that MOF-based composites prepared by an in situ growth strategy can also be used as organic anodes for LIBs. For instance, Wei and coworkers [110] used reduced graphene oxide (rGO) as a hard template for in situ growth of a fluorine-doped Co-based MOF with molecular formula $Co_2[F_x(OH)_{1-x}]_2(C_8O_4H_4)$, named as F-Co-MOF, using a facile solvothermal reaction. Owing to the synergistic effect of high conductive rGO networks and the F-Co-MOF structure, an outstanding reversible capacity of 162.5 mAh/g at a current rate of 200 mA/g after 300 repetitive cycles was achieved when this F-Co-MOF/rGO composite was adopted as an anode active material for a Li-ion full cell device assembled with LFP cathode material.

Based on the ultrahigh theoretical specific capacity of silicon (Si) and the large porosity of MOFs, some studies have focused on combining MOFs with Si through an in situ self-assembly technique to obtain higher cycling stability, relieve vigorous volume expansion of a silicon anode during the Li^+ insertion/extraction process, and achieve higher lithium storage capacity.

For instance, Zhou and coworkers [111] encapsulated Si nanoparticles with well-designed rod-like cross-linking Sn-based MOFs containing o-phthalic acid as organic linkers through an in situ self-assembly strategy. Thanks to the distinctive hybrid structure with remarkable synergistic effects based on Sn-based MOFs and Si nanoparticles, the Si@Sn-MOF composite delivered a discharge capacity of ca. 117.7 mAh/g after discharging/charging 150 cycles when tested as anode material for a Li-ion full (coin) cell device assembled with NCM622 cathode material at a fixed potential window from 2.8 to 4.3 V [111].

Likewise, Nazir and coworkers [112] reported a facile method for preparing a Si anode material coated with a conductive two-dimension (2D) Cu-based MOF named $Cu_3(HITP)_2$, in which HITP = 2,3,6,7,10,11-hexaiminotriphenylene, through an in situ growth. They evaluated this anode material in a Li-ion full cell with LCO cathode powder. Thanks to the improved ionic and electronic conductivity, as well as prospective volume expansion buffer, the full cell composed of LCO cathode powder and this Cu-MOF-coated Si anode material supplied high reversible capacities at various current densities: 1267 mAh/g at a 0.5C rate and 1105 mAh/g at a 1C charge–discharge rate.

It is worth noting that, unlike pristine MOFs, MOF derivatives possess better practical application potential in LIBs owing to their higher cycling stability and conductivity. Some studies have shown that the metals or other constituent heteroatoms of MOFs can be evenly distributed in the porous organic (carbon) matrix through the pyrolysis of the organic framework.

For instance, Yu and coworkers [113] synthesized MOF derivatives such as NiSb⊂CHSs, in which NiSb—embedded carbon hollow spheres. Firstly, they obtained the black Ni⊂CHSs precursor via the thermal treatment of a Ni-based MOF under an H_2/Ar atmosphere. Then, the as-prepared Ni⊂CHSs precursor was dispersed in the ethanol solution of $SbCl_3$ salt by virtue of ultrasonication. Finally, the black NiSb⊂CHSs product was collected by centrifugation, washed, and dried under a vacuum. The NiSb⊂CHSs were used for the first time as organic anode materials for LIBs, demonstrating remarkable rate capability, perfect cycling performance, and high specific capacity. When tested at a high voltage operation of nearly 3.5 V, the NiSb⊂CHSs||LMO full cell device displayed a high coulombic efficiency of ca. 99% and remarkable rate capability (210 mAh/g at a current density of 2 A/g) [113].

In 2021, Wang and coworkers [114] fabricated a Ni-Co-Sb/C nanosphere anode material for LIBs through a two-step "template sacrifice method" of calcination treatment of the Ni-Co-MOF nanosphere precursors. Due to the alloying mechanism, the Ni-Co-Sb/C nanospheres derived from Ni-Co-MOF delivered a high reversible capacity of 354 mAh/g at a current density of 100 mA/g even after 100 repetitive cycles when assembled with LCO cathode powder.

It should be noted that the sacrificial template method can also be utilized to fabricate Si-based anode material in which the in situ growth carbon possessing a porous structure generates a higher specific surface area. For instance, mesoporous silicon hollow nanocubes (denoted as m-Si HCs) were successfully produced by using such classic examples of MOFs as ZIF-8 as the sacrificial template [115]. The findings demonstrated that a full Li-ion cell consisting of LCO cathode material and m-Si HC-graphite anode material displayed outstanding cycle retention of 72% after 100 repetitive cycles at a 0.2C current rate. This excellent electrochemical performance corroborates that the applied template method is a facile and effective route to fabricate high-performance anode materials for LIBs [115].

Other than utilizing the sacrificial template method for producing Si-based organic anode material, using MOFs as a sacrificial template allows metal oxides with various properties and structures to be obtained by changing the calcination temperature and gas atmosphere.

For instance, Sun and coworkers [116] fabricated Co_3O_4 via both a two-step and one-step calcination of Co-based MOF precursor, respectively. Thanks to the 3D porous starfish-like structure, the Co_3O_4@N-C nanocomposite fabricated via a two-step technique delivered stable discharge capacity and could effectively buffer volume expansion during the charging/discharging process. Moreover, the Co_3O_4@N-C||$LiFePO_4$ full Li-ion battery showed stable capacity retention of 95.3% after 100 repetitive cycles [116].

It is worth noting that the calcination of Co-based MOFs that have heteroatoms or other additives allows cobalt oxide composites to be obtained that give full play to each component for achieving superior performance. For example, a Zn-doped hollow core–shell nano-sized Co_3O_4, named as Zn-Co-Oxide, derived from ZIF-67 via the template method exhibited a remarkable and stable specific capacity of 1600 mAh/g at a current density of 1 A/g for 700 repetitive cycles in the Li-ion full coin cell assembled with NCM532 cathode powder [117]. Fei and coworkers [118] produced a one-dimensional (1D) bunched Ni-MoO_2@Co-CoO-NC composite. Their synthetic strategy involved the room temperature growth of a ZIF-67 polyhedron onto $NiMoO_4 \cdot xH_2O$ nanowires followed by a carbonation treatment stage under Ar/H_2 atmosphere. Thanks to the synergistic effects of the variable metal and the well-designed morphology of Ni-MoO_2@Co-CoO-NC composite, a Ni-MoO_2@Co-CoO-NC||$LiFePO_4$ full Li-ion battery exhibited a high energy density of 329 Wh/kg and excellent cycling performance.

The electrochemical properties of rechargeable Li-ion full cells constructed with MOF-based anode materials and commercial cathode powders are summarized and compared in Table 3.

Table 3. Summary of the main electrochemical performance of Li-ion full cells assembled with MOF-based anode materials.

Organic Anode Material	MOF Template	Cathode Material	Current Density (C or mAg^{-1})/Specific Capacity (mAhg^{-1})/after Cycles/Capacity Retention (%)	Refs.
CoCOP	CoCOP	LFP	1C/69/300/83	[104]
Ni-IOHCs	Ni-IOHCs	LFP	0.1C/140/20/–	[105]
Fe-MIL-88B	Fe-MIL-88B	LFP	0.25C/86.8/100/73.7	[106]
Fe-MOF	Fe-MOF	LCO	500 mAg^{-1}/120/1000/–	[107]
Cu-CAT	Cu-CAT	NCM811	200 mAg^{-1}/371/200/–	[108]
[Zn$_2$(py-TTF-py)$_2$(BDC)$_2$]·2DMF·H$_2$O	(TTFs)-based Zn-MOF	NCM622	100 mAg^{-1}/131.9/70/–	[109]
F-Co-MOF/rGO	F-Co-MOF	LFP	200 mAg^{-1}/165.2/300/–	[110]
Si@Sn-MOF	Sn-based MOF	NCM622	20 mAg^{-1}/117.7/150/87.8	[111]
Si@Cu$_3$(HITP)$_2$	Cu$_3$(HITP)$_2$	LCO	0.1C/1038/50/46	[112]
NiSb⊂CHSs	Ni-based MOF	LMO	200 mAg^{-1}/228.2/100/–	[113]
Ni-Co-Sb/C	Ni-Co-MOF	LCO	100 mAg^{-1}/354/100/53.2	[114]
m-Si HC-graphite	ZIF-8	LCO	0.2C/–/100/72	[115]
Co$_3$O$_4$@N-C	Co–MOF	LFP	100 mAg^{-1}/266/100/95.3	[116]
Zn/Ni-Co-Oxide	Zn/Ni-ZIF-67	NCM532	1000 mAg^{-1}/1060/80/70	[117]
Ni-MoO$_2$@Co-CoO-NC	ZIF-67 polyhedron integrated with NiMoO$_4$·xH$_2$O nanowires	LFP	100 mAg^{-1}/130/60/92	[118]

As outlined in this section, the discharge/charge processes occurring in MOF-based anode materials are governed by conversion or intercalation charge-storage mechanisms. In the future, during the design of MOF-based anode materials, special attention should be paid to the following aspects:

1. Abundant redox active sites afforded by organic moieties or metal ion centers for the reversible electrochemical redox reaction;
2. Construction of conductive frameworks by importing heteroatoms and carbonization;
3. Adjustable porous frameworks for easier Li ions and electron transmission;
4. Utilization of the synergistic effect between various active sites of MOFs for next-generation LIBs with higher specific capacity.

6. Organic Compounds with Nitrogen-Containing Groups as Anode Active Material for LIBs

In recent years, to continue to investigate more organic materials that can be employed for rechargeable electrodes, the element N, which possesses an electronegativity close to that of oxygen, has attracted enormous attention.

Organic compounds with nitrogen-containing groups (OCNs) are regarded as some of the most promising organic electrodes because of their outstanding electrochemical properties, versatility, structural diversity, and adjustability [87].

To date, almost all organic anode compounds contain either an alicyclic ring group or a harmful aromatic ring, which leads to high toxicity. Such organic anode materials are undoubtedly not suitable for developing environmentally friendly and toxin-free LIBs. Zhu and coworkers [54] proposed the first linear chain-type organic compounds without benzene groups, 2,2′-azobis(2-methylpropionitrile) (AIBN) and 2,2′-azobis(2-methylpropionamidine) dihydrochloride (AIBA), as novel commercially available organic anode materials for LIBs based on the azo group. Another benefit of these low-toxic AIBN and AIBA anode materials is that they are renewable, sustainable, and inexpensive. Interestingly, AIBN-anode material contains one azo group which acts as a reversible electrochemical active site for the reversible insertion and extraction of lithium ions, whereas AIBA-anode has two active sites of azo (N=N) and imine (C=N) groups, confirmed with in situ Raman and ex situ Fourier transform infrared spectroscopy. The specific capacity of the AIBN anode in LIB was approximately 100 mAh/g at a current density of 10 mA/g from the 2nd to the 100th cycle [54].

Ye and coworkers [119] synthesized a highly conjugated polymeric Schiff base (PSB) through a facile solid-phase reaction. The obtained PSB anode exhibits a high specific capacity of 175 mAh/g at a current density of 10 mA/g and outstanding cyclic performance. Most importantly, due to the suppressed dissolution of the as-prepared PSB into the organic electrolyte, great stable cycling with a capacity retention of 90% was achieved after 100 repetitive cycles. Such success opened up new avenues for developing more Schiff base polymers and producing more promising and higher-capacity organic anode materials for LIBs.

Man et al. [120] synthesized a highly conjugated organic framework, poly(imine-anthraquinone) (PIAQ), through an environmentally friendly and green in situ solvothermal condensation reaction of 2,6-diaminoanthraquinone (2,6-DAQ) and p-phthalaldehyde (PPD) and adopted this material as the anode material of LIBs. Based on DFT calculations coupled with experimental investigations, Man and coworkers [120] proposed a 16 Li-storage mechanism of PIAQ. Intriguingly, the PIAQ-based organic anode material displays excellent reversible specific capacity (1231 mAh/g at a current density of 200 mA/g), outstanding long-term cycle stability (its specific capacity was 486 mAh/g after 1000 cycles at the high current density of 1 A/g), and perfect rate performance. Both the excellent charge–discharge performance and perfect structural stability of PIAQ material make it one of the most prospective organic anode materials for next-generation high-performance LIBs.

A summary of the electrochemical properties of OCNs which were reported as anodes for LIBs is presented in Table 4.

Table 4. Summary of the electrochemical properties of organic anode materials with nitrogen-containing groups employed as an anode for LIBs.

Organic Anode Material	Molecular Weight of Monomer	Anode Electrode Composition with the Mass Ratio (%)	Voltages of [Discharge]/[Charge] Platform or Average Voltages (V)	Initial Discharge (Lithiation) and Charge (Delithiation) Capacities (mAh/g)//at a Current Density (mA/g)	Discharge Capacity or Capacity Retention (mAhg^{-1} or %)/Current Rate (mAg^{-1})/After Cycles/Voltage Range vs. Li$^+$/Li	Discharge Capacities (mAh/g) with High Current Rate/Current Rate (mA/g)	Refs.
2,2′-Azobis(2-methylpropionitrile) (AIBN)	164	AIBN:super P carbon black:polyvinylidene fluoride = 80:10:10	[1.7, 1.2]/[-]	100/-//5	Unspecified	30/50	[54]
2,2′-Azobis(2-methylpropionamidine) dihydrochloride (AIBA)	198	AIBA:super P carbon black:polyvinylidene fluoride = 80:10:10	[1.7, 1.6, 0.6]/[-]	~1000/~160//5	110 mAhg^{-1}/10/200/0.01–3 V	~50/50	[54]
Free-standing nitrogen-rich graphene-like holey conjugated polymers (NG-HCP nanosheets)	1405	NG-HCP nanosheets:carbon nanotubes:poly(vinylidene fluoride) = 70:20:10	[1.4]/[-]	2497/1319//20	1015 mAhg^{-1}/100/230/0–3 V	237/2500	[90]
Bulk covalent organic framework (COF)	205	COF:acetylene black:polyvinylidene difluoride = 80:10:10	Unspecified	702/163//100	125 mAhg^{-1}/100/300/0.005–3.0 V	Unspecified	[91]
Two-dimensional covalent organic framework trapped by carbon nanotubes (COF@CNTs composite)	205	COF@CNTs composite:acetylene black:polyvinylidene difluoride = 80:10:10	[1.5]/[0.75, 1.6]	928/383//100	1536 mAhg^{-1}/100/500/0.005–3.0 V	217/5C (1C = 1Ag^{-1})	[91]
Polymeric Schiff base (PSB)	443	PSB:super P:polytetrafluoroethylene = 85:10:5	[0.15, 1.1, 2.3]/[0.3, 1.0, 3.2]	315/97//10	50.8%/10/More than 100/0.01–3.5 V	40.3/80	[119]
Poly(imine-anthraquinone) (PIAQ)	366	PIAQ:Ketjen black:Carboxymethylcellulose = 70:20:10	[1.0, 0.1]/[−1.3, −0.3]	1231/1130//20	89.1%/200/More than 100/0.01–3.5 V	259/2000	[120]

7. Conclusions, Challenges, and Outlooks on Further Developments of Organic Anode Materials for LIBs

7.1. Challenges and Strategies for Enhancing the Electrochemical Performance of Organic Anode Materials for LIBs

Although organic electrode materials possess so many advantages [36,121], there are also several challenges impeding their commercialization, such as their intrinsic low electronic conductivity at room temperature that hinders the reaction kinetics of organic rechargeable batteries, uncontrollable side reactions, and high solubility in organic electrolytes, which leads to fast capacity fading [87,122,123]. Ingenious strategies, such as molecular engineering, nanosizing [78,124], modification, and combination with other materials, have been applied to alleviate the stated intrinsic drawbacks of organic electrode materials [60,80,125–129]. Moreover, a deeper insight into the reaction mechanisms of this type of material is also essential.

Some of the typical strategies to overcome the poor electronic conductivity problem of organic electrode materials are introducing conductive carbon additives (10–60%) [127], substituent group design, and adjusting the length of the carbon chain in an organic molecule.

As for the problem of dissolving organic compounds in common organic electrolytes, it can be overcome by using a solid polymer electrolyte. Besides, electrolyte optimization, regulating the structure of organic electrode materials by molecular engineering, and polymerization emerge as the main effective methods for inhibiting the dissolution of organic small molecules and enhancing the electrochemical performance of organic anode materials [130].

The nanosizing of organic materials in different shapes has been one of the most popular techniques for improving the electrochemical performance of organic electrode materials. For instance, Wang and co-workers produced $Li_4C_8H_2O_6$ in nanosheets and reported that its electrochemical properties excel those in bulk particles or in nanoparticles [124]. A nanosheet $Li_4C_8H_2O_6$ electrode displayed superior electrochemical performance with a first discharge capacity of 358 mAh/g as an anode material. It should be noted that the nanosheet $Li_4C_8H_2O_6$ electrode noticeably outperforms its bulk counterpart, exhibiting a discharge capacity of 175 mAh/g as an anode active material even at a high rate of 5C, which is approximately two times higher than those of the $Li_4C_8H_2O_6$ bulk. This outstanding electrochemical behavior of the $Li_4C_8H_2O_6$ nanosheet was attributed both to the 2D particle morphology causing a rapid lithium diffusion rate and to the larger surface area making lithium ions more accessible to each organic molecule serving as the active material of negative electrode of rechargeable LIBs. Overall, the findings demonstrated that $Li_4C_8H_2O_6$ nanosheets having multifunctional groups and high-specific surface area are perspectives in the applications of all organic LIBs [124].

Using lithium salts of organic molecules is also one of the effective strategies for inhibiting the undesired dissolution of organic electrode materials in aprotic electrolytes widely employed in LIBs and enhancing the cycling stability of organic anode materials for LIBs [68,71,131,132].

7.2. Outlooks on Further Developments of Organic Anode Materials for LIBs

Considering the aforementioned challenges, the emphasis of upcoming research on the development of high-performance organic anode materials for LIBs should be placed on the following aspects:

- It is well-known that the theoretical capacity of an active cathode or anode material is ultimately dependent on the number of electrons transferred in each redox-active moiety and inversely proportional to the molecular weight of the organic molecule. Since organic anode materials exhibit redox activity based on redox-active centers, there is a large proportion of inactive mass which inevitably decreases the reversible capacity of the negative electrode. Therefore, enhancing the density of active centers in any redox-active organic molecule is required to boost the capacity of the negative electrode. Although the molecular design can enhance the theoretical capacity of

the organic anode material, several bulk properties of the anode material, such as crystallinity or particle size, influence the practical capacities.
- The search for unexplored redox-active functional groups in organic molecules continues to stimulate fundamental investigations. The early success of advanced porous materials, such as MOFs and COFs, towards rapid and stable cycling of Li-ions, is likely to accelerate the development of new organic anode materials for LIBs. Particularly, the optimal stability of MOFs or COFs in organic solvents may enable a robust pathway to utilize the features of organic redox into stable electrode materials.
- Apart from great efforts which have been made in molecular engineering to enhance the electrochemical parameters of organic anode materials, techniques to fabricate scalable organic anode materials are another aspect of further investigations. Expanding the scope of Li-ion battery technology will require reliable solutions for future energy needs. The investigations on such organic redox-active molecules are in their infancy and will require a comprehensive assessment of various classes of organic materials and compounds to fully realize their potential.

Thus, there is plenty of room to develop organic anode materials for LIBs with enhanced performance, such as long cycle life, high-output voltage, and high specific capacity.

Hopefully, this work provides a brief, useful introduction that outlines the representative organic anode materials in LIBs and gives new insight for the further application of organic anode materials for LIBs.

Author Contributions: Conceptualization, P.N. and K.P.; formal analysis, K.P., A.K. and A.A.P. (Alexander A. Pavlovskii).; investigation, resources; A.A.P. (Alexander A. Pavlovskii). and P.N.; data curation, K.P. and A.K.; writing—original draft preparation, A.A.P. (Alexander A. Pavlovskii), writing—review and editing, K.P., A.K. and P.N.; visualization, A.A.P. (Alexander A. Pavlovskii).; supervision, K.P., P.N. and A.A.P. (Anatoliy A. Popovich).; project administration, P.N.; funding acquisition, P.N. All authors have read and agreed to the published version of the manuscript.

Funding: The research was funded by the Ministry of Science and Higher Education of the Russian Federation by "Agreement on the grant in the form of subsidies from the federal budget for the implementation of state support for the creation and development of world-class scientific centers, those are performing research and development on the priorities of scientific and technological development" dated 20 April 2022 No. 075-15-2022-311.

Institutional Review Board Statement: Not applicable.

Informed Consent Statement: Not applicable.

Data Availability Statement: Not applicable.

Conflicts of Interest: The authors declare no conflict of interest.

References

1. Jyoti, J.; Singh, B.P.; Tripathi, S.K. Recent Advancements in Development of Different Cathode Materials for Rechargeable Lithium Ion Batteries. *J. Energy Storage* **2021**, *43*, 103112. [CrossRef]
2. Shen, Y.F.; Shen, X.H.; Yang, M.; Qian, J.F.; Cao, Y.L.; Yang, H.X.; Luo, Y.; Ai, X.P. Achieving Desirable Initial Coulombic Efficiencies and Full Capacity Utilization of Li-Ion Batteries by Chemical Prelithiation of Graphite Anode. *Adv. Funct. Mater.* **2021**, *31*, 2101181. [CrossRef]
3. Xu, N.; Mei, S.; Chen, Z.; Dong, Y.; Li, W.; Zhang, C. High-Performance Li-Organic Battery Based on Thiophene-Containing Porous Organic Polymers with Different Morphology and Surface Area as the Anode Materials. *Chem. Eng. J.* **2020**, *395*, 124975. [CrossRef]
4. Jiang, H.R.; Lu, Z.; Wu, M.C.; Ciucci, F.; Zhao, T.S. Borophene: A Promising Anode Material Offering High Specific Capacity and High Rate Capability for Lithium-Ion Batteries. *Nano Energy* **2016**, *23*, 97–104. [CrossRef]
5. Wang, X.; Tang, S.; Guo, W.; Fu, Y.; Manthiram, A. Advances in Multimetallic Alloy-Based Anodes for Alkali-Ion and Alkali-Metal Batteries. *Mater. Today* **2021**, *50*, 259–275. [CrossRef]
6. Sun, T.; Liu, C.; Wang, J.; Nian, Q.; Feng, Y.; Zhang, Y.; Tao, Z.; Chen, J. A Phenazine Anode for High-Performance Aqueous Rechargeable Batteries in a Wide Temperature Range. *Nano Res.* **2020**, *13*, 676–683. [CrossRef]
7. Chae, S.; Ko, M.; Kim, K.; Ahn, K.; Cho, J. Confronting Issues of the Practical Implementation of Si Anode in High-Energy Lithium-Ion Batteries. *Joule* **2017**, *1*, 47–60. [CrossRef]
8. Reddy, M.V.; Subba Rao, G.V.; Chowdari, B.V.R. Metal Oxides and Oxysalts as Anode Materials for Li Ion Batteries. *Chem. Rev.* **2013**, *113*, 5364–5457. [CrossRef]

9. Zhang, J.; Zhang, M.; Zhang, L. Binder-Free Three-Dimensional Porous Structured Metal Oxides as Anode for High Performance Lithium-Ion Battery. *Electrochim. Acta* **2013**, *105*, 282–288. [CrossRef]
10. Liang, C.; Gao, M.; Pan, H.; Liu, Y.; Yan, M. Lithium Alloys and Metal Oxides as High-Capacity Anode Materials for Lithium-Ion Batteries. *J. Alloys Compd.* **2013**, *575*, 246–256. [CrossRef]
11. Park, C.M.; Kim, J.H.; Kim, H.; Sohn, H.J. Li-Alloy Based Anode Materials for Li Secondary Batteries. *Chem. Soc. Rev.* **2010**, *39*, 3115–3141. [CrossRef]
12. Zhang, G.; Yang, Y.; Zhang, T.; Xu, D.; Lei, Z.; Wang, C.; Liu, G.; Deng, Y. Fe^{III} Chelated Organic Anode with Ultrahigh Rate Performance and Ultra-Long Cycling Stability for Lithium-Ion Batteries. *Energy Storage Mater.* **2020**, *24*, 432–438. [CrossRef]
13. Li, X.; Sun, X.; Hu, X.; Fan, F.; Cai, S.; Zheng, C.; Stucky, G.D. Review on Comprehending and Enhancing the Initial Coulombic Efficiency of Anode Materials in Lithium-Ion/Sodium-Ion Batteries. *Nano Energy* **2020**, *77*, 105143. [CrossRef]
14. Liang, Y.; Yao, Y. Positioning Organic Electrode Materials in the Battery Landscape. *Joule* **2018**, *2*, 1690–1706. [CrossRef]
15. Liang, Y.; Tao, Z.; Chen, J. Organic Electrode Materials for Rechargeable Lithium Batteries. *Adv. Energy Mater.* **2012**, *2*, 742–769. [CrossRef]
16. Lee, S.; Kwon, G.; Ku, K.; Yoon, K.; Jung, S.K.; Lim, H.D.; Kang, K. Recent Progress in Organic Electrodes for Li and Na Rechargeable Batteries. *Adv. Mater.* **2018**, *30*, 1704682. [CrossRef]
17. Lu, Y.; Chen, J. Prospects of Organic Electrode Materials for Practical Lithium Batteries. *Nat. Rev. Chem.* **2020**, *4*, 127–142. [CrossRef]
18. Cui, C.; Ji, X.; Wang, P.F.; Xu, G.L.; Chen, L.; Chen, J.; Kim, H.; Ren, Y.; Chen, F.; Yang, C.; et al. Integrating Multiredox Centers into One Framework for High-Performance Organic Li-Ion Battery Cathodes. *ACS Energy Lett.* **2020**, *5*, 224–231. [CrossRef]
19. Sieuw, L.; Lakraychi, A.E.; Rambabu, D.; Robeyns, K.; Jouhara, A.; Borodi, G.; Morari, C.; Poizot, P.; Vlad, A. Through-Space Charge Modulation Overriding Substituent Effect: Rise of the Redox Potential at 3.35 V in a Lithium-Phenolate Stereoelectronic Isomer. *Chem. Mater.* **2020**, *32*, 9996–10006. [CrossRef]
20. Lu, Y.; Zhang, Q.; Li, L.; Niu, Z.; Chen, J. Design Strategies toward Enhancing the Performance of Organic Electrode Materials in Metal-Ion Batteries. *Chem* **2018**, *4*, 2786–2813. [CrossRef]
21. Lee, M.; Hong, J.; Lopez, J.; Sun, Y.; Feng, D.; Lim, K.; Chueh, W.C.; Toney, M.F.; Cui, Y.; Bao, Z. High-Performance Sodium–Organic Battery by Realizing Four-Sodium Storage in Disodium Rhodizonate. *Nat. Energy* **2017**, *2*, 861–868. [CrossRef]
22. Banda, H.; Damien, D.; Nagarajan, K.; Raj, A.; Hariharan, M.; Shaijumon, M.M. Twisted Perylene Diimides with Tunable Redox Properties for Organic Sodium-Ion Batteries. *Adv. Energy Mater.* **2017**, *7*, 1701316. [CrossRef]
23. Yokoji, T.; Kameyama, Y.; Sakaida, S.; Maruyama, N.; Satoh, M.; Matsubara, H. Steric Effects on the Cyclability of Benzoquinone-type Organic Cathode Active Materials for Rechargeable Batteries. *Chem. Lett.* **2015**, *44*, 1726–1728. [CrossRef]
24. Shea, J.J.; Luo, C. Organic Electrode Materials for Metal Ion Batteries. *ACS Appl. Mater. Interfaces* **2020**, *12*, 5361–5380. [CrossRef]
25. Gannett, C.N.; Melecio-Zambrano, L.; Theibault, M.J.; Peterson, B.M.; Fors, B.P.; Abruña, H.D. Organic Electrode Materials for Fast-Rate, High-Power Battery Applications. *Mater. Rep. Energy* **2021**, *1*, 100008. [CrossRef]
26. Wang, Z.; Jin, W.; Huang, X.; Lu, G.; Li, Y. Covalent Organic Frameworks as Electrode Materials for Metal Ion Batteries: A Current Review. *Chem. Rec.* **2020**, *20*, 1198–1219. [CrossRef]
27. Zhao-Karger, Z.; Gao, P.; Ebert, T.; Klyatskaya, S.; Chen, Z.; Ruben, M.; Fichtner, M. New Organic Electrode Materials for Ultrafast Electrochemical Energy Storage. *Adv. Mater.* **2019**, *31*, 1806599. [CrossRef]
28. Schon, T.B.; McAllister, B.T.; Li, P.F.; Seferos, D.S. The Rise of Organic Electrode Materials for Energy Storage. *Chem. Soc. Rev.* **2016**, *45*, 6345–6404. [CrossRef]
29. Veerababu, M.; Varadaraju, U.V.; Kothandaraman, R. Improved Electrochemical Performance of Lithium/Sodium Perylene-3,4,9,10-Tetracarboxylate as an Anode Material for Secondary Rechargeable Batteries. *Int. J. Hydrog. Energy* **2015**, *40*, 14925–14931. [CrossRef]
30. Zheng, Z.; Zhu, J.; Yang, C.; Lu, C.; Chen, Z.; Zhuang, X. The Art of Two-Dimensional Soft Nanomaterials. *Sci. China Chem.* **2019**, *62*, 1145–1193. [CrossRef]
31. Zhang, C.; Lu, C.; Zhang, F.; Qiu, F.; Zhuang, X.; Feng, X. Two-Dimensional Organic Cathode Materials for Alkali-Metal-Ion Batteries. *J. Energy Chem.* **2018**, *27*, 86–98. [CrossRef]
32. Wu, Q.; Liu, J.; Yuan, C.; Li, Q.; Wang, H. Nitrogen-Doped 3D Flower-like Carbon Materials Derived from Polyimide as High-Performance Anode Materials for Lithium-Ion Batteries. *Appl. Surf. Sci.* **2017**, *425*, 1082–1088. [CrossRef]
33. Su, Y.; Liu, Y.; Liu, P.; Wu, D.; Zhuang, X.; Fan, Z.; Feng, X. Compact Coupled Graphene and Porous Polyaryltriazine-Derived Frameworks as High Performance Cathodes for Lithium-Ion Batteries. *Angew. Chem. Int. Ed.* **2015**, *54*, 1812–1816. [CrossRef] [PubMed]
34. Shi, R.; Jiao, S.; Yue, Q.; Gu, G.; Zhang, K.; Zhao, Y.; Kai Zhang, C. Challenges and Advances of Organic Electrode Materials for Sustainable Secondary Batteries. *Exploration* **2022**, *2*, 20220066. [CrossRef]
35. Sun, T.; Xie, J.; Guo, W.; Li, D.S.; Zhang, Q. Covalent–Organic Frameworks: Advanced Organic Electrode Materials for Rechargeable Batteries. *Adv. Energy Mater.* **2020**, *10*, 1904199. [CrossRef]
36. Chen, Y.; Wang, C. Designing High Performance Organic Batteries. *Acc. Chem. Res.* **2020**, *53*, 2636–2647. [CrossRef]
37. Poizot, P.; Gaubicher, J.; Renault, S.; Dubois, L.; Liang, Y.; Yao, Y. Opportunities and Challenges for Organic Electrodes in Electrochemical Energy Storage. *Chem. Rev.* **2020**, *120*, 6490–6557. [CrossRef]

38. Muench, S.; Wild, A.; Friebe, C.; Häupler, B.; Janoschka, T.; Schubert, U.S. Polymer-Based Organic Batteries. *Chem. Rev.* **2016**, *116*, 9438–9484. [CrossRef]
39. Song, Z.; Zhou, H. Towards Sustainable and Versatile Energy Storage Devices: An Overview of Organic Electrode Materials. *Energy Environ. Sci.* **2013**, *6*, 2280–2301. [CrossRef]
40. Zhu, L.M.; Lei, A.W.; Cao, Y.L.; Ai, X.P.; Yang, H.X. An All-Organic Rechargeable Battery Using Bipolar Polyparaphenylene as a Redox-Active Cathode and Anode. *Chem. Commun.* **2012**, *49*, 567–569. [CrossRef]
41. Min, D.J.; Lee, K.; Park, S.Y.; Kwon, J.E. Mellitic Triimides Showing Three One-Electron Redox Reactions with Increased Redox Potential as New Electrode Materials for Li-Ion Batteries. *ChemSusChem* **2020**, *13*, 2303–2311. [CrossRef] [PubMed]
42. Buyukcakir, O.; Ryu, J.; Joo, S.H.; Kang, J.; Yuksel, R.; Lee, J.; Jiang, Y.; Choi, S.; Lee, H.S.; Kwak, S.K.; et al. Lithium Accommodation in a Redox-Active Covalent Triazine Framework for High Areal Capacity and Fast-Charging Lithium-Ion Batteries. *Adv. Funct. Mater.* **2020**, *30*, 2003761. [CrossRef]
43. Aher, J.; Graefenstein, A.; Deshmukh, G.; Subramani, K.; Krueger, B.; Haensch, M.; Schwenzel, J.; Krishnamoorthy, K.; Wittstock, G. Effect of Aromatic Rings and Substituent on the Performance of Lithium Batteries with Rylene Imide Cathodes. *ChemElectroChem* **2020**, *7*, 1160–1165. [CrossRef]
44. Wang, Y.; Deng, Y.; Qu, Q.; Zheng, X.; Zhang, J.; Liu, G.; Battaglia, V.S.; Zheng, H. Ultrahigh-Capacity Organic Anode with High-Rate Capability and Long Cycle Life for Lithium-Ion Batteries. *ACS Energy Lett.* **2017**, *2*, 2140–2148. [CrossRef]
45. Lee, M.; Hong, J.; Lee, B.; Ku, K.; Lee, S.; Park, C.B.; Kang, K. Multi-Electron Redox Phenazine for Ready-to-Charge Organic Batteries. *Green Chem.* **2017**, *19*, 2980–2985. [CrossRef]
46. Lee, J.; Kim, H.; Park, M.J. Long-Life, High-Rate Lithium-Organic Batteries Based on Naphthoquinone Derivatives. *Chem. Mater.* **2016**, *28*, 2408–2416. [CrossRef]
47. Häupler, B.; Wild, A.; Schubert, U.S.; Häupler, B.; Wild, A.; Schubert, U.S. Carbonyls: Powerful Organic Materials for Secondary Batteries. *Adv. Energy Mater.* **2015**, *5*, 1402034. [CrossRef]
48. Han, X.; Qing, G.; Sun, J.; Sun, T. How Many Lithium Ions Can Be Inserted onto Fused C_6 Aromatic Ring Systems? *Angew. Chem. Int. Ed.* **2012**, *51*, 5147–5151. [CrossRef]
49. Zhang, C.; Chen, S.; Zhou, G.; Hou, Q.; Luo, S.; Wang, Y.; Shi, G.; Zeng, R. 3-Anthraquinone Substituted Polythiophene as Anode Material for Lithium Ion Battery. *J. Electroanal. Chem.* **2021**, *895*, 115495. [CrossRef]
50. Wang, H.; Yao, C.-J.; Nie, H.-J.; Wang, K.-Z.; Zhong, Y.-W.; Chen, P.; Mei, S.; Zhang, Q. Recent Progress in Carbonyl-Based Organic Polymers as Promising Electrode Materials for Lithium-Ion Batteries (LIBs). *J. Mater. Chem. A Mater.* **2020**, *8*, 11906–11922. [CrossRef]
51. Wang, Y.; Liu, W.; Guo, R.; Qu, Q.; Zheng, H.; Zhang, J.; Huang, Y. A High-Capacity Organic Anode with Self-Assembled Morphological Transformation for Green Lithium-Ion Batteries. *J. Mater. Chem. A Mater.* **2019**, *7*, 22621–22630. [CrossRef]
52. Gu, T.; Zhou, M.; Huang, B.; Cao, S.; Wang, J.; Tang, Y.; Wang, K.; Cheng, S.; Jiang, K. Advanced Li-Organic Batteries with Super-High Capacity and Long Cycle Life via Multiple Redox Reactions. *Chem. Eng. J.* **2019**, *373*, 501–507. [CrossRef]
53. Li, G.; Zhang, B.; Wang, J.; Zhao, H.; Ma, W.; Xu, L.; Zhang, W.; Zhou, K.; Du, Y.; He, G. Electrochromic Poly(Chalcogenoviologen)s as Anode Materials for High-Performance Organic Radical Lithium-Ion Batteries. *Angew. Chem. Int. Ed.* **2019**, *58*, 8468–8473. [CrossRef]
54. Zhu, Y.; Chen, P.; Zhou, Y.; Nie, W.; Xu, Y. New Family of Organic Anode without Aromatics for Energy Storage. *Electrochim. Acta* **2019**, *318*, 262–271. [CrossRef]
55. Wei, W.; Chang, G.; Xu, Y.; Yang, L. An Indole-Based Conjugated Microporous Polymer: A New and Stable Lithium Storage Anode with High Capacity and Long Life Induced by Cation–π Interactions and a N-Rich Aromatic Structure. *J. Mater. Chem. A Mater.* **2018**, *6*, 18794–18798. [CrossRef]
56. Banerjee, A.; Araujo, R.B.; Sjödin, M.; Ahuja, R. Identifying the Tuning Key of Disproportionation Redox Reaction in Terephthalate: A Li-Based Anode for Sustainable Organic Batteries. *Nano Energy* **2018**, *47*, 301–308. [CrossRef]
57. Maiti, S.; Pramanik, A.; Dhawa, T.; Mahanty, S. Redox-Active Organic Molecular Salt of 1,2,4-Benzenetricarboxylic Acid as Lithium-Ion Battery Anode. *Mater. Lett.* **2017**, *209*, 613–617. [CrossRef]
58. Deng, Q.; Xue, J.; Zou, W.; Wang, L.; Zhou, A.; Li, J. The Electrochemical Behaviors of $Li_2C_8H_4O_6$ and Its Corresponding Organic Acid $C_8H_6O_6$ as Anodes for Li-Ion Batteries. *J. Electroanal. Chem.* **2016**, *761*, 74–79. [CrossRef]
59. Janoschka, T.; Hager, M.D.; Schubert, U.S. Powering up the Future: Radical Polymers for Battery Applications. *Adv. Mater.* **2012**, *24*, 6397–6409. [CrossRef] [PubMed]
60. Zhu, L.; Ding, G.; Xie, L.; Cao, X.; Liu, J.; Lei, X.; Ma, J. Conjugated Carbonyl Compounds as High-Performance Cathode Materials for Rechargeable Batteries. *Chem. Mater.* **2019**, *31*, 8582–8612. [CrossRef]
61. Xie, J.; Gu, P.; Zhang, Q. Nanostructured Conjugated Polymers: Toward High-Performance Organic Electrodes for Rechargeable Batteries. *ACS Energy Lett.* **2017**, *2*, 1985–1996. [CrossRef]
62. Ogihara, N.; Yasuda, T.; Kishida, Y.; Ohsuna, T.; Miyamoto, K.; Ohba, N. Organic Dicarboxylate Negative Electrode Materials with Remarkably Small Strain for High-Voltage Bipolar Batteries. *Angew. Chem. Int. Ed.* **2014**, *53*, 11467–11472. [CrossRef] [PubMed]
63. Zhao, R.R.; Cao, Y.L.; Ai, X.P.; Yang, H.X. Reversible Li and Na Storage Behaviors of Perylenetetracarboxylates as Organic Anodes for Li- and Na-Ion Batteries. *J. Electroanal. Chem.* **2013**, *688*, 93–97. [CrossRef]
64. Renault, S.; Mihali, V.A.; Brandell, D. Optimizing the Electrochemical Performance of Water-Soluble Organic Li–Ion Battery Electrodes. *Electrochem. Commun.* **2013**, *34*, 174–176. [CrossRef]

65. Renault, S.; Geng, J.; Dolhem, F.; Poizot, P. Evaluation of Polyketones with N-Cyclic Structure as Electrode Material for Electrochemical Energy Storage: Case of Pyromellitic Diimide Dilithium Salt. *Chem. Commun.* 2011, *47*, 2414–2416. [CrossRef] [PubMed]
66. Senoh, H.; Yao, M.; Sakaebe, H.; Yasuda, K.; Siroma, Z. A Two-Compartment Cell for Using Soluble Benzoquinone Derivatives as Active Materials in Lithium Secondary Batteries. *Electrochim. Acta* 2011, *56*, 10145–10150. [CrossRef]
67. Nokami, T.; Matsuo, T.; Inatomi, Y.; Hojo, N.; Tsukagoshi, T.; Yoshizawa, H.; Shimizu, A.; Kuramoto, H.; Komae, K.; Tsuyama, H.; et al. Polymer-Bound Pyrene-4,5,9,10-Tetraone for Fast-Charge and -Discharge Lithium-Ion Batteries with High Capacity. *J. Am. Chem. Soc.* 2012, *134*, 19694–19700. [CrossRef]
68. Walker, W.; Grugeon, S.; Mentre, O.; Laruelle, S.; Tarascon, J.M.; Wudl, F. Ethoxycarbonyl-Based Organic Electrode for Li-Batteries. *J. Am. Chem. Soc.* 2010, *132*, 6517–6523. [CrossRef]
69. Han, X.; Chang, C.; Yuan, L.; Sun, T.; Sun, J. Aromatic Carbonyl Derivative Polymers as High-Performance Li-Ion Storage Materials. *Adv. Mater.* 2007, *19*, 1616–1621. [CrossRef]
70. Zeng, R.H.; Li, X.P.; Qiu, Y.C.; Li, W.S.; Yi, J.; Lu, D.S.; Tan, C.L.; Xu, M.Q. Synthesis and Properties of a Lithium-Organic Coordination Compound as Lithium-Inserted Material for Lithium Ion Batteries. *Electrochem. Commun.* 2010, *9*, 1253–1256. [CrossRef]
71. Armand, M.; Grugeon, S.; Vezin, H.; Laruelle, S.; Ribière, P.; Poizot, P.; Tarascon, J.M. Conjugated Dicarboxylate Anodes for Li-Ion Batteries. *Nat. Mater.* 2009, *8*, 120–125. [CrossRef] [PubMed]
72. Walker, W.; Grugeon, S.; Vezin, H.; Laruelle, S.; Armand, M.; Tarascon, J.M.; Wudl, F. The Effect of Length and Cis/Trans Relationship of Conjugated Pathway on Secondary Battery Performance in Organolithium Electrodes. *Electrochem. Commun.* 2010, *12*, 1348–1351. [CrossRef]
73. Wang, L.; Mou, C.; Wu, B.; Xue, J.; Li, J. Alkaline Earth Metal Terephthalates $MC_8H_4O_4$ (M=Ca, Sr, Ba) as Anodes for Lithium Ion Batteries. *Electrochim. Acta* 2016, *196*, 118–124. [CrossRef]
74. Liu, F.; Liao, S.; Lin, H.; Yin, Y.; Liu, Y.; Meng, H.; Min, Y. A Facile Strategy for Synthesizing Organic Tannic Metal Salts as Advanced Energy Storage Anodes. *ChemElectroChem* 2021, *8*, 2686–2692. [CrossRef]
75. Xiao, F.; Liu, P.; Li, J.; Zhang, Y.; Liu, Y.; Xu, M. A Small Molecule Organic Compound Applied as an Advanced Anode Material for Lithium-Ion Batteries. *Chem. Commun.* 2022, *58*, 697–700. [CrossRef] [PubMed]
76. Luo, C.; Ji, X.; Hou, S.; Eidson, N.; Fan, X.; Liang, Y.; Deng, T.; Jiang, J.; Wang, C. Azo Compounds Derived from Electrochemical Reduction of Nitro Compounds for High Performance Li-Ion Batteries. *Adv. Mater.* 2018, *30*, 1706498. [CrossRef] [PubMed]
77. Lakraychi, A.E.; Dolhem, F.; Djedaïni-Pilard, F.; Becuwe, M. Substituent Effect on Redox Potential of Terephthalate-Based Electrode Materials for Lithium Batteries. *Electrochem. Commun.* 2018, *93*, 71–75. [CrossRef]
78. Nisula, M.; Karppinen, M. Atomic/Molecular Layer Deposition of Lithium Terephthalate Thin Films as High Rate Capability Li-Ion Battery Anodes. *Nano Lett.* 2016, *16*, 1276–1281. [CrossRef]
79. Fédèle, L.; Sauvage, F.; Bécuwe, M. Hyper-Conjugated Lithium Carboxylate Based on a Perylene Unit for High-Rate Organic Lithium-Ion Batteries. *J. Mater. Chem. A Mater.* 2014, *2*, 18225–18228. [CrossRef]
80. Zhang, H.; Deng, Q.; Zhou, A.; Liu, X.; Li, J. Porous $Li_2C_8H_4O_4$ Coated with N-Doped Carbon by Using CVD as an Anode Material for Li-Ion Batteries. *J. Mater. Chem. A Mater.* 2014, *2*, 5696–5702. [CrossRef]
81. Walker, W.; Grugeon, S.; Vezin, H.; Laruelle, S.; Armand, M.; Wudl, F.; Tarascon, J.M. Electrochemical Characterization of Lithium 4,4′-Tolane-Dicarboxylate for Use as a Negative Electrode in Li-Ion Batteries. *J. Mater. Chem.* 2011, *21*, 1615–1620. [CrossRef]
82. Diercks, C.S.; Yaghi, O.M. The Atom, the Molecule, and the Covalent Organic Framework. *Science* 2017, *355*, eaal1585. [CrossRef] [PubMed]
83. Huang, N.; Wang, P.; Jiang, D. Covalent Organic Frameworks: A Materials Platform for Structural and Functional Designs. *Nat. Rev. Mater.* 2016, *1*, 16068. [CrossRef]
84. Beuerle, F.; Gole, B. Covalent Organic Frameworks and Cage Compounds: Design and Applications of Polymeric and Discrete Organic Scaffolds. *Angew. Chem. Int. Ed.* 2018, *57*, 4850–4878. [CrossRef]
85. Amin, K.; Mao, L.; Wei, Z. Recent Progress in Polymeric Carbonyl-Based Electrode Materials for Lithium and Sodium Ion Batteries. *Macromol. Rapid Commun.* 2019, *40*, 1800565. [CrossRef]
86. Molina, A.; Patil, N.; Ventosa, E.; Liras, M.; Palma, J.; Marcilla, R. New Anthraquinone-Based Conjugated Microporous Polymer Cathode with Ultrahigh Specific Surface Area for High-Performance Lithium-Ion Batteries. *Adv. Funct. Mater.* 2020, *30*, 1908074. [CrossRef]
87. Yu, Q.; Xue, Z.; Li, M.; Qiu, P.; Li, C.; Wang, S.; Yu, J.; Nara, H.; Na, J.; Yamauchi, Y. Electrochemical Activity of Nitrogen-Containing Groups in Organic Electrode Materials and Related Improvement Strategies. *Adv. Energy Mater.* 2021, *11*, 2002523. [CrossRef]
88. Yang, H.; Zhang, S.; Han, L.; Zhang, Z.; Xue, Z.; Gao, J.; Li, Y.; Huang, C.; Yi, Y.; Liu, H.; et al. High Conductive Two-Dimensional Covalent Organic Framework for Lithium Storage with Large Capacity. *ACS Appl. Mater. Interfaces* 2016, *8*, 5366–5375. [CrossRef]
89. Bai, L.; Gao, Q.; Zhao, Y. Two Fully Conjugated Covalent Organic Frameworks as Anode Materials for Lithium Ion Batteries. *J. Mater. Chem. A Mater.* 2016, *4*, 14106–14110. [CrossRef]
90. Lin, Z.-Q.; Xie, J.; Zhang, B.-W.; Li, J.-W.; Weng, J.; Song, R.-B.; Huang, X.; Zhang, H.; Li, H.; Liu, Y.; et al. Solution-Processed Nitrogen-Rich Graphene-like Holey Conjugated Polymer for Efficient Lithium Ion Storage. *Nano Energy* 2017, *41*, 117–127. [CrossRef]
91. Lei, Z.; Yang, Q.; Xu, Y.; Guo, S.; Sun, W.; Liu, H.; Lv, L.P.; Zhang, Y.; Wang, Y. Boosting Lithium Storage in Covalent Organic Framework via Activation of 14-Electron Redox Chemistry. *Nat. Commun.* 2018, *9*, 576. [CrossRef] [PubMed]

92. Lei, Z.; Chen, X.; Sun, W.; Zhang, Y.; Wang, Y.; Lei, Z.D.; Chen, X.D.; Sun, W.W.; Wang, Y.; Zhang, Y. Exfoliated Triazine-Based Covalent Organic Nanosheets with Multielectron Redox for High-Performance Lithium Organic Batteries. *Adv. Energy Mater.* **2019**, *9*, 1801010. [CrossRef]
93. Wu, M.; Zhao, Y.; Zhang, H.; Zhu, J.; Ma, Y.; Li, C.; Zhang, Y.; Chen, Y. A 2D Covalent Organic Framework with Ultra-Large Interlayer Distance as High-Rate Anode Material for Lithium-Ion Batteries. *Nano Res.* **2021**, *15*, 9779–9784. [CrossRef]
94. Wang, S.; Wang, Q.; Shao, P.; Han, Y.; Gao, X.; Ma, L.; Yuan, S.; Ma, X.; Zhou, J.; Feng, X.; et al. Exfoliation of Covalent Organic Frameworks into Few-Layer Redox-Active Nanosheets as Cathode Materials for Lithium-Ion Batteries. *J. Am. Chem. Soc.* **2017**, *139*, 4258–4261. [CrossRef] [PubMed]
95. Yang, D.H.; Yao, Z.Q.; Wu, D.; Zhang, Y.H.; Zhou, Z.; Bu, X.H. Structure-Modulated Crystalline Covalent Organic Frameworks as High-Rate Cathodes for Li-Ion Batteries. *J. Mater. Chem. A Mater.* **2016**, *4*, 18621–18627. [CrossRef]
96. Xu, F.; Jin, S.; Zhong, H.; Wu, D.; Yang, X.; Chen, X.; Wei, H.; Fu, R.; Jiang, D. Electrochemically Active, Crystalline, Mesoporous Covalent Organic Frameworks on Carbon Nanotubes for Synergistic Lithium-Ion Battery Energy Storage. *Sci. Rep.* **2015**, *5*, 8225. [CrossRef]
97. Li, X.; Cheng, F.; Zhang, S.; Chen, J. Shape-Controlled Synthesis and Lithium-Storage Study of Metal-Organic Frameworks $Zn_4O(1,3,5$-Benzenetribenzoate$)_2$. *J. Power Sources* **2006**, *160*, 542–547. [CrossRef]
98. Li, C.; Chen, T.; Xu, W.; Lou, X.; Pan, L.; Chen, Q.; Hu, B. Mesoporous Nanostructured Co_3O_4 Derived from MOF Template: A High-Performance Anode Material for Lithium-Ion Batteries. *J. Mater. Chem. A Mater.* **2015**, *3*, 5585–5591. [CrossRef]
99. Wu, R.; Qian, X.; Yu, F.; Liu, H.; Zhou, K.; Wei, J.; Huang, Y. MOF-Templated Formation of Porous CuO Hollow Octahedra for Lithium-Ion Battery Anode Materials. *J. Mater. Chem. A Mater.* **2013**, *1*, 11126–11129. [CrossRef]
100. Yang, S.J.; Nam, S.; Kim, T.; Im, J.H.; Jung, H.; Kang, J.H.; Wi, S.; Park, B.; Park, C.R. Preparation and Exceptional Lithium Anodic Performance of Porous Carbon-Coated ZnO Quantum Dots Derived from a Metal-Organic Framework. *J. Am. Chem. Soc.* **2013**, *135*, 7394–7397. [CrossRef]
101. Han, Y.; Qi, P.; Li, S.; Feng, X.; Zhou, J.; Li, H.; Su, S.; Li, X.; Wang, B. A Novel Anode Material Derived from Organic-Coated ZIF-8 Nanocomposites with High Performance in Lithium Ion Batteries. *Chem. Commun.* **2014**, *50*, 8057–8060. [CrossRef] [PubMed]
102. Serre, C.; Millange, F.; Surblé, S.; Férey, G. A Route to the Synthesis of Trivalent Transition-Metal Porous Carboxylates with Trimeric Secondary Building Units. *Angew. Chem. Int. Ed.* **2004**, *43*, 6285–6289. [CrossRef] [PubMed]
103. Wu, R.; Qian, X.; Rui, X.; Liu, H.; Yadian, B.; Zhou, K.; Wei, J.; Yan, Q.; Feng, X.Q.; Long, Y.; et al. Zeolitic Imidazolate Framework 67-Derived High Symmetric Porous Co_3O_4 Hollow Dodecahedra with Highly Enhanced Lithium Storage Capability. *Small* **2014**, *10*, 1932–1938. [CrossRef]
104. Song, H.; Shen, L.; Wang, J.; Wang, C. Reversible Lithiation–Delithiation Chemistry in Cobalt Based Metal Organic Framework Nanowire Electrode Engineering for Advanced Lithium-Ion Batteries. *J. Mater. Chem. A Mater.* **2016**, *4*, 15411–15419. [CrossRef]
105. Song, H.; Shen, L.; Wang, J.; Wang, C. Phase Segregation and Self-Nano-Crystallization Induced High Performance Li-Storage in Metal-Organic Framework Bulks for Advanced Lithium Ion Batteries. *Nano Energy* **2017**, *34*, 47–57. [CrossRef]
106. Shen, L.; Song, H.; Wang, C. Metal-Organic Frameworks Triggered High-Efficiency Li Storage in Fe-Based Polyhedral Nanorods for Lithium-Ion Batteries. *Electrochim. Acta* **2017**, *235*, 595–603. [CrossRef]
107. Sharma, N.; Szunerits, S.; Boukherroub, R.; Ye, R.; Melinte, S.; Thotiyl, M.O.; Ogale, S. Dual-Ligand Fe-Metal Organic Framework Based Robust High Capacity Li Ion Battery Anode and Its Use in a Flexible Battery Format for Electro-Thermal Heating. *ACS Appl. Energy Mater.* **2019**, *2*, 4450–4457. [CrossRef]
108. Guo, L.; Sun, J.; Zhang, W.; Hou, L.; Liang, L.; Liu, Y.; Yuan, C. Bottom-Up Fabrication of 1D Cu-Based Conductive Metal–Organic Framework Nanowires as a High-Rate Anode towards Efficient Lithium Storage. *ChemSusChem* **2019**, *12*, 5051–5058. [CrossRef]
109. Weng, Y.G.; Yin, W.Y.; Jiang, M.; Hou, J.L.; Shao, J.; Zhu, Q.Y.; Dai, J. Tetrathiafulvalene-Based Metal-Organic Framework as a High-Performance Anode for Lithium-Ion Batteries. *ACS Appl. Mater. Interfaces* **2020**, *12*, 52615–52623. [CrossRef]
110. Wei, R.; Dong, Y.; Zhang, Y.; Zhang, R.; Al-Tahan, M.A.; Zhang, J. In-Situ Self-Assembled Hollow Urchins F-Co-MOF on rGO as Advanced Anodes for Lithium-Ion and Sodium-Ion Batteries. *J. Colloid Interface Sci.* **2021**, *582*, 236–245. [CrossRef]
111. Zhou, X.; Yu, Y.; Yang, J.; Wang, H.; Jia, M.; Tang, J. Cross-Linking Tin-Based Metal-Organic Frameworks with Encapsulated Silicon Nanoparticles: High-Performance Anodes for Lithium-Ion Batteries. *ChemElectroChem* **2019**, *6*, 2056–2063. [CrossRef]
112. Nazir, A.; Le, H.T.T.; Kasbe, A.; Park, C.J. Si Nanoparticles Confined within a Conductive 2D Porous Cu-Based Metal–Organic Framework $(Cu_3(HITP)_2)$ as Potential Anodes for High-Capacity Li-Ion Batteries. *Chem. Eng. J.* **2021**, *405*, 126963. [CrossRef]
113. Yu, L.; Liu, J.; Xu, X.; Zhang, L.; Hu, R.; Liu, J.; Yang, L.; Zhu, M. Metal-Organic Framework-Derived NiSb Alloy Embedded in Carbon Hollow Spheres as Superior Lithium-Ion Battery Anodes. *ACS Appl. Mater. Interfaces* **2017**, *9*, 2516–2525. [CrossRef]
114. Wang, L.; Zhu, L.; Zhang, W.; Ding, G.; Yang, G.; Xie, L.; Cao, X. Revealing the Unique Process of Alloying Reaction in Ni-Co-Sb/C Nanosphere Anode for High-Performance Lithium Storage. *J. Colloid Interface Sci.* **2021**, *586*, 730–740. [CrossRef] [PubMed]
115. Yoon, T.; Bok, T.; Kim, C.; Na, Y.; Park, S.; Kim, K.S. Mesoporous Silicon Hollow Nanocubes Derived from Metal-Organic Framework Template for Advanced Lithium-Ion Battery Anode. *ACS Nano* **2017**, *11*, 4808–4815. [CrossRef] [PubMed]
116. Sun, Y.; Huang, F.; Li, S.; Shen, Y.; Xie, A. Novel Porous Starfish-like Co_3O_4@nitrogen-Doped Carbon as an Advanced Anode for Lithium-Ion Batteries. *Nano Res.* **2017**, *10*, 3457–3467. [CrossRef]
117. Han, Y.; Li, J.; Zhang, T.; Qi, P.; Li, S.; Gao, X.; Zhou, J.; Feng, X.; Wang, B. Zinc/Nickel-Doped Hollow Core–Shell Co_3O_4 Derived from a Metal–Organic Framework with High Capacity, Stability, and Rate Performance in Lithium/Sodium-Ion Batteries. *Chem.—Eur. J.* **2018**, *24*, 1651–1656. [CrossRef] [PubMed]

118. Fei, B.; Chen, C.; Hu, C.; Cai, D.; Wang, Q.; Zhan, H. Engineering One-Dimensional Bunched Ni-MoO$_2$@Co-CoO-NC Composite for Enhanced Lithium and Sodium Storage Performance. *ACS Appl. Energy Mater.* **2020**, *3*, 9018–9027. [CrossRef]
119. Ye, H.; Jiang, F.; Li, H.; Xu, Z.; Yin, J.; Zhu, H. Facile Synthesis of Conjugated Polymeric Schiff Base as Negative Electrodes for Lithium Ion Batteries. *Electrochim. Acta* **2017**, *253*, 319–323. [CrossRef]
120. Man, Z.; Li, P.; Zhou, D.; Zang, R.; Wang, S.; Li, P.; Liu, S.; Li, X.; Wu, Y.; Liang, X.; et al. High-Performance Lithium–Organic Batteries by Achieving 16 Lithium Storage in Poly(Imine-Anthraquinone). *J. Mater. Chem. A Mater.* **2019**, *7*, 2368–2375. [CrossRef]
121. Poizot, P.; Yao, Y.; Chen, J.; Schubert, U.S. Preface to the Special Issue of ChemSusChem on Organic Batteries. *ChemSusChem* **2020**, *13*, 2107–2109. [CrossRef] [PubMed]
122. Zhang, Y.; Niu, Y.; Wang, M.Q.; Yang, J.; Lu, S.; Han, J.; Bao, S.J.; Xu, M. Exploration of a Calcium-Organic Framework as an Anode Material for Sodium-Ion Batteries. *Chem. Commun.* **2016**, *52*, 9969–9971. [CrossRef]
123. Zhang, K.; Guo, C.; Zhao, Q.; Niu, Z.; Chen, J.; Zhang, K.; Guo, C.; Zhao, Q.; Niu, Z.; Chen, J. High-Performance Organic Lithium Batteries with an Ether-Based Electrolyte and 9,10-Anthraquinone (AQ)/CMK-3 Cathode. *Adv. Sci.* **2015**, *2*, 1500018. [CrossRef] [PubMed]
124. Wang, S.; Wang, L.; Zhang, K.; Zhu, Z.; Tao, Z.; Chen, J. Organic Li$_4$C$_8$H$_2$O$_6$ Nanosheets for Lithium-Ion Batteries. *Nano Lett.* **2013**, *13*, 4404–4409. [CrossRef] [PubMed]
125. Li, L.; Yin, Y.-J.; Hei, J.-P.; Wan, X.-J.; Li, M.-L.; Cui, Y. Molecular Engineering of Aromatic Imides for Organic Secondary Batteries. *Small* **2021**, *17*, 2005752. [CrossRef]
126. Liu, H.; Cheng, X.B.; Huang, J.Q.; Kaskel, S.; Chou, S.; Park, H.S.; Zhang, Q. Alloy Anodes for Rechargeable Alkali-Metal Batteries: Progress and Challenge. *ACS Mater. Lett.* **2019**, *1*, 217–229. [CrossRef]
127. Xie, J.; Zhang, Q.; Xie, J.; Zhang, Q. Recent Progress in Multivalent Metal (Mg, Zn, Ca, and Al) and Metal-Ion Rechargeable Batteries with Organic Materials as Promising Electrodes. *Small* **2019**, *15*, 1805061. [CrossRef]
128. Lee, M.; Hong, J.; Kim, H.; Lim, H.D.; Cho, S.B.; Kang, K.; Park, C.B. Organic Nanohybrids for Fast and Sustainable Energy Storage. *Adv. Mater.* **2014**, *26*, 2558–2565. [CrossRef]
129. Hu, P.; Wang, H.; Yang, Y.; Yang, J.; Lin, J.; Guo, L. Renewable-Biomolecule-Based Full Lithium-Ion Batteries. *Adv. Mater.* **2016**, *28*, 3486–3492. [CrossRef]
130. Zhao, Q.; Zhu, Z.; Chen, J. Molecular Engineering with Organic Carbonyl Electrode Materials for Advanced Stationary and Redox Flow Rechargeable Batteries. *Adv. Mater.* **2017**, *29*, 1607007. [CrossRef]
131. Chen, H.; Armand, M.; Demailly, G.; Dolhem, F.; Poizot, P.; Tarascon, J.M. From Biomass to a Renewable Li$_x$C$_6$O$_6$ Organic Electrode for Sustainable Li-Ion Batteries. *ChemSusChem* **2008**, *1*, 348–355. [CrossRef] [PubMed]
132. Chen, H.; Armand, M.; Courty, M.; Jiang, M.; Grey, C.P.; Dolhem, F.; Tarascon, J.M.; Poizot, P. Lithium Salt of Tetrahydroxybenzoquinone: Toward the Development of a Sustainable Li-Ion Battery. *J. Am. Chem. Soc.* **2009**, *131*, 8984–8988. [CrossRef] [PubMed]

Disclaimer/Publisher's Note: The statements, opinions and data contained in all publications are solely those of the individual author(s) and contributor(s) and not of MDPI and/or the editor(s). MDPI and/or the editor(s) disclaim responsibility for any injury to people or property resulting from any ideas, methods, instructions or products referred to in the content.

MDPI
St. Alban-Anlage 66
4052 Basel
Switzerland
Tel. +41 61 683 77 34
Fax +41 61 302 89 18
www.mdpi.com

Materials Editorial Office
E-mail: materials@mdpi.com
www.mdpi.com/journal/materials

www.ingramcontent.com/pod-product-compliance
Lightning Source LLC
LaVergne TN
LVHW070450100526
838202LV00014B/1695